THE ELDER JAMES WHATMAN

ENGLAND'S GREATEST PAPER MAKER
(1702–1759)

VOLUME II

State Papers 1747

A NOTE ON THE FRONTISPIECE TO THIS VOLUME

The Parish Church of All Saints', Maidstone, has served as a centre for the baptisms, marriages and burials of many Kent paper makers for a period spanning well over three hundred years. It is a fitting memorial, therefore, to incorporate an illustration of this fine church in a volume where so much space has been devoted to the genealogies of these paper makers. John Whichcord (who undertook this fifteenth century reconstruction, illustrated in this Plate, printed in 1849) said that "the form of this building presents none of those variations in arrangement that one finds in churches whose parts are of various dates, and where modifications were required to meet the newly introduced customs of successive generations. Perhaps few single buildings possess more completeness and uniformity than this church exhibited in its original state". The interior of All Saints' is surpassed by few parish churches in England. Its great and striking characteristic is its size and symmetry. From end to end it measures 227 feet, the nave being 91 feet wide and the chancel 64 feet. (Russell, J.M. Bib.41). During the 1970's–80's the church underwent extensive renovation and the screen illustrated in the Plate was removed, leaving the view of the nave from west to east unimpeded, thereby enhancing the great space of this interior.

Published by J.N. Balston
West Farleigh, Kent

ISBN 0 9519505 1 7 (Volume 2)
ISBN 0 9519505 2 5 (Set of 2 volumes)

Camera–ready copy by M. Notcutt, Oxford

Production consultancy by Laurence Viney, Berkhamsted

Printed in Great Britain by St. Edmundsbury Press,
Bury St. Edmunds, Suffolk

Printed on Acid free paper

The Collegiate Church of All Saints', Maidstone, restored to the date of the condition A.D. 1400 (John Whichcord jnr. Lithograph Day & Son, Lithographers to the Queen, 1849. Reproduced here by kind permission of the Churchwardens and Parochial Church Council).

Contents
The Appendices : Bibliography : Maps

Preface to Appendices I & II

Although the Hollingbourne mills, referred to in the text, have been given names and peopled with paper makers and otherwise discussed quite freely, almost as if they still existed, they have all in point of fact disappeared from to-day's landscape. An added complication is that relatively recent research[1] has shown the existence at various periods of at least a dozen watermills of one kind or another in the closely linked parishes of Hollingbourne, its opposite number Leeds together with its neighbour, Broomfield, the three parishes that encircle Leeds Castle. The assertion has been made that three of these mills were paper mills and that at least two of them played an important role in the development of papermaking in Kent from the early 18th C. onwards.

These claims have to be authenticated and the purpose of Appendices I and II is to reaffirm for the *specialist reader* (see FOREWORD n.8) the location of these paper mills; then to set out, using such documentary evidence as is available, the history of the mills and their various occupants in the form of synopses; and, finally, in Appendix II to identify some of the more important paper makers connected with them. It is perhaps appropriate to add at this point that much of what has been said above and that which follows applies equally to the paper mills and their occupants discussed in Appendices III and IV.

In recent years the most important work that has been done on the paper mills in this area is, first, that done by Dr. Alfred Shorter on the British Paper Industry as a whole[2] and on the mills in the Maidstone district in particular;[3] and, later, that done by Mr. R.J. Spain on the Watermills of the River Len.[4] Both authors have provided much of the information used here regarding the location of the three paper mills in Hollingbourne and the identity of many of their occupants. It might be asked what more was needed than this ? The author's own searches showed that an appreciable amount of significant detail was missing.

The object of this work has not been merely one of refining the previous accounts, but instead one of trying to present a more co-ordinated and sustained picture of these paper mills, treating them not as isolated units but as part of the expanding industry of the period and of this region.

Location

The question of locating the paper mills has not been a difficult one despite the fact that so little of them has remained. A millpond, its massive dam, the apron and foundations of the mill itself, buried deep in undergrowth, and the ruined sluices to the north is all that remains of Old Mill (in fact it would not be difficult to-day to pass it by without one knowing that there was anything there at all). Likewise, Spain has reported that "the small valley that

---

[1] Spain, R.J. Bib. 8 33.
[2] Shorter, A. Bib.12 (1957)
[3] Shorter, A. Bib. 7 (1960)
[4] Spain, R.J. Bib. 8 54–56

contained Grove (Eyehorne) millpond is now only a stream bed. A concrete cill that once held back the water is now broken and the mill house has long since gone"; as for Park mill, he writes that "it is now marked by a waterfall which can be seen from the Maidstone–Ashford road", which to–day traverses the valley at this point some 15 or more feet above the stream bed.[5] The mills, however, are all clearly marked as Corn mills on the 1864/5 6 inch to the mile O.S. map (Map 9); and from there it is a matter of tracing back the occupants until the evidence shows that they were paper makers and from surviving Excise Records one can quickly establish which mill was which; and when they ceased working (a much simpler operation than in the case of the mills described in App.III & IV).

Mill Names

Whereas the Ordnance Survey maps of the 19th C. (with the exception of the 1801 1 inch to the mile) name Park mill and Grove mill as such, no name is given for Old Mill. The 1839 Tithe Assessment also uses these names; but, if one goes back further to the 1816 Excise General List,[6] Grove mill is called "Eyehorne Mill" and Park mill "New Mill" (and both had Excise Numbers). Since Richard Barnard snr. and jnr., the proprietor and occupant respectively of Eyehorne mill, used the countermark "EYEHORN MILL" in some of their papers, this name has been used throughout this book in spite of the fact that later it came to be known as "Grove Mill". As the Excise Officers used the title "New Mill" for any mill that was added to their list, the name "Park Mill" has been retained. The name "Old Mill", however, seems to have been a matter of verbal tradition. The 1816 EGL described it as "Mote Hole", probably a mis–spelling of Moat meaning a pond; or even a steep–sided ditch with water in it because the River Len at this point flows through a steep–sided gorge. But, as it was by far and away the oldest of the Hollingbourne paper mills, it has been decided to use the name "Old Mill".

Occupants

In the case of the occupants the identification is less straightforward. It will be recalled that as a result of the Public Record Office Act of 1877 all Excise Records of the British Paper Industry prior to 1816 were destroyed (in fact the 1816 List is the earliest comprehensive list of mills and their occupants to have survived). So, in effect, it can be said that much of the history of the industry since the reign of Queen Anne went up in smoke. Shorter has described very fully the sources that he used to reconstruct the fabric of this history,[7] so there is no need to cover this ground again except to say that some of these sources have been re–examined by the author in greater detail. This last applies particularly to the Parish Records, to the genealogy and countermarks of the paper makers concerned, and to a lesser extent to what might be regarded as one of the corner stones of Shorter's monumental work, namely, his use of contemporary Insurance Records (many of these being in such a delicate state now that they are no longer accessible for inspection). A closer study of these sources has provided greater continuity and coherence to the history of these mills.

---

[5] Since this was written, the 1990 extension of the M20 motorway has obliterated one, if not both, of these sites. Ironically, Cowper has another apt line "where once we dwelt our name is heard no more".

[6] H.M. Customs & Excise Museum and Records, King's Beam House, Mark Lane, London EC3R 7HE.

[7] Shorter, A. Bib.12 20.

Parish Documents

Relevant extracts from the Hollingbourne and Leeds Ratebooks and Land Tax Returns are to be found in pages following the synopses of the histories of the three paper mills. Up till about 1800 these records provide solid evidence of their existence and occupants. In the 19th C. the picture becomes less clear as there are instances where some of the mills were sub–let to other paper makers than those entered in the official records (and this applies particularly to the occupants of the Sandling paper mills described in App.III). Examples of this are Crispe and Newman and, later, Stroud and Newman working Old mill in their names (countermarks) for Hollingworth and Balston as owners. It is most regrettable, however, that the vital Hollingbourne ratebook for the period prior to 1742 is missing; some of the ratebooks end prematurely and in other cases crucial Land Tax Returns have been lost. Resort had to be made to indirect evidence like genealogical data, Wills etc. Sometimes one is lucky in finding information in this way as, for instance, the entry of a multiple retrospective baptism of four of James Whatman's children in the Brenchley parish register in 1663;[8] but in many instances the searches were depressingly negative. For the sake of tidying up an account the facts included in the synopsis have occasionally been extended for this purpose.

Insurance Records

With regard to Insurance, the practice of insuring one's property or making a similar provision for adversities such as sickness, fire, shipwreck etc. dates back to at least mediaeval times and possibly earlier than this. Over and above this prevailing practice there was a noteworthy boom in Insurance undertakings at the beginning of the 18th C. It was at this time that, together with a number of other roughly contemporary ventures, the Sun Fire Insurance Company made its first appearance, between 1706–10.[9] The Sun's policies have survived from 1710[10] and are particularly useful in that this company seems to have been patronised by paper makers from an early date (at least, by those operating in the counties around London). William Gill, Whatman's and Harris' predecessor at Turkey mill, insured both this mill and his Forstal mill as early as 1716. Highly important for us also is the confirmatory evidence supplied by the Sun Fire Insurance Policy (SFIP) No.61845, which concerns the elder Whatman's investment in Old Mill, Hollingbourne, in September 1733.[11]

All the same the Insurance Policies do not provide the continuity of evidence that one might expect. Those policies that have been found carry a renewal date as, for instance, in the one for Old Mill cited above. In this case it was renewed (by Richard Harris) at the appointed

---

[8] Entry in the Brenchley PR (KAO P 45 1/2) for 8:Dec:1663. "Baptized upon the 8th day of this month four children of James Whatman and Lydia Waggoner his wife viz. Thomas, their son who was born upon the 6th day of September Anno Dom. 1654. Then James their son, who was borne upon the 20th day of January 1655 (1656 corr:). Then, Sarah, their daughter, who was borne upon the 6th day of November Anno Dom. 1657. Then Lydia their daughter, who was borne upon the 17th day November last passed viz. in this year of our Lord God 1663." For the full pedigree see App.II "The Whatman Paper Makers' Pedigree".

[9] Scott, W.R. Bib.43 Vol.III 363ff. and 480.

[10] Guildhall (London) MS. Dept. of Library MS.11936.

[11] Guildhall Library (London) MS 11936 Vol.39 Policy No.61845, dated 25th September 1733. Premium £1–10–0d. Renewal date Michaelmas 1734. Agent – Kirby.
    JAMES WHATMAN of Loose in the County of Kent Tanner. On a New built Dwelling house and paper mill being one entire building in the parish of Hollingbourne in the County aforesaid only Timber and Tiled now in the Occupation of Richard Harris paper maker exclusive of all other buildings whatsoever not exceeding Six hundred pounds. £600.

time. But in many other cases this is not so. Because of the consecutive nature of the Policy numbers it is clear that the renewals that one is looking for cannot be missing from the main record. One can only presume that either renewals were made at a local office and have since been lost; or else that the paper maker insured with another company (see below); or not at all. In the case of Old Mill there are some very substantial gaps, no renewals in the main record for example between 1736–39 and a huge gap between 1746 and 1770. It is no small undertaking to search through several hundred thousand policies and, apart from obtaining a few extra details from those already discovered by Shorter in his mammoth search, the attempt has not been made beyond a date of ca.1741 (with one or two later exceptions that were examined to see whether they threw any further light on a given situation). To try to cover the first gap in the SFIP records mentioned above a search was made through Hand in Hand Policies for the same period to see whether either Whatman or Harris had insured their mills through them; but no policies in their names could be found.[12] In this respect there are decided limitations to this source of evidence for detailed investigations on a year to year basis, although there can be no question of its value in a wider application provided that these records are accessible which in some cases for the time being is not so because of their fragile state.

COUNTERMARKS

In the author's experience the use of countermarks as a source of information for historical, and bibliographical, purposes is unequal. In some cases the powers that be keep meticulous records of their occurrence; others never look for them or even note general characteristics of the paper as, for instance, whether it is laid or wove. It is suggested here that more attention might be paid to these details and records kept of their incidence in institutions holding large collections of documents and drawings, dating mainly between the 17th and mid–19th C., where this is not already practised. Admittedly the subject is a difficult one and the interpretation of the findings frequently uncertain or unsolved. The comments below do not pretend to provide a solution to these problems but are given purely to assist people undertaking this work and, perhaps, provide a framework for a systematic record. These comments are confined to the countermarks themselves and do not embrace other features such as their position on the mould cover, the nature of the wiremarks that surround them etc.

Countermark is a difficult term to define; indeed it is doubtful whether there is any straightforward definition that would embrace all aspects of this subject satisfactorily especially when one tries to take into account the use of this kind of mark in different countries and for different periods in history. To simplify matters for the purpose of this book a new definition is given below which is aimed to cover, primarily, British papermaking of the 18th C., because once outside these limits matters become much more complicated.

---

[12] Hand in Hand Policies Guildhall Library (London) MS 8674 Vols.55–57 covering 1737–39. The Hand in Hand Records held by the Library date back to 1696.

The Royal Exchange Company were operating from 1721, but no records have survived before 1758. It is possible, therefore, that Whatman and Harris might have insured with this company during the missing years. But on balance seeing that they continued to insure with Sun Fire over a long period the notion of a local office is the solution favoured here.

Professor Alan Crocker (letter 5:Sept:89) has drawn my attention to his work on Ewell mill, Surrey, where the occupants, almost exclusively the Jubb family, insured their mills with both the Sun Fire and the Hand–in–Hand Companies between 1733–79, the latter providing useful information on the functions and dimensions of the buildings.

Labarre has defined the countermark as "the smaller or subsidiary watermark found in antique paper in addition to the main Watermark or mark proper. It served, as a rule, to indicate the name or initials of [the] maker and the date and place of making..... The countermark is generally placed in the second half of the sheet opposite the main mark, though border and corner positions are not infrequent".[13] It is not the intention to dispute this general definition here, a definition which has probably become as widely accepted as any definition is ever likely to become, but in the case of British papermaking of the 18th C., and even more so in the 19th C., it could be misleading to the uninitiated. A lot depends, for example, on what interpretation is given to the term "antique" papers. After all one has to consider the marking of wove paper, paper in which the "main mark" disappeared altogether and only the "smaller mark" (which in some cases became very large later) remained after 1760. One gets into even deeper water when one tries to disentangle all the other kinds of "watermarks" from one another, wiremarks, mould marks and in a more restricted sense cyphers, stereotypes and so on. The simplest answer is to consider what we usually mean when we say that a sheet of paper was "watermarked" or "had a watermark in it". The Oxford English Dictionary, for instance, defines it as "a distinguishing mark or device impressed in the substance of a sheet of paper during manufacture....", which is conveniently vague. If we are to re–define the term "countermark" for our use, then to avoid having to use awkward phrases like "the mark proper" we must have a definition that distinguishes the two terms from each other.

The term "Watermark", in its wider sense, embraces the "Countermark" as is implied in Labarre's definition. But it will be used in this book in a more restricted sense, namely, as Labarre's "Main mark". This will distinguish the countermark (re–defined below) and Stereotypes (also referred to below) from all those *other* Emblems or Mottoes that either denote different sizes of paper (Posthorn, Strasbourg bend etc.) or which may represent the Arms of the Sovereign, a City, State or Federation (Royal Arms, The Arms of Amsterdam, London, Pro Patria, Britannia etc.).[14]

The term "Countermark", as re–defined here, refers specifically to the personal mark, cypher, initials or name used by the "maker" of the sheet of paper concerned and *not* to imported or pirated stereotypes such as "IV" or "LVG" etc., widely used by British paper makers during the 18th C., supposedly as a symbol denoting a certain level of quality;[15] nor necessarily to a mark that is "in the second half of the sheet opposite the main mark".[16]

---

[13] Labarre, E.J. Bib.36 61.

[14] It should be noted that the use of these Royal or National devices by no means points to the paper originating from the countries indicated by these insignia.

[15] There are many other similar stereotypes whose significance is not properly understood such as " $w\ell$ " or the letters "HG", marks that were commonly used during the latter part of the 17th and early 18th Cs. (illustr:Gravell Bib.68 733/4). Then there are stereotypes such as "GR" with or without a crown, which originally may have been incorporated in papers intended for official documents; but who authorised their use and how widely it was pirated is not clear. One assumes that the earlier use of "ER", "AR" etc. would have resulted from special privileges granted to the paper maker.
Again, one finds slogans like "Pour Angleterre" in the early 18th C. (KAO P 241 12/12 d.o.d.1724); "Irish Lottery" (Susanna Whatman corresp. P.S.M. d.o.d. 22:Jan:1811); or "Warranted Not Bleached" of the 19th C. (Fitzwilliam Museum drawing 2676b attrib. J.J. Cotman) and so on, none of which can really be considered as countermarks in the author's definition although they may have been used by identifiable paper makers.

[16] Frequently the countermark (author's definition) may be found within the "Main mark" e.g. within the frame or shield surrounding a Britannia or similar watermark; or as the subscript to the "Main mark" while the mark in "the second half of the sheet" may be a stereotype such as "IV" or "GR".

Plate 16 – Examples of Early 19th C. Forged Whatman Countermarks

The examples of the forged countermarks illustrated below were sent to the author very kindly by the late Mr. E.G. Loeber who had an extensive collection of them.

1.

A typical specimen of a genuine "J WHATMAN 1831" countermark (P.S.M.)   x 0.7

2.

A forged Whatman countermark in a Hague document dated 1828   x 0.7
Note (i) full–stop after "J"; (ii) the "A"s with dipped bridges; (iii) the centre portion of the "M" does not reach the base line; (iv) faulty "N".

3.

A good imitation of a genuine Whatman mark in a Karlsruhe document of 1828   x 0.7
Note once again the faulty "M" and "N". (See Note 19 opposite).

4.

Two Bremen documents of 1822/23. (i) "M" is the wrong way round; (ii) in one of the specimens the "A"s have dipped bridges.   x 0.7

5.

A Gottingen document of 1828 (i) full–stop after the "J" (sometimes there is a full–stop after the "N"); (ii) "A"s dipped bridges; (iii) a faulty "M".   x 0.7

6.

A Gottingen document of 1812. Note the reversed "A" and the poor formation of the mark.   x 0.7

7.

A Hague document of 1835 shows a DARK countermark in a frame. This kind of mark (as far as the author knows) was *never* used by the Balston family when making paper.   x 0.7

Having re-defined the term "Countermark" for the purpose of this book, the definition, unfortunately, has to be qualified immediately in several respects.

(i)   First, the word "maker" usually refers to the Master Paper Maker (or to a partnership) and with one or two exceptions this applies to all countermarks mentioned in the text and appendices.[17]

However, sometimes it may also refer to the person who commissioned the paper as, for instance, a Stationer, who in turn may have owned the mill perhaps; or to a Printer for whom the paper was destined. It is as well then to check this point before ascribing the mark to the Master Paper Maker at the mill.

(ii) Towards the end of the 18th C. and increasingly in the 19th C. the situation is complicated by:-
      (a) the perpetuation of a maker's name (countermark) by successors at a given mill; or even by its acquisition and transference elsewhere.
      (b) by the forgery of established countermarks (to be distinguished from the earlier practice of pirating well-known continental countermarks).

Perhaps the best known example that demonstrates both of these cases conveniently is that of the countermark "J Whatman". In the first case (a) Hollingworth and Balston continued to use this countermark after the younger Whatman had sold his business to them and after his death also. In 1807, when they had parted company, Hollingworth used "J Whatman Turkey mill" and Balston used, among others, "J Whatman" on its own, transferring the latter to his new mill, Springfield.[18] In the other case (b) the Whatman countermark was widely forged in the 19th C., (PLATE 16) particularly by German and Austrian makers[19] (and, indeed, examples of similar forgeries have been found in the 20th C.).

(iii) There are a number of problematic countermarks, clearly belonging to Master Paper Makers, which can only be accounted for satisfactorily, if one postulates the sub-letting of a given mill owned by one Master and operated by another. Several examples of this occur in the Maidstone area, one of these being the countermark "Crispe & Newman 1797", which is discussed later in App.I under the Park mill countermarks.[20]

---

[17] After 1794 and before they parted company "J Whatman" would have been used by Hollingworth & Balston (see also under (ii)a in text above). Miss P.M. Pollard (private communication) mentioned that a Dublin surgeon, Bartholomew Mosse, had his own personal mark inserted in paper made ca.1753.

[18] Further details and variants of the Whatman countermarks after 1806 may be found in Balston, T. Bib.74 164-7. These include the "J Whatman W Balston" marks etc.
      As examples of the perpetuation of other makers' names "W King" of Alton may be cited here as used by John Edward Spicer and descendants from 1828 and possibly earlier since King had left Alton or had died by 1796.
      As an example of the acquisition and transference of another well-known countermark to another, unconnected mill, there is "R Barnard" of Hollingbourne to Wookey Hole in Somerset (see App.I under Eyehorne mill countermarks).

[19] Common mistakes found in 19th C. forgeries of the Whatman countermark include (a) a full-stop after the "J" (although this was used by the makers in the 20th C. for a special purpose); (b) an unusual "A"; and (c) the use of Dark Watermarks (filigraine ombre). Schulte cites a "J Whatman Turkey Mill 1833" chiaroscuro countermark (dark centre with a light outline) which is also more than suspect.
      A 19th C. forged Whatman countermark received from Germany (Oct.1990) stylistically identical to PLATE 16 (3) but with a full stop after the "J" and most unusual wiremarks in the paper necessitated a re-examination of the comment that example (3) was "a good imitation" where the "M/N" only is considered to be uncharacteristic. How certain can one be of this claim? The author has compared 17 examples of (3), dimensions differ, obtained from 6 sources in Germany, 8 from Holland and 3 from Antwerp d.o.d.'s 1824-1836 with a wide range of genuine Whatman marks in papers made at both Springfield (Balston) and Turkey mills (Hollingworth) from 1801-1865 in which all the "M/N"s are identical in style with example (1). In other words (3) is still considered to be "a good imitation". Gravell, T.L. Bib.68 (769) d.o.d.1828 illustrates a similar mark with a subscript "L". Two other similar marks with a subscript "L" have been found in a Hanover document (1836) and one from Gottingen (1841), the former with a full-stop after the "J". The provenance of Gravell's example has yet to be determined, but on the face of it another forged mark. The author has also seen suspect "J LARKING" marks with a dot over the "I" from the Continent.

[20] Other examples mentioned in App.I are "Stroud & Co."; and in App.III "Green & Young" and "John Young".

(iv) Certainly from the early 19th C. (and possibly in the late 18th C. as well) Stationers sometimes applied their own embossed marks to certain kinds of paper already bearing the maker's countermark.[21]

(v)   In some cases where a paper mill had become well-known or where the paper maker perhaps wanted to distinguish one of his mills from another or else advertise it for some reason, one finds the name of the mill in the place of the maker's personal mark. In this sense it can be regarded as a countermark. Examples of this are "Eyehorn Mill", "SM" (Springfield mill), "Valleyfield", "Lower Tovil" etc.[22]

(vi) One sometimes finds a client's name or initials or even insignia coupled with the maker's name and these have to be distinguished from the countermark, the watermark and the stereotype as redefined here.[23]

One might include under this heading the enigmatic use of certain types of script "W" marks which are clearly quite different to the single line "W" countermarks used by the elder James Whatman. It has been suggested that these marks might have been used to designate a class of paper eg."writing" paper ? For further discussion of these "W" marks see App.V.

It will be seen that in spite of a simplified and restricted definition of the term countermark it is not an easy term to wrap up neatly in a single phrase.

Earlier it was said that countermarks could be used more advantageously as a source of information than is often the case. Apart from providing one with confirmation that a given paper maker existed and was engaged in White paper manufacture,[24] to be useful one must obviously be able to relate this information to something else. For instance, it may be useful in dating an edition of a printed book where the countermark or watermark dates conflict with the date shown on the title page; engravings are a rich field for this kind of discrepancy; and many other examples may be found in the case of paintings copied or forged long after the artist concerned had died or in works to which incorrect dates have been attributed. In the context of this book and in papermaking history in general countermarks can be useful in several other ways.

---

[21] These embossed marks usually seem to occur in fairly large sheets of paper e.g. Drawing paper. Early examples include Whatman papers embossed by an anonymous Stationer "Bristol Paper"; another, "Drawing Paper : Rough : Thomas Creswick" who was both a paper maker and Stationer; another, "Bath Superfine" with surrounding ostrich plumes and countermarked "W KING 1828", clearly used by Spicers in their capacity as Stationers.

[22] Examples of this are "Eyehorn Mill: (see App.I Eyehorne mill countermarks); "SM" (see Balston, T. Bib.74 167 fig.12); "Valleyfield 1816" (Turner Bequest CLXXXIX 5 d.o.d.1819) representing Charles Cowan's mill in Midlothian, a mill active before 1750; and "Tovil Mill", "Ivy Mill" etc. (see Chap.VII note 234).

[23] "VEIC" in a quartered diamond-shaped frame (United East India Company) associated with various makers' countermarks (V = U). Other interesting examples are 3 "J Whatman Turkey Mill" countermarks incorporating a script "N" (1853); and another plus a capital "B" (1868). The author's interpretation of these is that the "N" stood for James Newman, the Artists' Colourman of Soho Square and represented a special order. As Hollingworths gave up making handmade paper (and Whatman papers) in 1858 and as Balstons had acquired the sole rights to use the "J Whatman" mark in 1859, but not "Turkey Mill" (except for use in one kind of paper, Antiquarian, where the special "JAMES WHATMAN TURKEY MILL KENT" mark was used), it seems likely that Balstons were asked to repeat (1868) this special order and were licensed to use "Turkey Mill" for it and that they incorporated a "B" to distinguish it. Other Artists' Colourmen were doing much the same thing at the time. (These sheets were discovered by Mr. Peter Bower in a collection of papers).

[24] Generally speaking (no exceptions have been found yet) prior to the 19th C. "coloured" papers would have been regarded as inferior and would thus have carried no countermarks; colour would have been added to the furnish to mask the low colour of the rags and dirt. From ca. 1820 onwards some coloured papers were made for a special purpose e.g. Artists' papers, where dated countermarks (not easy to detect) are found. Examples of these are "S & A" (Smith & Allnutt see App.III p.110); and "B E & S" (Bally, Ellen & Steart, de Montalt mill, Bath), the former found in a painting of James Stark's (BM 1911-4-12-5); the latter in drawings by J.S. Cotman (Norwich NCM 381.235.951).

Firstly, the presence of a countermark is a fairly positive indication that the paper maker designated was engaged in making White paper, although there may have been instances where he might have had paper made for him, bearing his countermark, at another mill. It is not thought that this would have been practised much, if at all, during the 18th C. when, as previously noted, the use of countermarks became increasingly more a matter of the prestige of the maker and his products than one of identification for Excise purposes. In circumstances like these he would not have made paper for others too readily if they bore their mark rather than his.[25]

The converse does not necessarily follow; the absence of a countermark might have signified that White paper either was never intended for export and the associated drawback of Duty or that it was of a lower quality such as demi-fine or seconds not meriting the use of a mark. So though a countermark can tell us something useful about the existence and pretensions of a paper maker, one cannot say that because no Quelch or Terry marks have been discovered before 1780 they did not make White paper before this date.[26]

Secondly, countermarks can relate a paper maker to a locality in a specific way that is informative as well as point to his success in the market place. With regard to the first, examples will be found in these appendices of marks whose occurrence in the normal way might have remained undiscovered, turning up in documents (scraps of paper) belonging to the villages or locality where the paper mill was situated. These discoveries tell one something about the early history of newly commissioned mills that quite clearly had not yet established any notable reputation for themselves.[27] Then there are (watermark) dated countermarks belonging to paper makers who cannot be accommodated satisfactorily at that date anywhere so that an alternative explanation for their papermaking activities has to be sought such as the sub-letting of a mill by those who from ratebook or other evidence one might have expected to have been the occupants.[28] In much the same way a series of dateable countermarks will illustrate the continuity or else the changes that may have taken place in the management of a mill e.g. the formation or the dissolution of a partnership.[29] In Kent, if not equally elsewhere, this kind of information is most readily found in the County

---

[25] As allowed in the original Act (10 Anne, c.18), first operative in 1712, the use of a countermark in theory permitted the paper maker to claim drawback of Duty on his paper under certain circumstances including exportation. But this drawback was by no means easy to recover as witness the Master Paper Makers' Memorial to the Commissioners delivered by the younger Whatman 2:Dec:1764 (see Balston, T. Bib.1 App.III pp.149–151). Their complaint made little difference and the fact that countermarks were so widely used after the 1760's confirms the view that the makers saw their countermarks as a mark of the perfection of their products compared to those of others (Chap.VII note 357).

[26] Although this conclusion appears to be almost axiomatic, one is constantly faced with the problem of trying to penetrate problematic situations such as knowing whether Edward, or Mrs. Russell, was at Forstal mill during the 1770's (in the absence of a ratebook) or that Munn and Sweetlove were at Ford mill, Little Chart. The discovery of countermarks enables a distinction to be made between two such mills and the movements of their occupants that otherwise might not be evident.

[27] Examples of these are the William Avery (Eyehorne mill) and Munn & Sweetlove (Ford mill, Little Chart) countermarks found in Leeds village documents (see App.I under Eyehorne mill countermarks and App.II the Table of Lewis Munn marks p.90).

[28] In addition to the Crispe & Newman marks mentioned earlier other examples include those belonging to seemingly homeless paper makers like Stroud, Charles Brenchley and Green & Young (see App.III pp.110,112).

[29] A good example is the case of the various Lewis Munn countermarks (see App.II Table p.90).

Archives; one might almost say that a great part of the history of the county's papermaking lies hidden in the paper used for local documents now held in Record Offices.[30]

The other aspect mentioned above can tell one something about the success paper makers had, indirectly through their Stationers, in penetrating markets outside their district. Apart from an element of chance, an achievement of this kind would in the end have depended on other factors such as a reputation for quality; and in that cut–throat age price. But in the longer term consistent quality would have been the determining factor. Obviously, it would be difficult, if not impossible, to follow the fortunes of most paper makers in a market as large and diverse as London, although in the earlier part of the 18th C. it might be easier to discern the progress made by paper makers like the elder Whatman. If one takes a provincial market instead, one has a much greater chance of success.[31] Export markets could also yield interesting information about the distribution of papers found there from the home market.[32]

Finally, countermarks in addition to the nature of the paper can help in questions of dating. For example, the discovery of the "IW" mark, discussed in Chapter III, compelled one to reconsider very carefully accounts of the elder Whatman's papermaking career; had it not been for this discovery it might not have been thought necessary to add to Thomas Balston's brief biography of him. There are other examples of a similar nature.[33] Conversely, one comes across examples of countermarks, dated or otherwise, or the use of certain kinds of paper, e.g. Wove paper, which show that the date assigned to an undated document on other grounds is incorrect. The author has seen drawings, for example, executed on Wove paper to which a date has been assigned earlier than that of the invention of Wove paper itself, in this case it was later declared a copy; or, as mentioned earlier, cases of books or engravings that have been reprinted on paper bearing a much later countermark than that of the original printing.

---

[30] Unless one is undertaking a search specifically aimed at recording countermarks in a given collection (e.g. Gravell's Bib.68) the findings of the individual historian will inevitably be limited. A helpful approach, however, would be for people using archives to note marks in the documents they were searching and that by recording these systematically an invaluable corpus of marks would be accumulated in time (perhaps a rather idealistic thought ?).
The early land tax returns, usually the first 3 or 4 years, are often a fruitful source of marks attributable to paper makers working in or near that parish. A good example of this was the discovery of Hugh Bennett countermarks in the 1780 Foots Cray Land Tax Returns (see Chap.VII note 182).

[31] An indication of this kind of penetration was noted by the author in documents originating in Norfolk and held by the Norfolk R.O; Sessions Order Books and Consistory Court Wills from 1750–91 were examined. Up till the early 1780's the marks are anonymous (e.g. Fleur–de–Lys/IV : Dutch Lion/GR : Pro Patria/Crown GR in circle); but from 1784/5 onwards 5 different Kent countermarks appear (d.o.d. order –– T FRENCH : J CRIPPS : R WILLIAMS : EDMEADS & PINE : and I TAYLOR). Only 1 Surrey mark (CURTEIS & SONS d.o.d.1788) and 3 Norfolk (J HAMERTON of Taverham d.o.d.1780 : monogram believed to belong to Mackglashan & Co., Swanton Morley : "TAVERHAM" by then under Anstead, 1794). This limited search showed quite an impressive invasion by Kent paper makers, an invasion by no means limited to records of this kind; and, significantly, mostly Maidstone paper makers.

[32] Numerous examples of British countermarks found in documents originating in America have been discovered and recorded by T.L. Gravell (Bib.16 and 68). Others are to be found in documents originating in India (and elsewhere) in the 18th C; and, more recently, in early 19th C. Europe (I am indebted to the late Mr. E.G. Loeber of Hilversum for sending me a wide selection of British countermarks borne in documents found in Belgium, Holland and Germany between ca.1806–36 including many genuine and forged Whatman marks).

[33] Another example is that of PORTAL & BRIDGES countermarks found in drawings by William Daniell made in India in 1788 (RIBA Drawings Colln). The papers in question must have been made as early as 1786, perhaps before this even, much earlier in fact than the date usually given for the formation of this partnership, the first known evidence of this being ca.1791 (Universal Directory). Earlier Portal countermarks appear to be uncommon, so it is perhaps worth recording here an "I PORTAL" mark in a Smeaton Drawing in the Library of the Royal Society on Laid paper [Box V/1 No.91], an undated document possibly early 1780's and belonging to Joseph Portal (act.1747–93).

Watermark dates were obligatory for drawback of Excise Duty on Exports made after 1794 (34 Geo.lll,c.20)[34] though this clause was repealed in 1811. To begin with dates were not always altered annually, though probably by the end of the first decade of the 19th C. this became an annual event; earlier than this dating can only be judged on the basis of accumulated experience of their occurrence; the irregularity in the sequence of dated Whatman marks, for instance, between 1794 and 1804 has by now been firmly established, sufficiently so that an uncertain mark found to be dated in one of the gaps would immediately be suspect.[35] (There seems to be no good reason to doubt the accuracy of the watermark dates recorded in this book).

If one has the experience, expertise and eyesight like the late Mr. E.G. Loeber, one can extend the subject of watermarks and countermarks a great deal further by examining the construction of the marks as seen from their impressions in the sheet of paper.[36] He was able to dissect many marks into their component lengths of wire, the bends made in these, the avoidance of or the flattening of potentially proud crossing points and the method used to sew them onto the cover. Using this technique requires very real expertise and enormous experience. Using techniques of this kind one might be able to establish, for instance, whether paper makers in an area like Hollingbourne shared the services of a common mould–maker or not. The technique is also useful in detecting forgeries.[37]

Although some of the illustrations used here really lie outside the subject of this book, they have been included to show how the countermark can provide the historian, archivist, curator etc. with useful information about:–

(i)   The identity of the paper maker and partners.
(ii)   Their location; and the mill name.
(iii) The early history and status of a mill from the discovery of unusual specimens in collections of parish documents, bills etc. relating to the locality of the mill in question.
(iv)   The pretensions of the paper maker.
(v)   Mould–makers; and Forgeries.

---

[34] The practice of watermark dating paper was introduced on the Continent in the 17th C. (allegedly on the instructions of Colbert); but these dates are usually irregular and unreliable. The occurrence of dates in British papers before 1794 is very rare. Two examples are known to the author (i) paper made by Nicholas Dupin in Dublin dated "1692" (Pollard, M. Bib.19 233); and (ii) the younger James Whatman dated his Antiquarian paper certainly as far back as 1784 (Drawing by John Smeaton in the Library of the Royal Society Box IV/1 No.79).

[35] See Notes on "W" marks App.V.

[36] Loeber, E.G. Bib.45 Text pp.31 ff; and Plates 79–81, 117. As a Filogranist he prefers the term "Wire Profile" to that of "Watermark" for the kind of marks discussed in these papers. This is in order to distinguish this kind of mark, made by bits of wire sewn on to the mould cover and thus making an impression on the wet sheet of paper, from various other methods of watermarking introduced in the 19th C.

[37] Loeber discovered recently moulds in a French mill mould store bearing certain very well–known British countermarks together with both 19th and 20th C. watermark dates. The moulds and the marks were constructed in a different manner to the genuine articles but deceptively similar to the inexpert eye.

THE SITE OF OLD MILL, HOLLINGBOURNE

Before coming to the Synopsis of the history of Old Mill as a paper mill, it might be appropriate to comment on the site as it is to-day. Some of the remarks will have a personal note in them since my experience of the place spans a period of well over 50 years. There could not possibly be a greater contrast in my response to it, however, between the time when I first knew it and now.

To-day there is nothing left of the mill-house and though it has been described as having been a fine old house there is no ruin of it left of which one could even say "plus belle que la beauté est la ruine de la beauté". Nevertheless the setting has a haunting air about it and the size and tranquillity of the millpond creates a landscape of considerable grandeur and beauty (see Frontispiece Vol.I).

It ceased working as a paper mill in 1848, 115 years after it was built as such. It continued to operate, however, as a corn mill until 1913 at least but had definitely been abandoned by 1922.[38] Spain mentions that Coles Finch visited the site in 1928 and again in 1929 in order to take photographs which unfortunately have been lost. It is worth quoting here a few lines from Coles Finch's description as given by Spain:-

> "The mill-wheel is idle, the mill-race choked with water plants and coarse growths. The ivy is claiming the fine old house and running riot over it. A few of the upper rooms shelter a poor family, the basement being entirely deserted and sadly neglected...........
>
> The mill, a huge five-storey erection, once busy night and day....is now silent in decay. Not a window is intact and the near apple trees thrust their trusses of pink and white blossom into the very openings, as if to cover the scars neglect and decay have made. I could find no one to grant permission to enter the old mill, no one had a care for it, not even to adjust the weir. When the winter floods came, the pond overflowed and cut a gully and thus gained the mill-race, moving the wheel in spasmodic jerks, which, as it slowly revolved, dragged round the rusty shaft and gearing."

During the great Depression, when so many fine rural buildings of one kind or another fell into decay, the prospects for their restoration were non-existent. By 1934 the deterioration of Old Mill was such that the building was no longer habitable by families however poor.

In 1933 I had completed my first stint in the laboratories at Springfield mill before going up to University so that, when my father who had no personal connection with the paper industry (only a family one) bought Caring House at the western extremity of Leeds parish (1934/35), I had already received my baptism in papermaking; but it was to be another 30 years or so before I was to profess any interest in the history of the industry.

Caring, which had been one of the principal mediaeval farmsteads of Leeds, looked down on orchards at the foot of which flowed a stretch of the River Len. We were surrounded, although we did not appreciate the fact at the time, by old watermill sites. Less than a mile downstream Thurnham mill[39] was still working and I passed it every day on my way to

---

[38] Spain, R.J. Bib.8 62/63.
[39] Spain, R.J. Bib.8 68.

work. It had replaced Ballard's mill, a little way upstream, which had fallen into disuse about 1829.[40]

Looking in an upstream direction from Caring one could see "Fulling Mill"[41] only 600 yards away. It had been a corn mill belonging to Leeds Abbey in the 12th C., had become a fulling mill by the beginning of the 17th C. but had quite possibly been reconverted to some other function during the 18th C. because it appears to have been working as a watermill as late as the 1820's.[42] In 1934 the mill-house was empty and in a derelict state (but restored since and inhabited). Looking beyond this, but in the same direction, the terrain surrounding Old Mill could be seen just over three-quarters of a mile away but not the mill itself, which was hidden from view by the gorge in which it lay. Even closer, along the road to Leeds village, lay Spout House occupied by one of the Hollingworths between 1810-1817; and Caring House itself had been the home of the Saxby family from an early date. They figure prominently in the parish records at a time when Whatman was building his paper mill.

In the 1930's I and my younger brother were great walkers and frequently we would follow a route through the orchards below Caring, across the Len, along the farm road and over the river again past Fulling Mill and up the hill and thence by a footpath that led down to Old Mill. In those days the site, in contrast to to-day, was almost treeless. This had a decidedly flattening effect on the landscape. Although the site lies in a ravine, the absence of trees, which now accentuate and dramatize the valley at this point, gave the place a prosaic appearance. From memory, the millpond was choked and bordered with rushes so that one hardly noticed it. The only prominent feature was the mill itself and, though it had the ruinous appearance described by Coles Finch, it is conceivable that even at this late stage it could have been restored; the rusty waterwheel was still there beside it. But the scene lacked a Humphry Repton simulation of an attractive setting that might have encouraged someone to restore the building.

In the event the mill, its contents and the waterwheel all disappeared sometime during World War II. All that remains now is the massive dam with the waste sluices at its northern end (see Map 6 p.xvi), the top of the dam being nearly 16 feet above the level of the water as it issues below. Of the mill only the foundations, covered with matted undergrowth and nettles, may be discerned; the mill-race leading to an apron (8'8" in width : Spain has estimated a wheel width of ca.7 feet); a fall of some 14' 3" (Spain estimating a 12 ft. diameter overshot waterwheel); and the island below between the tail-race and sluice water a carpet of periwinkle. (See PLATE 17 overleaf).

Above, the millpond about 70 yards wide at its widest point extends eastwards for nearly 500 yards.[43] At the eastern end the Len enters through marshy ground beyond which may be seen what are perhaps the descendants of the osier beds that figure in the Poor Rate Assessments against the mill in the 18th C.

---

[40] Ibid. 67.

[41] This was the mill-house, the mill apart from the remains of some foundations had disappeared.

[42] Spain, R.J. Bib.8 64.

[43] See frontispiece (1982) Vol.I. The great hurricane of 1987 has altered this view felling many of the tallest trees.

Plate 17 – Old Mill, Hollingbourne (1984)

(photographs by author)

Fig. iii – THE EXIT TO THE NORTHERN SLUICE.

width of sluice at knees 6ft. 6in. (1.98m.)
height of dam ca. 16ft. 10in. (5.13m.)

Fig. ii – THE FALL

14ft. 3in. (4.35m.) high

Fig. i – THE APRON

8ft. 8in. (2.64m.) wide

It is a scene of great tranquillity with little more than the cry of wildfowl to disturb the silence. In terms of the history of papermaking it is a place of exceptional significance for it was here that the elder Whatman made his first and important investment in the industry just over two hundred and fifty years ago. In my view, it is a place that should be carefully preserved (it is already in danger of becoming an unofficial rubbish tip), but not publicized in a way that might destroy the peacefulness of its rural state. It could be maintained without too much difficulty as a fitting memorial to the once famous White paper industry of this country.

A note on the 2 Sketch—maps of the Old Mill site (Map 6 overleaf)

No early maps of the Old Mill site have come to light. The earliest map discovered to date showing Old Mill is the 1844 Tithe map for Leeds (sketch—map No.2) and this is clearly diagrammatic. The paper mill, which was still operating at this date, is shown as a single building; and the parish boundary as running along the northern bank of the mill pond. Just how much one can read into this is uncertain. The 1864/5 6 inch O.S. map (see Map 9) shows two buildings one on either side of the tail—race and the parish boundary running down the centre of the mill pond. As will be seen in App.I the Poor Rate Assessment against Old Mill switched from Hollingbourne to Leeds in 1808 and, possibly, for a period the parish boundary was regarded as lying to the north of the mill site. All later maps show the mill as lying in the parish of Leeds, but prior to 1808 the mill had always been assessed under Hollingbourne with the exception of the Osier beds, which came under Leeds. It is interesting to note, but confusing to the issue, that the Land Tax Assessments for the mill remained under Hollingbourne throughout.

Sketch—map No.1 is based on the 25 inch O.S. map of 1877–1885 (Sheets 43/6 &.10). The parish boundary, which is shown as running down the centre of the mill pond, across the dam and down the tail—race, has been omitted from this. As in the 1864/5 O.S. map it shows two buildings one on either side of the tail—race (A & B), but the northern one is indicated only in outline (not shaded) as if to indicate that it was perhaps roofless, a ruin or just a storage yard. The little structure shown just to the north represents steps leading down from the top of the dam. Old Mill cottages, just to the west of the mill site, are shown on the north side of the track; to—day they lie on the south side. The only other extant buildings shown on the 1968 25 inch O.S. map are the Outhouse (C) and the Shed.

Map 6 – TWO SKETCH-MAPS OF OLD MILL SITE, HOLLINGBOURNE

1. ca. 1880   (approx. to scale)
2. 1884   (not to scale)

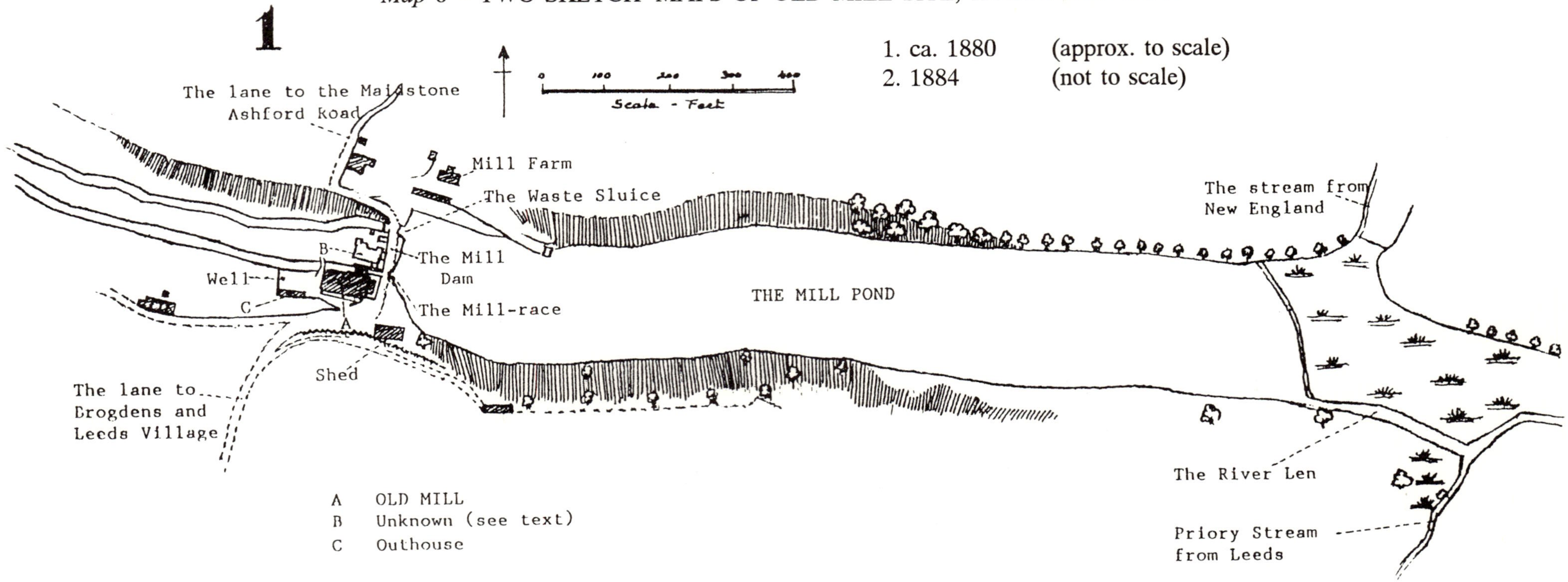

Scale - Feet

0   100   200   300   400

1

The lane to the Maidstone Ashford Road

Mill Farm

The Waste Sluice

The Mill Dam

The Mill-race

The stream from New England

THE MILL POND

Well

Shed

The lane to Brogdens and Leeds Village

A   OLD MILL
B   Unknown (see text)
C   Outhouse

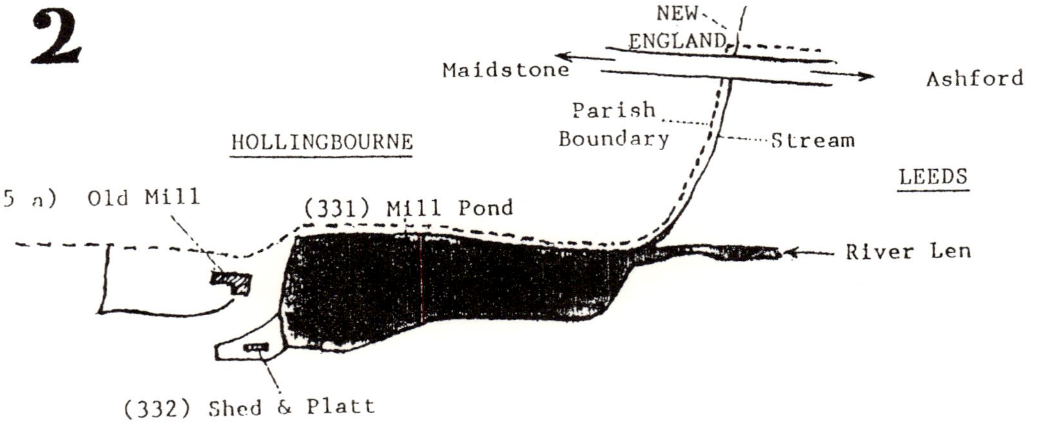

The River Len

Priory Stream from Leeds

2

Maidstone

Ashford

NEW ENGLAND

Parish Boundary

Stream

LEEDS

River Len

HOLLINGBOURNE

(335 n)   Old Mill

(331)   Mill Pond

(332)   Shed & Platt

SKETCH MAP based on the LEEDS Tithe Map for 1844 (Ref.296 & Apportionments L 4) Cathedral, City and Diocesan Record Office, Canterbury.

Owner of Land
Charles Wykeham Martin
Occupier
John Hollingworth

APPENDIX I

The Hollingbourne Paper Mills

OLD MILL, Hollingbourne, Kent MR TQ 820541 (Excise No. 302)

This synopsis of the history of Old Mill, Hollingbourne (also sometimes referred to as "Old Mill" Leeds and Mote or Moat Hole) is based on a number of existing accounts of the mill (listed below) as well as on a re-examination of local records and various other archives, corrected and brought up-to-date where necessary.

Principal sources of Information (other sources given in the tables below)

Spain, R.J.	Bib.8 60.
Shorter, A.	Bib.12 194/5.
Shorter, A.	Bib.7 (May) 54.
Balston, T.	Bib.1 Chapt.XV.
Kent County Archives	Various.
Balston Archives	PSM.
Guildhall Library Manuscript	Sun and Hand in Hand Insurance Records.
Greater London Record Office	

Sum Insured	Dates		References
		Spain considers it to have been a Domesday mill site. The last mill there had fallen into disrepair and had disappeared by the beginning of the 18th C. It was a magnificent site for generating power.	
	1654	Deeds describe this as a Fulling Mill; a reference also to a Pole Field. Confirming this there is an earlier reference to Wm. Dann, a Fuller of Leeds (1640).	Spain
	1723	The mill had disappeared. Owners of the site Roger and Hugh Meredith.	Spain
	1733	*Richard Harris* (James Whatman) Recital to a Tripartite Indenture of 1762 shows that James Whatman "did act and *build*... a dwelling house and water mill... on land leased from Sir Roger Meredith of Leeds (30:July:1733)". An Insurance Policy of the same year covers "a new built dwelling House and *Paper* Mill being one entire building in the Parish of Hollingbourne... now in the occupation of *Richard Harris,* Papermaker" exclusive of all other buildings.	KAO U 289 T 34 (see below)
£600		*James Whatman* insured house and mill Sept. 1733.	MS 11936/39 SFIP 61845 25:Sept:1733
		The significance of this act of building a paper mill has been discussed in the main text of this book; as also the importance of the Whatman/ Harris relationship in this context. (See respective Whatman and Harris Genealogies).	App.II
		Thomas Carter of Leeds, Richard Harris' father-in-law (and, in the future, James Whatman's also) may have supplied the timber for the building of this mill and advised on the choice of site.	App.II
		The Hollingbourne RB is missing for this period, nothing to indicate whether Richard Harris was living in the dwelling house or not. The Insurance Policy states that it was in his occupation. That he was living in the parish is supported by other evidence, the entry in the	GLRO P 92/ SAV/3039.

Sum Insured	Dates		References
		Marriage Register, St. Saviour's Southwark and his daughter's baptism in Hollingbourne 5:Oct:1735; and from 1733–1739 (first half) he paid dues on the osier beds that lay in the parish of Leeds as an Outdweller.	KAO P 187 1/1 KAO P 222 12/4
£600	1734	*Richard Harris* insured house and mill.	SFIP 65053 7:Oct:1734
£800	1736	*James Whatman* insured house and mill including Stables and Loft in the occupation of *Richard Harris*. The next (*consecutive*) Policy shows that *Richard Harris of Hollingbourne* insured his stock in trade in his Papermills and Store Rooms all in one building in *Boxley* and not elsewhere (£350) and on his Stock in a drying shed thereto but not adjoining (£50).	SFIP 72264 27:Dec:1736
(£400)		(Turkey Mill. SFIP 72265 27:Dec:1736).	
		No further policies have been found for either Old Mill or Turkey Mill until Dec. 1739. (The suggestion has been made that the insurance for these mills might have been made through the Royal Exchange Company whose records are missing for this period).	MS 11936 vols. 48–53 SFIP's. MS 8674 vols. 55–57 H. in H.
		The question arises as to who worked Old Mill during this period. In 1736 Whatman had insured the house and mill in the occupation of Harris and although the latter is not shown as an Outdweller in Boxley records neither is a dwelling house covered in the Turkey mill policy (no mention of Turkey Court built ca. 1680–1690).	
		Not until he made his Will in Aug. 1739 that Harris refers to "all that now built messuage or tenement wherein I now dwell." At some point between 1737–1739 he must have moved from Hollingbourne, though he continued to pay his dues (as shown in the Leeds RB) up to 1739 (first half). With the pulling down and reconstruction of Turkey Mill on his hands after 1738 it raises the intriguing question as to whether Whatman had had a hand in the running of Old Mill.	KAO PRC 17/91 Part I fo.4 KAO P 222 5/1
£1300	1739	*William Quelch* of Loose and *John Terry* of Hollingbourne insured house and mill in the occupation of Terry (£700) and their household goods and stock in trade (£600). (For the movements of Quelch and Terry see TABLES XXII & XXIII).	SFIP 82644 25:Dec:1739 App.II
£800	1740	*James Whatman, Paper Maker* insured Old Mill in the tenure of *John Terry*... house and mill. It is difficult to account for this action because within 10 days Quelch and Terry insured the mill again (See SFIP 86467 below : note, Shorter gives the old, uncorrected calendar date as 8:Jan:1740. Evidence supporting this correction is provided by the higher number of Quelch's Policy). One can only assume that a mistake had been made. (vide Policy for 1739).	SFIP 86382 29:Dec:1740.
£600	1741	*Quelch* (Loose) and *Terry* (Hollingbourne) house, mill and stock. (Turkey mill in the meantime was insured for £1700).	SFIP 86467 8:Jan:1741.
£500	1741	*John Sanders* insured the mill.	SFIP 89143 15:Oct:1741
	1742–5	Mr. Saunders was assessed in the Hollingbourne RB. (bankrupted in 1748). (See Extracts from Hollingbourne RB for Saunders' occupation, pp.28–29).	KAO P 187 5/1
£300	1746	*James Whatman* insured the mill.	SFIP 105423 30:June:1746.

Sum Insured	Dates		References
	1746–8	*Mr. Whatman* was assessed for "the pepare mill"; latterly as Mr. James Whatman *or* Occupier, but none of these three entries as an Outdweller. These last entries coupled with the low insurance may indicate that the mill was not being worked.	KAO P 187 5/1
		There are no records of Old Mill being insured again until 1770. Possible reasons for this have been advanced above.	
	1749	*Mrs. French* (£43–10) Hollingbourne (see separate notes on Extracts). This entry is of possible interest, though not as yet explained. The recital to the Tripartite Indenture (KAO U 289 T 34) referred to above and dated 1762 shows that in *1751* Whatman "did demise the said dwelling house and Water Mill and all other the said premises to *Mary French*, aforesaid Papermaker's executor etc. to hold from the feast of St. Michael then last past for the term of nineteen years at and under the yearly rent of Eighty Pounds". The original leasehold agreement between Sir Roger and Whatman had been for 60 years, uncertain why 19 years was specified as indeed the reason for Mrs. Mary French being involved.	KAO P 187 5/1
	1750–70 (1760– 70)	*Henry French* (snr.) was assessed in the Hollingbourne RB, particularly in this style during the later years, whereas in actual fact after 1759 the mill was run by *Henry French jnr.* Henry French snr. moved to Gurney's mill in Loose in 1760 and insured it (SFIP 176721 on 10:July:1760). Later he insured Gurney's mill in his son Thomas' tenure and a William Greenway (SFIP 293987 31:Jan:1771). Henry snr. died in 1775 and was buried in Loose[1]. Henry French jnr. described as a Paper Maker was buried in Hollingbourne in 1776.	(The Calendar was adjusted after 1752). For French's genealogy see App.II.
£600	1770	*James Whatman* (II) insured utensils and stock (note not the mill). The unanswered question is why did he not insure the mill ? He had inherited the freehiold[2] from his father in 1759 (although by the terms of his father's Will not directly while his mother remained life tenant : in the event the Estates were made over to him in 1762. Balston, T. Bib.1 129 and 16/17).	SFIP 290474 25:Oct:1770.
	1771–4	*Whatman* is *not* entered as an Outdweller in the Hollingbourne RB, the assessment remaining unchanged at £40 indicating that the mill was being worked by him again. This is confirmed by the fact that in 1771 he added a new Whole–Stuff Engine to his 2 Half–Stuff ones and carried out experiments with his improved rag preparation equipment. It is worth noting this point as it might be the key to understanding some of the changes and events that follow.	For an account of these see App. V.

[1]   Shorter, A. Bib.12 195 noticed that the Leeds RB entries between 1749–60 refer to *Mr.* French and thereafter to Henry French. That this difference is significant is shown by the Hollingbourne and Loose Parish Registers (see French Genealogy, App.II).

[2]   In the first place James Whatman I leased the site of Old Mill from Sir Roger Meredith in 1773 for 60 years; but at some later date, not known but evidently in his lifetime, he must have acquired the freehold of that part of the site that lay in the parish of Hollingbourne (see Land Tax Extracts for the 2 parishes, pp.28–33).
Under the terms of his Will (PRO/PROB./11/848), drawn up on 9:June:1759 only 20 days before his death on the 29th June, he left instructions to his widow that when his son reached the age of 23 "I do recommend and so desire my said Wife to surrender up to and put him [James II] in the Trade or Business of paper making in the aforesaid Mills [Turkey and Hollingbourne, the latter not specified in the Will]..." In the normal course of events this would have been in 1764; but in the event Ann, his mother, handed over the business to him in 1762.

Sum Insured	Dates		References
£1400	1774	*Clement Taylor jnr.* (Proprietor of Upper Tovil Mill, Maidstone[1] from 1772) insured Old Mill.	SFIP 341155 24:June:1774
	1775	Mr. Taylor not assessed as an Outdweller.	
	1776–93	*Mr. Taylor* assessed as an O.D. for the mill (see Extracts from RBs and Land Tax Records for both Hollingbourne and Leeds).	
£9500	1791	*Clement Taylor* insured the mill again.	SFIP 591378 3:Nov:1791.
		The dramatic increase in the Insurance Premium should be noted rising from £1400 to £9500 for the 17 years of his occupation. Does this reflect Whatman's improvements to the rag preparation equipment installed in 1771 or an enlargement of the paper mill undertaken by Taylor ? In 1739 (see above) Quelch and Terry had paid a premium of £1300 which covered the house and mill as well as household goods and stock in trade. Taylor's £1400 might have covered much the same and not included the new Whole–Stuff Engine (see next para.).	
		Without any of the intervening premiums it is impossible to say whether Taylor undertook installations and improvements of his own progressively during this period or all in one. All that can be said here is that (a) by 1848, when the mill was laid still, Old Mill had 5 vats compared to possibly 2 vats at most in 1774. Shorter (Bib.12 191) pointed out that Clement Taylor jnr. had enlarged Upper Tovil mill in 1772–3, probably to a 5–vat mill. With a general expansion of the industry in the period that followed he could have started enlarging Old Mill as well; and (b) he might either have paid Whatman for the use of this new Whole–Stuff Engine or enlarged the mill through services provided by Whatman, either of which would account for the mysterious £400 paid annually by Clement Taylor to Whatman, to which Thomas Balston refers, between 1780–91.	Balston, T. Bib.1 54 & 130.
		The Land Tax assessments from 1780–1820's remain constant, so there does not seem to have been any increase in the size of the property. In contrast to this, whereas the Ratable value was doubled in 1776 for all properties, that for Old Mill increased dramatically in 1794, rising from £80 to £150 with no comparable increase for Eyehorne mill. Under Hollingbourne this situation remained constant up to 1809 and, though complicated by other factors, later under Leeds. It would seem that this rise points to a significant increase in the facilities and size of Old Mill during the period 1771–1791.	
	1791 ff.	With an enlarged Upper Tovil mill, probably an enlarged Old Mill and as MP for Maidstone Taylor must have required assistance in running an outlying mill like Old Mill. As pointed out in Chap.VII and later in this Appendix and App.II a good case can be made out for *Daniel Newman* acting as manager from 1791 (and possibly some association also of Lewis Munn and Henry Salmon with this mill at an earlier date). Certainly, later in this decade and by 1797 Newman had gone into partnership with a member of the Crispe family, evidenced by a CRISPE & NEWMAN 1797 countermark, a date 3 years before Park mill came into being. As Newman had been living in Hollingbourne since 1793 (and earlier in Leeds), apart from Eyehorne under Robert	See Table XIII

[1] For information on Clement Taylor snr. (William Quelch's son–in–law) and his family see App.II

Sum Insured	Dates		References
		Williams, there was nowhere else but Old Mill where these two could have made their paper.[1]	
		Appropos of the above Shorter had noticed that Newman's name came next to Clement Taylor's in the Leeds RB from 1791 and assumed from this that Newman was the Master Paper Maker at Old Mill (which he undoubtedly was later), but produced no evidence for this. The entries in fact show that Newman followed Robert Salmon (father of Henry & Thomas) in 1791 and that this assessment, with a small addition in 1808, continued to 1809. Although seemingly continuous in 1810 this is replaced by a much larger one against Stroud & Newman, Stroud being Newman's current partner at Park mill (q.v.). Later, in 1813 when Stroud had left the area, this assessment was made alternately against Daniel Newman and his son, Denny, up to 1823 indicating that this was not only a different property but one associated with Park mill.	Bib.12 195 See pp.32–33; and App.II for Salmon
		The interpretation of this data (given in greater detail in the preamble to Park mill) is that the changes in the ratable values given above, though seemingly a continuous sequence, refer in fact to assessments made against 2 different properties, the confusion arising from a change in the parish boundaries which must have taken place in 1808. The ratable value of Old Mill which had been £149 under Hollingbourne (1795–1807) fell to £110 under Leeds in 1808/9. This figure was complicated in 1810 by further assessments made against the Hollingworths for "cottages", "Spout House" and an unexplained reference to "part of New Mill at New England" discussed later on p.26. All this points to the property formerly occupied by Salmon and, later, Newman, as belonging to Old Mill; soon after the boundary change this assessment must have been absorbed under the "cottages" and its place taken in the RB by a totally different property linked with Park mill (in Hollingbourne) but lying on the Leeds side of the parish boundary. All this ties in with the current occupants of Park mill, Stroud & Newman, who were still working Old Mill, with Hollingworth clearly resuming responsibility for running what was his mill in 1810. It is suggested that the Old Mill cottage had originally been occupied by Salmon's 2 sons as trainees, with Henry possibly as manager from 1786 followed by Newman as manager in 1791 but not occupying it as a Master until his partnership with Crispe. This could have taken place when Balston & Hollingworth took over the mill from Whatman in 1794/5.	See PREFACE Map 6 and p.xv See Map 7. App.II
	1793–4	*Whatman* may have temporarily occupied the mill again but only in name. There is confusion in the evidence regarding the transition from Taylor's to the Hollingworth/Balston "occupation". The RBs. for Hollingbourne and Leeds together with the respective Land Tax Returns are at sixes and sevens. Hollingbourne RB shows Whatman 1794, the Land Tax 1793–4; Leeds RB implies 1793–4 and LT 1795. The confusion is increased by entries in Whatman's accounts, Thomas Balston believing that Whatman may not have owned the mill after	Bib.1 130/1

[1] Shorter (Bib.7 (May) 55) argued that entries in the Leeds RB against Newman and "Crispe" pointed to Park Mill having been assessed for a time under this parish. The current search contradicts this assertion (see preamble to PARK MILL and App.II under CRISPE). The only mill assessed at this time in Leeds is made against Edward Crispe (£10 "more for the mill") which clearly refers to Abbey Corn mill (Spain, R.J. Bib.8 58 cites an indenture of 1789 KAO U 855 T 1/4 to this effect).

Sum Insured	Dates		References
		1788. He suggests that after 1791 Calcraft had somehow acquired the freehold and that Whatman had leased the mill for these final years.	
	1795	*Hollingworths & Balston* (actual entries, RB Messrs. Baldstone & Hollingworth; LT Bostone Finch & Co. for Hollingbourne; all the Leeds entries are for Hollingworth only). Whatman's Account book (1796) puts the date for this transfer as 23:Aug:1794, thus ending a 61-year association with this mill which his father built 1733, an event that was to prove of great moment not only to his own career but to the development of the White Paper Industry in Kent.	
		Although it has been suggested above that by this date Calcraft may have acquired the freehold, the fact is that it is not known from whom Hollingworth and Balston acquired the freehold themselves. Calcraft continued to own the Leeds portion of the property until well into the 19th C. (see Leeds Land Tax extracts, pp.32–33).[1]	Bib.1 130/1
	1795– 1804	*Hollingworths & Balston.* Although assessed as such, it has been demonstrated that the mill during this period was worked by *Newman* et. al. Reference has already been made to a CRISPE & NEWMAN 1797 countermark and others have been found prior to 1800.	See TABLE XVIII (p.25)
	1805– 1848	*Hollingworths.* The partnership between Balston and Hollingworth came to an end and the latter retained control of the mill for this period.	
	(1805– 1848)	William Balston, contrary to Thomas Balston's account, did *not* operate Old Mill in 1805; he occupied Eyehorne mill (see next Synopsis) and was followed there by the Hollingworths for a further 3 years (up to and including 1808). Old Mill was not vacant or available to either Balston or the Hollingworths for this period, the evidence pointing to Newman and his current partner (Stroud in 1805) still working it.	Bib.1 131
		At some point, as yet undetermined, the Hollingworths resumed the control of this mill, probably some time between 1810 and the end of 1812. In 1810 (Leeds RB) Finch Hollingworth was assessed for Spout House, just over half a mile from Old Mill. By the end of 1812 the Newman and Stroud partnership came to an end, Stroud moving to Pratling Street, Aylesford. He continued, however, to live in Leeds parish as a farmer. The fact that Stroud left at the end of 1812 must indicate that Hollingworth had already resumed control of Old Mill and that it was no longer worth his while making paper at Park mill without the resources of Old Mill to back this up.	See App. II & III

---

[1] After 1804 the Hollingbourne Records no longer mention Hollingworth and Balston in association with Old Mill; the partnership came to an end in 1805. Balston (Bib.1 120) states that the partnership had bought 8 vats from Whatman (1 Loose, 2 Poll and 5 Turkey mills) and he suggests that Old Mill had passed to them on Mortgage (Bib.1 131); but Balston's reference here to 9 vats is far from clear, particularly as he goes on to quote a letter of Susanna's to William written in 1810 referring to the "profits from the three vats which I had from Hollingworth". The questionable validity of these statements is discussed further under Eyehorne mill (p.14[1806] para 3).

With reference to Calcraft's ownership of the freehold, the Leeds part of the property had passed to Charles Wykeham Martin by 1844 (Tithe Map).

Sum Insured	Dates		References
	1848	The Excise General Letter for May 1850 states that Mill No.302 ceased working; but the Original Papermakers Recorder for May 1902 gives a more precise date – 14th:Nov:1848 – when the vats were laid still. Four apprentices working there went to Turkey mill, J. & T. Green (brothers) and J. & W. Proctor (brothers). W. Proctor died in 1904 and he laid the last post at Hollingbourne Old Mill, No.3 Stamps, which were afterwards made on the machine at Turkey mill. Enlarged probably by 1791, it was certainly a 5–vat mill when it closed.	Bib.7 (May) 55.
	(1850)	Old Mill was converted to a Flour mill.	

Mill Numbers

Unlike the case of EYEHORNE and PARK MILLS there is no evidence that OLD MILL was ever reconverted to a paper mill. When it shut down in 1848 it lost its Excise Number 302.

The Excise Authorities had a way of re–allocating disused numbers to mills sometimes in totally different Collections and consolidating them in others.

It is worth noting (in connection with the other two mills) that at some point before 1861, when the Excise Duty on paper was repealed, Eyehorne and Park mills must have been allotted Nos.301 and 302 respectively, which suggests that Park mill at least may have been operating after 1848/50 (for details see the Synopses of the history of these two mills). No.301 formerly belonged to Poll mill, which had been pulled down in 1836.

Countermarks (Old Mill)

Only one set of Countermarks can be attributed with absolute certainty to a paper maker who worked at this mill. These belong to Henry French (see below). With reasonable certainty one can also attribute the Crispe & Newman countermarks appearing before 1800. Otherwise most of the occupants of this mill were operating other mills at the same time, sometimes their principal mills. Under these circumstances it is unlikely that their countermarks could be identified positively as emanating from Old Mill. These paper makers include Whatman, Clement Taylor and the Hollingworths.

Apart from the more positive attributions the search has posed one or two questions and these are discussed in chronological order below:–

(1)   An "R H" countermark illustrated below found in a document dated Feby.19th 1733 (corrected date 1734) and originating in England, was discovered by Mr. T.L. Gravell (Delaware) in a document in an American Archive (Bib.16. 19:Feb:80 – 6:June:80).

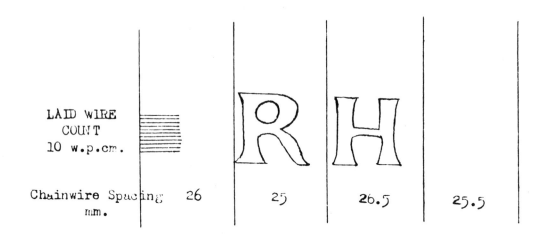

Traced from a light sensitive proofing paper copy of the countermark kindly supplied
by Mr. T.L. Gravell and with whose permission it has been reproduced above.

This countermark has been mentioned because it is just possible that it might have been a Richard Harris countermark, although there are several reasons which throw doubt on this suggestion but do not condemn it outright.

First, if one assumes that the mill was operating by September 1733 (the first insurance date), though possible it would not have left much time (4–5 months) in which to make the paper and for it to have reached the user in February 1734 (cf. the case of Whatman/Walsh in Chap.III).

Second, there are three other possible candidates amongst English paper makers of the time, that is to say if the paper was made in England. (The information below is taken from Shorter, A. Bib.12).

(a) Richard Hinton	Down mill, Cobham (Bib.12 237). The mill was burnt down in 1733; and some evidence that Hinton was the paper maker there.
(b) Richard Heritage	Coltnett mills, Wraysbury, Buckinghamshire (Bib.12 147). He was there in 1699 but gone by 1725. The Company of White Paper Makers issued a licence to the master paper maker, Mercier, in 1687 which suggests that the mill may still have been a White mill in Heritage's time.
(c) Richard Heath	Hurcott mill, Worcestershire (Bib.12 246). The only mention of Richard is in 1715; by 1756 it is Charles Heath. In view of the fact that it was producing watermarked paper by 1743 (Bib.9 57) it might have produced countermarked paper also.

The letter is signed by a Mr. Lionel Lyde. Lyde or Loyd may be a West country name and, since the goods with which the letter is concerned were shipped on the Eagle, a Liverpool ship, Heath may be a more likely candidate than either of the others or Richard Harris.

Third, the lettering of this countermark has a continental appearance; but it must be said that, so far, it has not been possible to identify "RH" with any contemporary continental paper maker.

(2)   *The Henry French countermarks, a shared Mould Maker ?*

The subject of "W" countermarks, and in particular the single line "W" examples, is discussed further in App.V. In the present context attention is drawn to these single line marks because several of the Henry French marks, found within watermarks of a shield under crown type (e.g. Britannia or Pro Patria ejusque Libertate), are similarly formed in single line lettering.

Those discovered to date are found in documents dated between 1760–69. There are in addition 2 rather later Thomas French marks with a single line "F" placed above the tail of the Lion within the Pro Patria watermark and exactly similar to the Whatman examples discovered by Mr. Gravell (App.V Fig.14). The "F" marks are found in 2 documents, a 1781 Maidstone Land Tax/Loddington Ward Return and in a Croydon Parish document also of 1781 (for details and further discussion see App.V).

The occurrence of the rather crudely formed single line "W" marks, on present evidence, seems to terminate in the early 1770's. They appear to have been superseded by more professionally formed "W"s found so far in documents between 1765–1777 (illustrated left and App.V). It is possible that the pressures arising from increased production (later manifested in his occupation of Hollingbourne and, even later, Upper Mill, Loose) may have required the younger Whatman to employ the full–time services of a Mould Maker to cope with the increased demand; his output by this time must have been equivalent to 5–6 times that of the French family (though this disparity would have lessened after 1760). It is known, for example, that by the 1780's Whatman employed Thomas Harris, as a mould maker.

To account for the close similarity of the early single line countermarks used by Whatman and French, it is suggested that they may have shared a mould maker, and that, later, French continued to employ him while Whatman engaged a new one. This would certainly help explain the positioning and use of the much later single line "F" used by Thomas French, though there is an alternative explanation to this, namely that Thomas French may have acquired a pair of moulds discarded by Whatman (App.V).

(3)   *Crispe & Newman Countermarks before 1800*

A number of these countermarks with watermark dates of 1797 and 1798 have been discovered, and discussed already in the synopsis of the history of this mill (see also Park mill).

TABLE XVI – TABLE OF COUNTERMARKS FOR OLD MILL

d.o.d.			Sources
1760	H F	Under a Crown within a frame surrounding a Britannia Watermark with Pro Patria Ejusque Libertate (as in Balston, T. Bib.1. 163 fig. 22 illustrating a similar single line "JW" used by Whatman from 1750). Single line.[1]	Leeds KAO P 222 5/9
1761, 63, 69.	H F	As above, single line HF. Truncated in the 1763 doc.	P 222 5/10
1763/4	H F	As above, single line.	Harrietsham P 173 5/2 Pt.I
1965	H∘F	Lion Passant and Vryheit : single line.	P 173 5/2 Pt.I
(1766) (1770)		(In the same archive, 2 single line Whatman countermarks were discovered, illustrated below. Both were in the surround of Britannia watermarks. In the first example the single line "W" was slightly off centre and a faint mark to its left suggested that there might once have been a "J" there too; but *not* in the other example).	
1779	HF	Large double line lettering (~1 cm. or more high).[2]	Leeds KAO P 222 5/11.
(1769)		A truncated "H" in the same style might have belonged to a similar "H F".	P 222 5/10.
before 1800		CRISPE & NEWMAN countermarks see under PARK MILL.	

[1] The paper used for the document dated 1760 must have been made in the 1750's and thus by Henry French snr. at Old Mill. It is probably safe to assume that the other single line "H F" countermarks, found in documents belonging to the neighbouring villages of Leeds and Harrietsham, were also in papers made at Old Mill.
The provenance of the paper used for the documents containing the large double line "H F" countermarks might have been, from either Gurney's (Loose) or Old Mill. As they belong to the same series of Leeds village documents as the others (and since Thomas French had started using his own marks), it has been assumed that they too belong to Old Mill papers. (See also Note 2 below).
Gravell, T. Bib.68 13 (No.336) records a single line "HF" mark. d.o.d. 1781.

[2] Shorter, A. Bib.12 303 Fig.62 records a large double line countermark "H FRENCH" of 1768.

Map 7 – A ROUGH SKETCH OF EYEHORNE & PARK PAPER MILLS, HOLLINGBOURNE

(Based on the Hollingbourne Tithe map of 1839 [ref. TO 1117] Cathedral,
City and Diocesan Record Office, Canterbury).

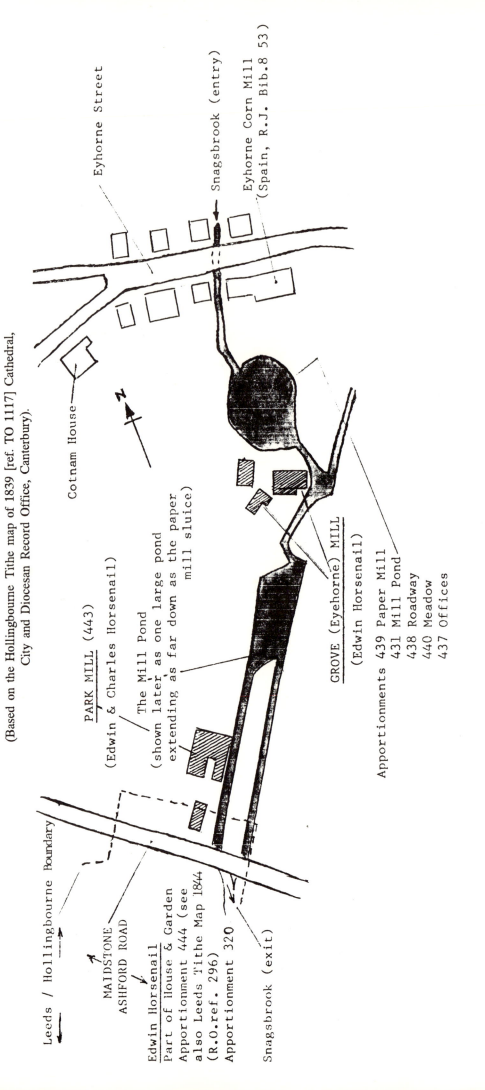

Eyhorne Street

Snagsbrook (entry)

Eyhorne Corn Mill
(Spain, R.J. Bib.8 53)

Cotnam House

N

PARK MILL (443)
(Edwin & Charles Horsenail)

The Mill Pond
(shown later as one large pond
extending as far down as the paper
mill sluice)

GROVE (Eyehorne) MILL
(Edwin Horsenail)

Apportionments 439 Paper Mill
431 Mill Pond
438 Roadway
440 Meadow
437 Offices

Leeds / Hollingbourne Boundary

MAIDSTONE
ASHFORD ROAD

Edwin Horsenail
Part of House & Garden
Apportionment 444 (see
also Leeds Tithe Map 1844
(R.O.ref. 296)
Apportionment 320

Snagsbrook (exit)

The Tithe Map does not include OLD MILL.

Apportionments at NEW ENGLAND are 366 House & Garden
371)
370) surrounding land.

(Not to Scale)

EYEHORNE MILL, Hollingbourne MR TQ 835545 (Excise No.306)

Principal Sources of Information (Other sources given in tables below)

Spain, R.J.	Bib.8 53.
Shorter, A.	Bib.12 195/6.
Shorter, A.	Bib.7 (May) 55.
Kent County Archives	Various.
Balston Archives	PSM.
Guildhall Library MSS.	Sun Fire Insurance Records MS 11963.

The reason for using the name Eyehorne mill (known also as Grove mill) is given in the countermark section following the Table below and distinguishes it from the Corn mill illustrated in the sketch−map opposite. The latter is the earliest (rough) map known of the two mills illustrated and differs in some respects from later maps e.g. in the shape of the Park mill pond and the fact that the house adjoining Park mill lies entirely within the Leeds parish boundary.

Sum Insured	Dates		References
£1500	1762	*James Austen* of Chatham insured his *new−built* paper mill. The Austens were proprietors of this mill up to 1780. James Austen 1762−68 : Mrs. Austen 1769−80	SFIP 192741 28:Sept:1762.
	1763	*Mr. Austin* assessed for the paper mill in the Hollingbourne RB (first mention).	KAO P 187 5/1.
	1764−74	*William Avery.* First RB entry for him is 1764. William Avery insured his utensils and stock; Mr. Austin insured the mill in the tenure of Avery.	SFIP 208828 SFIP 208829 28:Apr:1764.
	(1769)	Mary Austen of Maidstone insured the mill in the tenure of William Avery.	SFIP 271812 8:July:1769.
	(1774)	William Avery appears in the RB for the last time. A "WA" countermark found in a Leeds village document dated 1779 probably indicates that he was the paper maker at Eyehorne mill during the 1770's. This is mentioned because Shorter suggested that Williams might have succeeded Avery earlier than 1775.	KAO P 222 5/11
£400	1775− 1803	*Robert Williams* was the occupant of this mill to the time of his death in 1803. He insured his stock and utensils.	SFIP 355470 29:June:1775
		It is possible that before he became the occupant of this mill Robert Williams may have been working for Avery. In 1764 "Robert Williams" insured a house, of which he was tenant, in Eyhorne Street (SFIP 205630 24:Jan:1764). For the years 1765−67 there is an entry in the RB under Outdwellers which reads "late Williams". In 1768 (1st entry)− 1774 Robert Williams is assessed for £2−10.	(See note on Robert Williams p.19)
		The earliest "R WILLIAMS" countermark discovered so far is in the Boxley RB for the year 1780. To indicate the early distribution of his paper an "R WILLIAMS" countermark has been found in Consistory Court Wills in the Norfolk R.O. (d.o.d. 1787). Four (R) WILLIAMS countermarks have also been	KAO P 40 12/4

Sum Insured	Dates		References
		discovered by Gravell in American documents, 1 dating from 1785. (See also Table p.18.)	Bib.68 23.
	(1782)	Richard Barnard had become the proprietor of this mill by 1782. (The 1781 Tax Record is missing).	KAO QRP1 180
	1804	"Heirs of Robert Williams" RB entry; Williams died the previous year.	KAO P 187 5/1
	1805	*William Balston*. The RB entry is "Baldstone, late Williams". The assessment remained virtually unaltered. Although there are no Land Tax Returns between 1803–1810, there is no reason to suppose that there had been any change in the ownership of the mill. Richard Barnard was proprietor in 1803 and in 1810. Balston and his successor were tenants.	
		Thomas Balston, in his chapter on Hollingbourne Old Mill, states that the Hollingworths let Old Mill to Balston in 1805 after the breaking up of their partnership. But there is no evidence to support this assertion. On the contrary the evidence in the Hollingbourne and Leeds parish records contradicts this.	Bib.1 131
	(1806)	In both the correspondence between Catherine Vallance and William Balston during their engagement and while the building of Springfield mill was in progress and the correspondence between the Hollingworth brothers and Balston, the references are always to Hollingbourne, nothing more explicit than this. The anxieties that Catherine felt for William riding out there and the enquiries the Hollingworths made about "Stroud's mill" make no reference to Old Mill or Mote Hole. (See also notes on Thomas Stroud). This state of affairs continued until July 1806 (just before William and Catherine's wedding in August).	PSM. PSM. App.II.
		In May 1806 the Hollingworths wrote to Balston saying "We have been in expectation of meeting you in order to arrange settling the business of your Inventory of the effects at Hollingbourne...it will be necessary for Mr. Burr to look over the lease". It is highly improbable that the Hollingworths would have wanted Mr. Burr to look over the lease of *their own mill*. The next RB entries make it clear that the lease and the inventory referred to Eyehorne mill (the inventory has been reprinted in Bib.1 131/132).	PSM.
		Now that the assertion that William Balston leased Old Mill in 1805 is considered incorrect (Bib.74.41 no evidence given) it is not valid to transpose the claim that Old Mill was a 1–vat mill (Bib.1 131) to Eyehorne where it is known for certain that Balston had the lease in 1805 despite the fact that the latter may only have had 1 vat. Thomas Balston's evidence presented for Old Mill (Susanna Whatman's ref. to William's 3 vats in 1810) is conflicting (cf. Bib.1 131 1 vat v. Bib.74 41 5 vats no evidence given for either of these claims).	
		Under Countermarks (see p.18) it will be noticed that there is a "J WHATMAN W (Cursive) 1805" example, which might have originated from Eyehorne mill under Balston (also discussed in App.V. on p.276).	Springfield Collection.

Sum Insured	Dates		References
	1806–8	*Hollingworth*, late Williams, is the entry for the Paper Mill. The assessment remained unchanged at £80. Hollingworth's occupation adds further confirmation to the proposal that Old Mill was being managed by Stroud and Newman. Again, under Countermarks, it will be noticed that there is an unusual example of a "Hollingworth" mark dated 1807, which may well have emanated from Eyehorne mill.	
	1809–36	*Richard Barnard snr. and his relations* worked this mill. The Barnard connection with this and, from 1819, Park mill (q.v.) is a complicated one.[1] Richard snr., hitherto owner of this mill, mortgaged it ca.1810, his nephew Richard jnr. remaining as tenant and clearly the Master Paper Maker as indeed later at Park mill also. Richard snr. was a tanner by trade but must have retained an element of control over	
	(1819)	this situation because the June 1819 EGL. shows Richard snr. and jnr. as occupants of both Eyehorne and Park mills. Richard snr. died in 1825 leaving his estate, to be held in trust by his widow during her lifetime, to his nephew. John Maxsted, Richard snr.'s brother-in-law was by now the mortgagee (see below). The Land Tax Returns show Richard Barnard, "Proprietor".	KAO/QRP1/180
	(1831)	Richard Barnard, Paper Maker of Hollingbourne, was bankrupt. This must refer to the operations at Park mill since (i) the Land Tax Returns for Eyehorne mill (1828–32) show "Richard Barnard/Himself" *Paper Mill* taxed at £40 as it had been since 1780 (in Mrs. Austen's day); and (ii) "RB" countermarks continued over this period.	See Countermarks section
	(1833) (1835) (1836)	Richard Barnard re-instated at "New Mill" (Park Mill), but he was made bankrupt again in 1835 and the mill was temporarily laid still in 1836.	EGL Feb.1836 re Mill 307
	1836	At the same time as the above Eyehorne mill (No.306) "left off work".	
		This seems to have been the end of the "Richard Barnards" paper-making activity in this area although "Richard Barnard" appears to have remained (in name) *owner* of the mill up till 1839, if not as late as 1841 (an 1841 map of Grove mill and surrounding land shows it in his name though in fact it may have belonged to Maxsted as Mortgagee; Richard jnr. appears to have left (1836) and his aunt died in 1837.[2] Richard jnr.'s eldest son, Richard (b.1812), could have been the new "owner" but not a paper maker). This, however, was not the end of the name Barnard here (see below).	Shorter Bib.7 (May) 55. Hollingbourne Tithe Map 1839. KAO U 442 P69 Inf.C.Boas.

[1] I am indebted to Mr. Clifford Boas (of Loose) who has been carrying out extensive research into the history of the Barnard family for this new and recent information. Only items that are relevant to papermaking at these Hollingbourne mills have been included. But in passing it may be noted that Mr. Boas also has information to the effect that steam operated papermaking machinery had been installed at Eyehorne mill sometime prior to 1831. The steam raised was probably used to power the beaters and plate glazing machinery but *not* a papermachine.

[2] In spite of Mr. Boas' detailed work the Barnard situation has not been fully resolved. It was thought that Richard jnr. might have moved to Priddy near Wookey Hole (see under Countermarks). A Richard Barnard died there in 1836, but does not appear to have been Barnard jnr. though possibly related ? His aunt, who had a life interest in the Estate died in 1837 and, again, it is not clear who succeeded her though Maxsted must have remained the Mortgagee.

Sum Insured	Dates		References
	1836–7	*Edwin Horsenail* took over the papermaking at "Grove" mill (and in the following year that of Park mill also). His brother Charles joined him in 1839.	
	1840	*James Barnard*, according to Pigot's 1840 Directory, was the paper maker at Eyehorne mill. (It is not known who this "James" was, but clearly another member of the large Barnard family).	Spain Bib.8
	1842–9	*Timothy Healey* was the paper maker at Eyehorne mill; the EGL for Mar.1849 records that the mill had "left off".	
		An advertisement (Newspaper not identified) in the possession of the Maidstone Museum refers to the sale by order of the trustees of the Will of Mr. John Maxsted on the 25th November 1847 of a valuable Paper Mill situate at Hollingbourne. It is thought that this must refer to Eyehorne mill from the description of the property.	
		Both the 1864/5 6–inch O.S. and the 1877 25–inch O.S. maps of Hollingbourne show Grove and Park mills as *Corn mills.*	
		Spain, quoting from Bagshaw's 1847 Directory, mentions a William Clay as Mill board and Hollingworth as Paper manufacturers in Hollingbourne as well as 3 Corn mills. It is not known which mill Clay worked, probably Park mill (see under next synopsis).	
	1880's	In the 1880's it appears that papermaking was revived at both Grove and Park mills. Writings, Banks, Drawings and Ledger papers were made at the former; and similar papers including Blottings at the latter. This "New Hollingbourne Paper Mills Co.", which used the countermark "H B Barnard", was wound up in 1887. At that time they were using mill numbers 301 and 302. (See further comments under Countermarks).	Shorter Bib.7 (May) 55

COUNTERMARKS

The spelling of the mill name adopted here is Eyehorne (initially used by Shorter Bib.12 195) instead of the place name EYHORNE, but altered by him later to Eyhorne (Bib.7 55). The latter is the generally accepted spelling but in the past this has been variable (see examples below).

Until more countermarks have been discovered, it has been decided to retain the spelling EYEHORNE for the paper mill partly to distinguish it from the Corn mill further upstream and partly because this may also have been the form used early in the 19th C.

EYHORNE	Harris, Hasted and Phillipot Histories. Ordnance Survey, modern and 19th C. (with the exception of the 1801 O.S. when the clearly phonetic "Iron" Street was used).
EYEHORNE	Excise General List for 1816 and Shorter, A. Bib.12 195.
EYEHORN	4 paper mill countermarks (see next page).
EYHORN	1 paper mill countermark (see next page).

Table XVII sets out a selection of the countermarks discovered (see comments on these below):–

R WILLIAMS	These are very numerous in local records. An "R WILLIAMS 1794" (Leeds LT Return of 1795) shows paper put to use almost immediately.
R BARNARD	There is a sprinkling of these and EYEHORN countermarks in the 1820's not shown in this Table e.g. in the Hollingbourne Land Tax Returns (KAO Q/RP1 180) R BARNARD for 1823–4; "EYEHORN 1824" for 1825. The "R BARNARD" countermark appears in Moulds (at the time of writing in the possession of Madame Tussauds) at Wookey Hole, Somerset. Formerly, the paper mill at Wookey Hole had been run by W.S. Hodgkinson & Co. Ltd., who took over the mill ca.1850. A disastrous fire there in 1858 destroyed buildings and invaluable records of the earlier history of this ancient paper mill (which dates back to 1610, when it was converted). At some point during their ownership, Hodgkinsons must have acquired the rights to use this countermark.[1]
J WHATMAN W 1805	The paper that the Hollingworths used to write to William Balston in 1806 is countermarked J WHATMAN plus date. A draft letter to Hollingworth written by William on the 25th May 1806 is countermarked "J WHATMAN 1805". The inventory of Eyehorne mill is on paper marked "J WHATMAN 1804" all normal countermarks. The "J WHATMAN W (cursive) 1805" might have been used to distinguish the paper made that year at Eyehorne mill (see Hollingworth countermark below). (An alternative explanation is discussed in App.V).

---

[1] In a letter written by Mr. Guy Hodgkinson (28.Dec.1949) to the late Mr. Alex Robertson (formerly a member of the Springfield mill staff) he said "We have no record that "R BARNARD" paper was ever made anywhere else, but you may be right in thinking that it was originally made at Hollingbourne", which shows how careful one must be in relying on verbal tradition.

Recently Mr. Clifford Boas (see p.15 n.1) discovered a Will of a Richard Barnard of Priddy, Somerset, about 2 miles from Wookey Hole, who died in 1836 (Will registered in 1837 in the Wells Court). On present evidence it seems doubtful that this was Richard jnr., but it is conceivable that he might have been a kinsman whom Richard jnr. had gone to live with (?) and that either he or a member of his family sold their rights to use this countermark to Golding & Snelgrove or their successors at Wookey Hole. Churchill, W.A. (not always reliable) Bib.3 48 records an "R BARNARD 1854" countermark, indicating that it might already have been put to use there (certainly not in Hollingbourne) by this date.

HOLLINGWORTH 1807   From 1794–1806 the Hollingworth/Balston partnership made use of the "J WHATMAN" countermark in their papers. From 1807–1859 Hollingworths used "J WHATMAN TURKEY MILL". The HOLLINGWORTH 1807 mark is unusual (in fact it is the only one known to the author) and because of its date it may well indicate that the paper had been made under unusual circumstances e.g. at a mill, such as Eyehorne, leased by the Hollingworths for a period of 3 years and not at one of their own.

TABLE XVII – COUNTERMARKS EYEHORNE MILL

d.o.d.	Countermark	Provenance	References
1779	W A [truncated] (William Avery)	Leeds Parish Document	KAO P 222 5/11
1780–	R WILLIAMS	Boxley RB starting 1780	KAO P 40 12/4
1781	R WILLIAMS	Leeds Parish Document	KAO P 222 5/12
1783	R WILLIAMS	Sutton Valence Tax	KAO Q/RP1 374
1787	R WILLIAMS	Norfolk Record Office	Consistory Court Wills
1799	R WILLIAMS 1795	Papers Springfield Mill	MPM Correspondence
prob. 19th C.	R WILLIAMS KENT	Springfield Collection	
1803	R W KENT	Leeds Parish Document	KAO P 222 12/24
?	(POSTHORN & SHIELD) : J WHATMAN : 1805	Springfield Collection	(see note above)
?	HOLLINGWORTH 1807	Library Royal Society	Drawing 188 Smeaton Collection, d.o.d incorrect
?	R BARNARD 1812	Turner Watercolours	T.B. CCIV, CCXI CCXIII, CCXXVI, CCXXXVIII, CCXXXIX CCLXXVIII
1828	EYEHORN o MILL	West Farleigh Parish Document	KAO P 143 13/1
1829	EYEHORN MILL	Broomfield 2 Parish Documents	KAO P 222b 5/8
?	RB 1830	Hewlett Colln. B.P. & B.M.A.	Shorter, A. Paper-Maker & Brit. Paper Trade Jo. 1959 Mar. 41.
?	RB 1831	Turner Watercolour (noted 1977)	(not recorded).
1834	R BARNARD 1832	Birmingham Alabama Document	Gravell, T. Bib.68 5 (No. 80).
1880's	H B BARNARD KENT (not seen)	Unknown	Shorter, A. Bib.7 (May) 55 under "New Mill"

ROBERT WILLIAMS AND LEWIS MUNN SNR.
(see also Appendix II)

One of the by–products of this work has been to explore possible relationships within the paper industry where signs of this are thought to exist and the case discussed below is just such an example. (The genealogical data associated with this case will be found in Appendix II).

It has not been discovered yet where Robert Williams, the occupant of Eyehorne mill (1775–1803), came from. The name Williams is not unusual. It may be, therefore, nothing more than a coincidence that another Robert Williams who, in terms of age, could have been the father or uncle of the Hollingbourne Robert is described as a paper maker of Croxley Hamlet near Rickmansworth as early as 1758. In 1773 a Robert Williams, paper maker of Rickmansworth, insured the utensils and stock in his paper mill called Mill End mill, also near Rickmansworth.[2] This Robert Williams died in 1779. From the evidence presented in Appendix II it seems safe to assume that the Croxley Hamlet and the Mill End mill Robert Williams were one and the same person; and it is also quite clear that the Eyehorne paper maker was a different Robert. From the facts known about the Croxley Hamlet Robert, it is possible that the Hollingbourne one could have been his son by an earlier marriage had there ever been one; or a nephew or cousin.

There is nothing particularly remarkable about the fact that there were two Robert Williams paper makers. What heightens this coincidence is that some 20 years after the Mill End mill Robert had died (1779) this mill was occupied by Lewis Munn jnr. (1799). During the years ca.1774–1776 Lewis Munn snr. was resident in Hollingbourne, quite possibly learning how to make paper or more probably learning the business of running a paper mill, because in 1776, together with Thomas Sweetlove of Leeds village, he set up Ford Paper Mill at Little Chart, Kent (a mill that continued to operate until 1987).[3] Munn is unlikely to have started a new paper mill without training and experience; and it is equally unlikely that he would have stayed in Hollingbourne for two years doing nothing. If he had been training or working, there were only 2 paper mills where he could have obtained this experience, Old Mill under Clement Taylor or Eyehorne mill in a stage of transition between the occupation of Avery and Robert Williams. Whatever the case Munn would almost certainly have met Williams, where the latter had been resident certainly since 1768. Munn, who had 2 children baptised in Hollingbourne, may, for example, have lived in Eyhorne Street from 1774 and assisted Williams in taking over from Avery in 1774-5. There are no facts to substantiate this assumption. Nevertheless this curious concurrence of paper makers' names in Kent and Hertfordshire suggests that there may have been some link between the two Robert Williams and the Munn family.[4]

The alternative, that Lewis Munn snr. might have worked or trained at Old Mill, is equally speculative though there are pointers in this direction and these are examined in App.II p.89.

---

[2] Shorter, A. Bib.12 178 quoting SFIP 333134 (23:Nov:1773).

[3] See Chap.VII p.325.

[4] The fact that the Lewis Munn jnr. (Hertfordshire) was the son of Lewis snr. has not been established yet with absolute certainty. Nevertheless the evidence presented in App.II about his relationship with Lewis snr. and the fact that he took precedence over the other sons at Little Chart after Lewis snr.'s death strongly suggests that he was the son and heir.

PARK MILL, Hollingbourne   MR TQ 832542   (Excise No.307)

This synopsis of the history of Park Mill, Hollingbourne (known also, in the Excise Lists, as NEW MILL; and in correspondence as "Cotterams") is based as before on a framework taken from existing accounts re-examined, and corrected and brought up-to-date where necessary.

Spain deals with both Grove (Eyehorne) and Park Mill together in one section; Shorter in 1957 gives Park Mill very summary treatment and in 1960, particularly in respect of the early history of this mill, reverses his first account on what is considered here to be incorrect assumptions. In view of this not only has the evidence been re-examined, but before the results can be set out in the form of a synopsis it will be necessary to disentangle the various ratebook and Land Tax entries referring either to the mill or to other property owned by the occupants; it is perhaps the failure to do this that has led to the earlier confusion.

The only clearly perceptible evidence for the early years of this mill is to be found in the Hollingbourne RB and this parish record ends with the year 1815. After 1815 it is possible to trace the history of the mill by means of Excise Letters; but it is the beginnings of this mill about which there has been some uncertainty. The first Ratebook entry, indeed the first mention of it as a *Paper Mill*, is for the year 1800.

There would be no need to question this date as a starting date for the mill were it not for the fact that evidence exists for the papermaking activities of Crispe and Newman, the new mill's first occupants, prior to it. For example, a CRISPE & NEWMAN 1797 countermark can be found in the paper used for the 1804 Land Tax Return for Leeds village;[5] and Shorter has shown that these two paper makers attended a General Meeting of the Master Paper Makers in June 1799.[6]

The question as to where Crispe and Newman might have been operating before 1800 has already been discussed under Hollingbourne Old Mill.[7] Shorter believed that evidence for the existence of Park mill before 1800 might be found in the Leeds Parish Records;[8] but the results of the present search do not support this proposal. For one thing Park mill has always lain within the Hollingbourne Parish Boundary. An earlier theory of his[9] that first Newman and later both of them may have been working at Old Mill accommodates these awkward facts much better than his later suggestion.

There is, however, another puzzling feature about Park Mill; it does not feature overtly, if at all, in the Land Tax Returns for either Hollingbourne or Leeds. It has been thought that the entry for this tax might have come under an umbrella covering the tax for the Castle Estate, a Return which has not yet been located. Since the Hollingbourne RB finishes in 1815, the absence of this Land Tax entry creates difficulties in confirming with absolute certainty some details of the history of Park mill.

---

[5]   KAO Q/RP1 214 (1804).
[6]   Shorter, A. Bib.7 (May) 56. Document in the PSM.
[7]   See this Appendix, pp.4-6.
[8]   Shorter, A. Bib.7 (May) 55.
[9]   Shorter, A. Bib.12 195.

To solve this question four of the recorded properties are re-examined (i) Park Paper Mill; (ii) a property designated "Cotnams"; (iii) a house lying for a time on the Hollingbourne/Leeds parish boundary next to the paper mill; and (iv) a house occupied by Thomas Stroud. So far as the mill itself is concerned the common factor from 1800-1818 is its occupation first by Daniel and, later, by Denny Newman, after which the Barnard family took over. From this last then one must assume that, unless the mill had an earlier existence than 1800, other entries in either the Hollingbourne or Leeds records referring to Newman must be associated with other properties than the paper mill.

One of these other properties was the house called "Cotnams" situated in Eyhorne Street (see MAP 7 ; and also in the 1864/5 6 inch O.S. [MAP 9]). To have been marked on the map it must have been an important property. It is suggested here that this is one of "the two houses mentioned [by Hasted], one belonging to Robert Salmon Esq. was only rented by him of Lord Fairfax, who demised it in 1793 to Mr. Daniel Newman, who now resides in it....".[10] Both the Hollingbourne RB and Land Tax returns confirm this (see Extracts p.29).[11] In short the assessments relating to Newman in Hollingbourne prior to 1800 can be shown to refer to his house, Cotnams, and not to the mill.

If the entries for Cotnams, referred to above, are followed through, it will be noticed that in the RB under 1809 another Newman entry appears, that for Denny Newman, Daniel's son, who would have been aged 23 by then. In 1813 both the Cotnam and "Denny's property" are lumped together with the paper mill assessment, although in later years they were separated again. We know that the assessment against Daniel refers to Cotnam House, so it is clear that Denny's assessment must have referred to the house that lay between the paper mill and the main Maidstone/Ashford Road and which up till that date, certainly, had been divided by the Hollingbourne/Leeds parish boundary.

If we now examine the Hollingbourne Land Tax Returns, we can identify Cotnams and, if followed through to 1832, the same entry will be seen to refer to *House and Land* as distinct from the L.T. entries for Hollingworth and Richard Barnard, which mention, specifically, *Paper Mill*. Entries here for Denny Newman start in 1814 (as opposed to 1809 in the RB). Stroud appears here to have been both proprietor and occupant of this property up to 1812, but it clearly does not refer to his domicile (see below) and, further, if the entries are followed through it must be seen as the house adjoining Park mill. In any case Stroud had left Hollingbourne by 1813.[12]

There are *no* Newman entries in the Leeds Land Tax Return. There are, however, RB entries to which reference has already been made under Old Mill against Daniel Newman which start

---

[10] Hasted, E. Bib.42 (1798) Vol.VI 566.

[11] The Hollingbourne RB indicates that this property was assessed under the name of Newman as early as 1785. This must refer to John and Sarah Newman, who had two children baptised in Hollingbourne, 1786 and 1790. Daniel Newman did not move to Hollingbourne until sometime between 1791/1793; he appears in the Land Tax Retaurns for the first time in 1792. The RB entries, after Mrs. Newman (1785/1790) become "Mr. Newman" from 1791. It looks as if Robert Salmon, mentioned by Hasted, had sub-let Cotnam House to the Newmans until it became Daniel's in 1793 (For further details of the Newmans see App.II under Daniel and Denny Newman).

[12] The Leeds RB and LT returns show that from 1810 Stroud was assessed for some major property there, certainly farm land and probably the Abbey Corn Mill as well. Also by 1812/13 Stroud had become the occupant of Pratling Street paper mill in Aylesford (see App.II p.83 and App.III under Pratling Street Paper mill).

long before Park Mill came into being. It has been suggested that changes in the parish boundary, or some other similar adjustment, around 1808 led to the transfer of these Old Mill assessments to Leeds and that this property, which belonged to Old Mill, was absorbed under Hollingworth. A close study of the Hollingbourne and Leeds records confirms this. In Leeds Newman following Robert Salmon, is assessed from 1791–1809 for 5d., later consolidated to 10d. plus 4d. "more for the Mill House" for 1808/9. This entry then disappears at a time when the Old Mill assessments were transferred from Hollingbourne to Leeds and at a time when, it has been suggested, the Hollingworths resumed control of Old Mill (see p.6). On this basis these Newman assessments represent Newman's association with Old Mill.

After 1809 the Leeds RB assessments are much larger (£8–10) and relate to Stroud & Newman (and later Newman) going on as late as 1823 long after Newman had had any association with Old Mill; and because of Stroud's name being linked with Newman earlier they must refer to the Leeds portion of the house associated with Park Mill. These dates tie in with the start of the Denny Newman assessments (£10) in 1809 in the Hollingbourne RB.

In contrast to this, Stroud was assessed in Hollingbourne in his own name for property (£8/RB) from 1805–1812, which presumably related to the house in which Stroud lived, a house not connected with either Old or Park mills.

Lastly, there is the Crispe of Crispe and Newman to be accounted for. Apart from the name appearing in connection with Park mill, there are no Crispe assessments (at least none of any significance) in the Hollingbourne Records; but the Crispe family feature prominently in the Leeds and Broomfield Records. Shorter believed (a) that a John Crispe was associated with Newman; and (b) that he came from Leeds. The Crispe genealogy is a complicated one and details of it can be found in App.II. It can be shown from this that the only nearly contemporary John Crispe linked with Leeds was buried there in 1793, too early for the partnership. It is by no means certain that Newman's partner was John Crispe but, if he was, it seems likely that it was John Crispe of Sutton Valence, an opposite number to Richard Barnard and a man in his fifties at the time of the partnership and also known to have backed another papermaking venture (see under Salmon in App.II). Edward Crispe of Leeds, another man in his fifties, is also a possible though less likely candidate. In any case all the Crispe assessments in the Leeds Records can be shown to be connected with either Abbey Corn mill or other farming interests and *not* with Park mill.

Summarising this complicated preamble we have the following:–

(a) Unequivocal evidence for the existence of Park mill in the Hollingbourne RB as from the year 1800; it is clearly specified as a Paper Mill from the start and there is *no* evidence that any other mill had existed on that site before.[13]

(b) Suggestions that the mill may have been operating before 1800 either on the grounds of unexplained countermarks e.g. CRISPE & NEWMAN 1797 or because of the existence of certain puzzling RB or LT assessments in the names of either Crispe or Newman, may be discounted.

---

[13] Spain, R.J. Bib.8 45–63

(c) A perfectly satisfactory explanation for the earlier Countermarks can be provided by assigning the operation of Old Mill at this time to the partners clearly under licence from Hollingworth and Balston, the owners.

(d) In the foregoing preamble all the entries for properties under the names of Newman, father and son, Crispe and Stroud can be accounted for separately from those of the paper mill. In short there is no evidence that Park Mill was working before 1800.

(e) The absence of any reference to Park Mill in any of the local Land Tax records suggests that the paper mill may have been connected for a time with Leeds Castle; the demise of Cotnam House to Daniel Newman by Lord Fairfax supports this proposal.

Dates		References
1800	*Messrs. Crispe & Newman* for the *Paper Mill*; £70 (RB entry).	KAO P 187 5/I
1801–2	*Crispe & Newman*. This is the RB entry; but it seems likely that during 1802 Thomas Stroud succeeded Crispe in this partnership. He was certainly there by 1803, but the proposal made above would accommodate an otherwise inexplicable countermark "STROUD & CO 1802".	See App.II. for notes on Crispe and on Thomas Stroud (1767–1815)
1803–4	*Messrs. Newman & Co.* It can be shown that by 1803 this RB entry represented the partnership of Stroud & Newman. Stroud & Co. (with no reference to Crispe) appear in a printed list of Master Paper Makers dated 13:June:1803; and duplicate MS notes enclosed with a printed resolution of 19:Sept:1803, giving lists of employees, refer to Stroud & Newman at "Cotterams Mill". Cotterams clearly refers to Daniel Newman's house in Hollingbourne, "Cotnams". (For explanation, see preamble to Park Mill and App.II).	PSM/MPM. PSM/MPM.
1805–12	*Newman & Stroud.* This again is the Hollingbourne RB entry. Stroud also appears, separately, for the first time in the Hollingbourne and Leeds RB and LT Records (not till 1810 in Leeds).	(See Extracts).
(1806)	A letter dated May 1806 from Hollingworth to Balston points to a "walk–out" at this mill (for further details see under Stroud).	App.II.
	The order of the names varies in these entries e.g. in 1810/11 it is Stroud & Newman. There is a STROUD & CO 1812 Countermark indicating that the partnership still existed; and Stroud's continued presence in Hollingbourne is confirmed by the fact that he was Churchwarden in 1812, but by 1813 he had moved his papermaking activities to Pratling Street, Aylesford, though he continued to live and farm in Leeds village.	KAO P 173 5/2 App.III
1813–18 (1816) (1818)	*Daniel Newman* or Newman & Son were the occupants for these years (the Hollingbourne RB ends with 1815). This is confirmed by the 1816 EGL. and by Countermarks, the former "Messrs. Daniel Denny Newman at New Mill" Excise No. 307. There is also a D NEWMAN 1818 countermark.	KAO P 173 5/2 Part II.
1819–31	*Richard Barnard Snr. & Jnr.* The EGL for June 1819 shows these two as the proprietors of both Eyehorne and Park mills.	

Dates		References
	From now on the history of this mill is tied almost inseparably to that of Eyehorne mill, under which heading the subject has already been discussed in some detail (q.v.). Almost all the information henceforward comes from Shorter's (1960) paper on it. The details for completeness are summarised below.	Bib.7 (May).
1831	*Richard Barnard* jnr. was made bankrupt and it seems clear the the reference was to this mill. Eyehorne mill continued to work under Richard Barnard (Land Tax Returns and countermarks confirm this).	
1833	*Richard Barnard* re–instated at NEW MILL (No.307) EGL July 1833.	
1835–6	*Richard Barnard* was made bankrupt again in 1835 and the Feb.1836 EGL shows that the mill was closed temporarily.	
1837–9	*Edwin Horsenail* took over the mill (EGL Oct.1837). He was joined later by his brother, Charles, and they worked this mill and Eyehorne jointly (see sketch-map based on 1839 Tithe map).	See Map VII
	Shorter states that no further reference to the mill could be found as a paper mill until the 1880's, but there does seem to be some evidence that it might have re-opened.	Bib.7 (May) 55.
(1847)	(William Clay ?). Spain quotes Bagshaw's 1847 Directory as saying that, in addition to Hollingworth's Paper Mill, William Clay was operating a Millboard Mill in Hollingbourne. Only 2 paper mills are mentioned in this reference, so either Timothy Healey must have finished at Eyehorne and Clay taken over there; or else Park mill had re-opened. As mentioned under Eyehorne mill, it is not known which of these mills was still working. The EGL for 1849 states that mill *No.306* had left off; but Shorter refers to a Paper Mill Directory of 1860 showing Healey as a manufacturer of paper and millboard at Eyehorne mill and yet somehow before 1861 Eyehorne mill had acquired Poll Mill's Excise number and Park mill Old Mill's Excise number (302) and Old Mill according to the Excise did not leave off until 1850 (although in fact the vats were laid still in 1848).	Spain Bib.8. Bib.7 (May)
1880's 1887	*The New Hollingbourne Paper Mills Company Limited, Kent* re–opened both Eyehorne and Park Mills as paper mills. They shut down in 1887. In the Maidstone Museum there is a Ream Label referring to *PARK MILLS* (Mill Number 302). The mill is shown as a Corn mill in the 1864/5 6–inch O.S. Map.	Map.9

COUNTERMARKS

The Table that follows shows a selection of countermarks, some of which may *not* be connected with Park mill but have been included for reasons which will be obvious. The Table has been set out in 5 bands which are comprised of the following countermarks:-

Band 1   CRISPE & NEWMAN countermarks with a watermark date earlier than 1800, the date when Park mill started working. It has been proposed that the papers concerned were made at Old Mill.

Band 2   CRISPE & NEWMAN countermarks with a watermark date of 1800 or later. The papers in which these were found may have been made at either mill. Since Newman produced White countermarked papers later in his career, when only Park mill was available to him, it is safe to assume that White had been made there from the outset. One cannot therefore assign these marks definitely to one mill or the other.

Band 3   The synopsis has shown that Stroud partnered Newman from 1802-12. Although Daniel Newman must have been older than Stroud, the latter (as with Crispe) appears to have been the senior partner. This probably accounts for the 2 "STROUD & CO" countermarks shown here. To date these are the only two found with a watermark date. Other undated marks have been seen, but only in 1 case has it been possible to assign it to Park mill, STROUD (only) in a document dated 1809.
This band also includes the first DANIEL NEWMAN and NEWMAN & SON marks of 1813.

Band 4   A STROUD 1814 countermark has been included here to show that the paper must have been made at his Pratling Street mill, which he was working by 1813.

Band 5   This includes the later NEWMAN & SON marks up to 1818, after which Richard Barnard took over the mill. In the case of the latter it is not possible to assign his marks to one mill or the other except where he used "EYEHORN MILL" as his mark.

TABLE XVIII – COUNTERMARKS PARK MILL, HOLLINGBOURNE

	d.o.d	Countermark	Provenance	References
1	1804 1801	CRISPE & NEWMAN 1797 C & N 1798	Leeds Parish Land Tax 1804 2 Drawings : Plans and Specifications Joseph Patience	KAO Q/RP1 214 Drawings Collection RIBA L/6 171, 172.
2	1800 n.d. 1803 n.d.	CRISPE & N......18.. CRISPE & NEWMAN 1800 CRISPE & NEWMAN CRISPE & NEWMAN	Leeds Parish Document Shorter, A Leeds Parish Document William Balston memo	KAO P 222 5/13 Bib.12 286 Fig.31 KAO P 222 12/24 PSM (ca.1805)
3	1804 1809 n.d. 1818 n.d. before 1816	STROUD & CO 1802 STROUD ....UD & CO 1812 D NEWMAN 1813 NEWMAN & .... 1813 NEWMAN & SON 1813	Susanna Whatman/William Balston correspondence Broomfield Document Harrietsham Parish Document Hollingbourne Land Tax Harrietsham Parish Document Boxley Ratebook (1816–1824)	PSM. KAO P 222b (bundle) 5/5 KAO P 173 5/2 Part II KAO Q/RP1 180 KAO P 173 5/2 Part II KAO P 40 12/7
4	n.d.	STROUD 1814	Harrietsham Parish Document	KAO P 173 5/2 Part II
5	1816 1818 n.d.	NEWMAN & SON 1815 (also N & S under Watermark) D NEWMAN 1815 D NEWMAN 1818	Hollingbourne Land Tax Hollingbourne Land Tax Harrietsham Parish Document	KAO Q/RP1 180 KAO Q/RP1 180 KAO 173 5/2 Part II

NOTE: Gravell, T. Bib.68 20 (No.651) records a "STROUD & Cº 1802" in a Boston, Mass. document of 1807; and Bib.68 17 (No.526) a script "N" under Shield also in a Boston, Mass. document of 1818 which could be a Newman countermark.

A NEW MILL AT NEW ENGLAND ?   (Approx: MR TQ 825545)

In 1808 there is an enigmatic entry in the Hollingbourne RB which follows directly underneath the entry for the property identified here as EYEHORNE MILL and which the Hollingworths had leased in 1806 (following William Balston's occupation). 1808 was their last year there, after which Richard Barnard took over. This curious entry reads "£90 Hollingworth, part of New Mill at New England" and the assessment continues up till 1815 when the Hollingbourne RB comes to an end. In contrast to the RB there is nothing in either the Hollingbourne or Leeds Land Tax Returns to indicate any increase in the Hollingworth liabilities beyond those already discussed under Old Mill. (In 1808 the RB entries for Old Mill were switched from Hollingbourne to Leeds, but *not* the Land Tax). This entry refers to "part of New Mill at New England" and not to cottages or other houses.

New England is an outlier of Hollingbourne village about equidistant between Park mill to the East and Old Mill to the South–West (see Map 9) and lies just to the North of the main Maidstone/Ashford road (this particular stretch was part of the old road). About 50 yards to the North–East of the road at this point there is an area of land that looks as if it might once have been a mill pond[14] (or at least had the makings of one); it adjoins a stream that flows into Old Mill mill pond at its eastern end.

Only one NEW MILL is listed in the 1816 EGL that can be identified with Hollingbourne and that was PARK MILL. Whatever this project might have been it never materialised; it might have been intended as a mill to prepare "half–stuff" to increase the capacity of Old Mill, a practice that the Hollingworths were to adopt later at Otham mill.

In an attempt to throw some light on this subject, entries for New England have been noted from 1791 onwards in the Extracts from the Hollingbourne RB. The spelling seems to have varied over the years. The entries pass from Daniel Daniel to Thomas Rowler in 1799 and the place–name changes to "Newlands". In 1805 the RB value is increased from £10–£30 and in 1806 it passed to the "Home of Industry" (the Workhouse). This assessment continues until 1808 and then disappears as "Newlands" in 1809, presumably because it had been taken over by the Hollingworths. The enigma remains unsolved.

Dates				References
1808	*Hollingworth* part of New Mill at New England		£90	KAO P 187 5/1
1809	*Thos. & Finch Hollingworth*		£90	
		Cottages	£11	
1810/11	*Thos. & Finch Hollingworth* New England		£90	
			£10	
	T. Hollingworth (not clearly legible but resembles "late grant sat."?)		£100	
1812	(as above)		£94	
			£10	
			£100	
1813	(as above)		£90	
			£4	
			£116	
1814/15	*Thos. & Finch Hollingworth* New England		£90	
			£4	

---

[14] This can be clearly seen in the 1877/85 25 inch O.S. Map 43 series. The site suggested for the mill pond is an area marked "472" just to the NE of New England.

A Note on the HOLLINGBOURNE and LEEDS Ratebook and Land Tax Records

These extracts have already been referred to in the Preface to these Appendices. Until 1816 they supply the principal evidence of the existence of the three paper mills in Hollingbourne and their occupants, evidence that is confirmed by other sources such as Insurance Records. It is important to note, however, that the names of the paper mills do not appear in the original records; they have to be deduced from information about the occupants obtained from other sources. The Hollingbourne RB at one point or another does confirm in each case that the respective entries do in fact refer to a paper mill. The only "Mill" not confirmed in this way is the enigmatic one at New England.

A number of other entries are included in the Tables that follow. These come under 2 headings, Miscellaneous Entries and Main Entries, such as those under Leeds, referring to other kinds of mill. The Miscellaneous entries provide details that throw light on people or property where their omission might have led to unnecessary confusion e.g. the earlier Newmans in Hollingbourne and the Crispes in Leeds.

The mills identified in these records are:–   (Map Reference)

(a)	*Hollingbourne Old Mill* built by James Whatman in 1733, but because the Hollingbourne RB is missing for this period the first entry is not until 1742. The first mention of it as a "pepare" mill is in 1746.	TQ 820541
(b)	*Eyehorne mill* built by James Austen in 1762; but the first reference to him and the "paper mill" is in 1763.	TQ 835545
(c)	*Park mill* makes its first appearance as a "paper mill" in 1800, but who built it is not known.	TQ 832542
(d)	*New Mill at New England* (?). Nothing is known about this "Mill" beyond the RB entries, which start in 1808.	TQ 825545

Spain describes two Corn mills in Hollingbourne during this period, Manor Mill (MR TQ 843552) and Eyhorne Corn Mill (MR TQ 835546).[15] The RB and Tax Records refer, for instance, to Edward Laws "More for the mill" in 1791; this refers to Eyhorne Corn mill. There are also mentions of Thomas Clark (£9) and Thomas Weller (£5) "House and mill" in 1826, the former definitely at Manor mill. A Charlton was assessed in 1809 for a windmill. These have been mentioned to show that the sums involved are small by comparison to those for paper mills.

Other references show that Edward Crispe and John Calcraft both had small interests in Hollingbourne at the turn of the century. William Balston was also assessed separately, that is from the Hollingworths, for some property from 1800–1825 at least for Land Tax.

In the village of Leeds, apart from Thomas Stroud's farming interests (1810–17), references to paper makers are confined mostly to Outdwellers. Edward Crispe's "Abbey Corn Mill" can be identified (MR TQ 823531). There is another entry for John Hills "more for the mill" in 1809; but this mill has not been identified here. The other mills that had once existed in Leeds village had ceased working long before this period (except possibly "Fulling Mill", MR TQ 812540; was this Hill's ?). There had been Priory Mill (MR TQ 823530) in the 16th C. and Le Nethertoune (MR TQ 823534) even earlier.[16]

---

[15] Spain, R.J. Bib.8 50,53

[16] Spain, R.J. Bib.8 56, 59. (Abbey Mill 57, Fulling Mill 64; and there were 2 ancient mills associated with Leeds Castle, Castle Mill 45 and Le Mille 47).

EXTRACTS 1742–1799 FROM:

1. Hollingbourne Ratebook (1742–1799)   KAO P 187 5/1

Identified as	1742/3 1744/5	1746/7/8	1749/50/1/2	1761/2	1763 1764	1768/9/70	1771/2/3	1774
OLD MILL	£40 £43-10 Mr.Saunders	£43-10 not Outdweller 6 Mr.Whatman for the pepare Mill 7 Mr.James Whatman 8 do. or Occupier.	£43-10 £50 Mrs.French 49 50+ Mr.French	£40 Henry French	£46 do. do.	£40 Henry French snr. (in fact Henry jnr. after 1760).	£40 not O.D. James Whatman	£40 not O.D. James Whatman
EYEHORNE (Grove) MILL				No mention of Mr.Austin but he insured his new-built paper mill in Sept.1762.	£35 £35 Mr. William Austin Avery for for paper paper mill. mill.	£35 William Avery (July 1769 Mary Austen of Maidstone insured mill in tenure of Wm.Avery).	£35 William Avery	£35 William Avery
MISC. ENTRIES.						£2-10 Robert Williams	£2-10 do.	£2-10 do. £2 Henry French

2. Hollingbourne Land Tax Returns (from 1780–1798)   KAO Q/RP1 180

Identified as				1780		1782–90	
OLD MILL		Propr:	James Whatman		James Whatman	
		Occup:	Clement Taylor	£48	Clement Taylor	£48
EYEHORNE (Grove) MILL		Propr:	Mrs. Austin		Richard Barnard [1]	
		Occup:	Robert Williams	£40	Robert Williams	£40
Property possibly insured by Robert Williams in 1764 in Eyhorne St.		Propr:			Robert Williams [2]	
		Occup:			himself	£3
		Occup:			himself late Bottle	£5

Footnotes

[1] Unable to confirm whether Austin or Barnard was proprietor of Eyehorne in 1781; LT return missing.

[2] Robert Williams LT entries included as this property seems to have been associated with Eyehorne mill (see next sheet) and might refer to property insured by him in 1764 (cf. misc. RB entries from 1768–1774).

[3] (Eyhorne RB 1791/2) The next entry does not seem to relate, refers to Mrs. West who reappears in 1809.

1775	1776	1785/90	1791/2	1793	1794	1795	1796/7	1798/9
£40 not O.D. Mr.Taylor for the paper mill.	£80 O.D. Clement Taylor	£80 O.D. Clement Taylor	£80 C.D. Clement Taylor	£80 O.D. Clement Taylor	£150 O.D. James Whatman	£149 O.D. Messrs.Baldstone & Hollingworth late Mr.Whatman.	£149 O.D. do.	£149 O.D. Baldstone & Hollingworth or vice versa.
£2 Robert Williams £35 more for the paper mill	£6 Robert Williams £70 more for the paper mill	£6 £70 Robert Williams more for late Mr.Swan £8. 3	£6 £70 £8 Robert Williams £2-10 for the other house.	£78 £8 £2-10 Robert Williams	£78 £8 £2-10 do.	£78 £8 £2-10 do.	£78 £8 £2-10 do.	£78 £8 £2-10 do.
£2 Henry French	£4 Mrs.French	£40 Mrs.Newman	£40 Mr.Newman £2 Danl.Daniel £5 more for New England	£30 Mr.Newman late Salmon £2 Danl.Daniel £5 more for New England.	£30 do. £2 do. £5 do.	£30 do. £2 do. £5 do.	£30 Daniel Newman £2 do. £5 do.	£24 Daniel Newman £2 do.£6 Thos. Rowler £5 £10 for Newlands

Identified as	1791		1792		1793		1794		1795		1796/7/8	
James Whatman			do.		do.		do.		Messrs.Bostone Finch & Co.		Messrs. Hollingworth & Co.	
Clement Taylor	£48		do.	£48	himself	£48	do.	£48	themselves	£48	themselves	£48
Richard Barnard			do.		do.		do.		do.		do.	
Robert Williams	£40		do.	£40	do.	£40	do.	£40	do.	£40	do.	£40
Robert Williams			do.		do.		do.		do.		do.	
himself x 2	£8		do	£8	do.	£8	do.	£8	do. x 3	£10	do.	£10
Lord Fairfax			do.		do.		Daniel Newman		do.		do.	
himself	£20		himself for Dl. Newman	£20	do.	£20	himself	£20	do.	£20	himself 96/97 / himself 98	£20 £17

COTNAMS

EXTRACTS 1800–1815 FROM:
1. Hollingbourne Ratebook (1742–1815)   KAO P 187 5/1

Identified as	1800	1801/2	1803	1804	1805	1806	1807
OLD MILL	£149 Outdweller Hollingworth	£149 O.D. (H&B) or (B&H)[1]	£149 O.D. do.	£149 O.D. do.	£149 O.D. Messrs.Hollingworth	£149 O.D. do.	£149 Not O.D. Hollingworth "more for old Mill"
EYEHORNE MILL	£78 £8 £2-10 Robert Williams	£78 £8 £2-10 do.	£78 £8 £2-10 do.	£78 £8 £2-10 Heirs of Robt. Williams	£80 Baldstone late Williams for Paper Mill.	£80 Hollingworth late Williams for Paper Mill.	£80 Hollingworth
NEW ENGLAND							
PARK MILL	£70 ( ) Messrs.Crispe & Newman for the Paper Mill	£70 O.D. Crispe & Newman	£70 Not O.D. Messrs.Newman & Co.	£70 Not O.D. do.	£70 Not O.D. Newman & Stroud	£70 etc. Newman & Stroud	£70 do.
COTNAMS (NEW ENGLAND) (STROUD)	£28 Daniel Newman £ 6 Thos.Rowler £10 more for Newlands	£28 do. £6 do. £10 do.	£28 do. £6 do. £10 do. (No Stroud)	£28 do. £6 do. £10 do. (No Stroud)	£27 do. £25 Home of Industry £30 more for Newlands late Rowler £8 Thomas Stroud	£27 do. £25 do. £30 do. £8 do.	£27 do. £25 do. £30 do. £8

2. Hollingbourne Land Tax Returns (from 1780–1832)   KAO Q/RP1 180

Identified as		1800		1801/2/3		(1804-9 missing)	1810		1812	
OLD MILL	Propr:	Messrs. Hollingworth & C?		Hollingworth			Hollingworth		do.	
	Occup:	themselves	£48	themselves	£48		themselves	£48	do.	£48
(unknown)	Propr: Occup:	Balston / himself	£1	do.	£1		Balston/himself	£1	do.	£1
EYEHORNE MILL	Propr:	Richard Barnard		do.			Richard Barnard		do.	
	Occup:	Robert Williams	£40	do.	£40		himself	£40	do.	£40
Property possibly associated with Eyehorne mill ?	Propr:	Robert Williams		do.		COTNAMS - - -	Daniel Newman		do.	
	Occup:	himself x 3	£10	do.	£10		himself	£16	do.	£16
COTNAMS	Propr:	Daniel Newman		do.			Thomas Stroud [5]	£1	do.	£1
	Occup:	himself	£17	do.	£17					
							Thomas Green Thomas Stroud [6]	£3	do. himself	£3

[1] To reduce the space required for names abbreviations have been used here and elsewhere in the Table, in this case they stand for Hollingworth & Balston or vice versa.

[2] This entry follows directly below the entry identified as Eyehorne mill.

[3] "Newlands" gone.

[4] Thomas Stroud was Churchwarden for this year.

[5] This line is interpreted as the assessment for the house adjoining Park mill, part in Hollingbourne and part in Leeds.

[6] This line might represent Stroud's domicile; he must have moved to Leeds sometime after 1810.

1808	1809	1810/11	1812	1813	1814	1815
NO MORE ENTRIES FOR OLD MILL IN THE HOLLINGBOURNE RB : TRANSFERRED TO LEEDS FROM 1808.						
£80 — Hollingworth late Williams	£80 — Rd. Barnard P'Mill £1 do. late West	£88 — R.Barner (sic)	£88 — Richard Barnard	£120 do.	£120 do.	£120 do.
£90 — Hollingworth part of New Mill at New England [2]	£90 — Thos. & Finch Hollingworth £11 do. Cottages	£90 — Thos. & Finch Hollingworth £10 New England £100 T.Hollingworth late grant sat. (not clearly legible)	£94 do. £10 do. £100 do.	£90 do. £4 do. £116 do.	£90 do. £4 New England	£90 do. £4 do.
£70 — do	£70 do.	£70 — Stroud & Newman	£70 do.	£113-10 — Daniel Newman	£70 — Denny Newman	£70 — Daniel Newman & jnr.
£27 do. £25 do. £30 do. £8 do.	£27 do. £10 Denny Newman £25 Home of Industry[3] £8 Stroud	£27 do. £10 do. £25 do. £12 do.	£27 do. £10 do. £25 do. £12 do. [4]	(gone into mill assessment see 1814 entry) (no Stroud)	£31-10 do. Daniel Newman £12 Denny Newman do. house	£31-10 do. £12 do.

1814 1815	1816/17/18/19/20		1821		1822	1823/4/5		Last Returns	1826[7] 1828-32	
Holl./Co.	do.		do.		do.	do.			Hollingworth	
themselves	do.	£48	do.	£48	do.	do.	£48		themselves Paper Mill[8]	£48
do.	do.	£1	do.	£1	do.	do.	£1			
do.	do.		do.		do.	do.			Richard Barnard	
do.	do.	£50	himself Jo.Coulter	£40 / £10	do.	do.	£50		himself, Paper Mill[8] / Matthews, House & Land	£40 / £10
do.	do.		Denny Newman		do.	do.			Daniel Newman	
do.	do.	£16	Rd. Barnard	£16	do.	do.	£16		Thomas Samway, House & Land[9]	£16
Denny Newman	do.	£1						?	Daniel Newman/Rd.Barnard[10]	£2
	Hollingworth		do.		do.	Holl.				
	Robt.Larking	£7	do.	£7	do.	himself	£7			
(from 1820)	Hope / Robt.Larking	£51	Hollingworth (late Hope)/ Robt.Larking	£51	Holl./ h'self	do.	£51		Finch Hollingworth/ James Honey,	£51

[7] The 1826 Return includes Thomas Clark, House and Mill £9 and Thomas Weller, House and Mill £5.

[8] For the first time the entries refer specifically to "Paper Mill". Richard Barnard also appears to be Proprietor to property occupied by Richard Orford, House and Land £2.

[9] Throughout this entry has been identified at "COTNAMS"; there is no reference to a paper mill in the Land Tax Returns under Lord Fairfax, The Rev. Dr. Fairfax; and no references to the Rev. Denny Martin, who succeeded Lord Fairfax.

[10] It has been assumed that this is a continuation of the earlier entries on this line.

EXTRACTS 1777–1823 FROM:

1. The Leeds Parish Poor Ratebook (1777–1823) KAO P 222 5/2

Identified as	1777-1788	1789-90	1791	1793/4	1795/6	1798/99	1800	1801/2/4
ABBEY CORN MILL 1	£10 Edward Crispe	£50 Edward Crispe £10 more for the mill £7 more Mr. Taylor u.s.d. ?	£68-10 do. £10 do. £7 do.	£68-10 do. £10 do. £7 do.	£68-10 do. £10 do. £7 do.	£68-10 do. £10 do. £7 do. £19 more Mr. Fullagar u.s.d. ?	£68-10 do. £10 do. £7 do. £19 do.	£134 Messrs & W.Cr £20 more f the mi £14 more Taylor £38 more M Fullag
OTHER MILLS								
MISC. CRISPE ENTRIES	£12-10 more Crispe (1788) £12-10 more Crispe £11 more Castle Land	£12-10 do. £11 more the Castle land.	£12-10 do. £11 do.	£12-10 do. £11 do.	£12-10 do. £11 do.	£12-10 do. £11 do. £2-10 Edw. Crispe	£12-10 do. £11 do. £2-10 do.	£25 do. £22 do. not no
HOLLINGBOURNE OLD MILL & OTHER 2 OUTDWELLERS	10d Clement Taylor 5d. Robert Salmon	10d do. 5d do.	10d do. 5d Mr.Newman	10d late Clement Taylor 5d do.	£3-15 late Whatman 5d Daniel Newman	£3-15 Hollingworth ye mill 5d Daniel Newman	£7-10 do. 10d do.	£7-10 do. 10d do.

2. The Leeds Parish Land Tax Returns 1781–1828 KAO Q/RP1 214³

Identified as			1781/84/86	1790		1791/93/94	1795		1797/8 1800		1802/3/4/5	
HOLL. OLD MILL ABBEY CORN MILL	Propr:		John Calcraft 4	John Calcraft jnr.		do.	do.		do.		do.	
	Occup:		Clement Taylor	do.	£1	do.	Whatman	£1	Hollingworth	£1	do.	£1
	Occup:		Edward Crispe	£8	do.	£65 do.	do.	£65	do.	£64	Thos.& Wm.Crispe	£65
		Propr:	Edw.Crispe⁵	do.		do.			Thos.& Wm.Crispe			
		Occup:	3 x Occupants	do.		do.						

---

1 From 1789 until at least 1908 Edward Crispe and sons are assessed "more for the mill" and as there does not seem to have been any other mill than Abbey Mill available, it has been assigned to them. There is a Deed (KAO U 855 T1/4) concerning John Calcraft and this mill dated 1789 and the letting of it. The other names shown here seem to be associated with Crispe, perhaps as tenants.

2 The whole of this block is devoted to Outdweller entries considered to be relevant.

3 Only the Leeds Tax Records for the years shown, 1780–99, examined; after 1800 all extant ones.

4 John Calcraft was the Proprietor of many properties figuring in the Tax Records. Only two of these have been selected. For the 1784 Return the entry is "Heirs of John Calcraft".

5 This block has only been included to show Crispe's interests as a Proprietor.

6 RB 1809, since "more the mill" appears twice for this year and since the assessments were consolidated in 1801, these entries must surely refer to 2 separate mills. It has been proposed that Hills may have been working "Fulling Mill" MR TQ 812540.

1808	1809	1810/11	1812/13	1813/14	1814/16	1816/17	1817-1823
£134 Thos. Crispe	£134 do.	£220-10 Thomas Stroud [7]	£197-10 Thomas Stroud (Crispe gone).	£197-10 do.	£197-10 Thomas Stroud (widow)	£197-10 Widow Stroud	(gone)
£20 more the mill	£20 do.	£50-15 Thomas Crispe					
£4 more Taylor	£4 do.						
£38 more Watt's farm	£38 do.						
	£40 John Hills [6]						
	£25 more for the mill						
(not noted further)							
£110 Hollgwth.	£110 do.	£216-10 Hollgwth.	£120 T & F. [10] Hollgwth.	£120 do.	£120 do.	£120 do.	£120 T & F Hollgwth.
10d Daniel Newman	10d do.	£66-10 more for Spout House	£10 more Cottages	£10 do.	£10 do.	£10 do.	£10 more the Cottages
4d more for the mill house.	4d do. [8]		£96-10 Holl.Esq.	£96-10 Geo. Selby late Holl. Esq.	£96-10 do.	£96-10 do.	£100 Geo.Selby £157 Philip Martin Esq.
			£66-10 More Spout Ho.				
				£66-10 Holl.Esq.	£66-10 do.	£66-10 do.	
		£8-10 Stroud & Newman [9]	£8-10 Stroud & Newman	£8-15 Denny Newman	£8-15 Daniel Newman	£8-15 do.	£8-15 Daniel Newman

1806	1810	1814	1815	1816	1817	1818-1825		1826-1828	
do.		do.		do.		do.		do.	
do.		Finch Hollingworth	£1	do.	£1	do.	£1	(not noted).	
Thos.(only) Crispe		Thomas Stroud	£65	Elizabeth Stroud	£70	William Crispe	£70	do.	£14

(not noted further)		Thomas Hollingworth	£18			(not noted further)
		Hollingworth/ himself	£11	Robert Hollingworth	£11	

Reference has been made earlier to the fact that there are no obvious entries in the Tax Returns of either Hollingbourne or Leeds for Park Mill. Mention should be made here that there are two other large assessments in the Leeds Return not shown here. Since both date from the 1780's they cannot refer to Park mill; both show Owlett as Occupant. One is between John Calcraft/Owlett for £110 rising to £147 after 1795; and the other between Lord Fairfax/Owlett in 1789; in 1795 Dr. Fairfax/himself & Owlett £43; later, 1803 General Martin/himself & Rugg £43 (see App.II under Salmon).

[7] It is not clear whether Stroud's very large assessment for 1810 and after included Abbey Mill or not ? His assessment is reflected in the Land Tax where he appears to replace Crispe.

[8] RB 1809, this would seem to be Newman's last connection with Old Mill.

[9] RB 1810 and after, this would appear to be property associated with Park Mill.

[10] RB 1812/13, the £216-10 for 1810 has been broken down here into Old Mill, Cottages and some unspecified property seemingly connected with Old Mill, but later assigned to Selby and Philip Martin. The latter appears to have taken over Spout House from Hollingworth by 1817.

APPENDIX II

Identification of the Occupants of and others connected with the Hollingbourne Paper mills

TABLE XIX – THE WHATMAN / HARRIS RELATIONSHIP

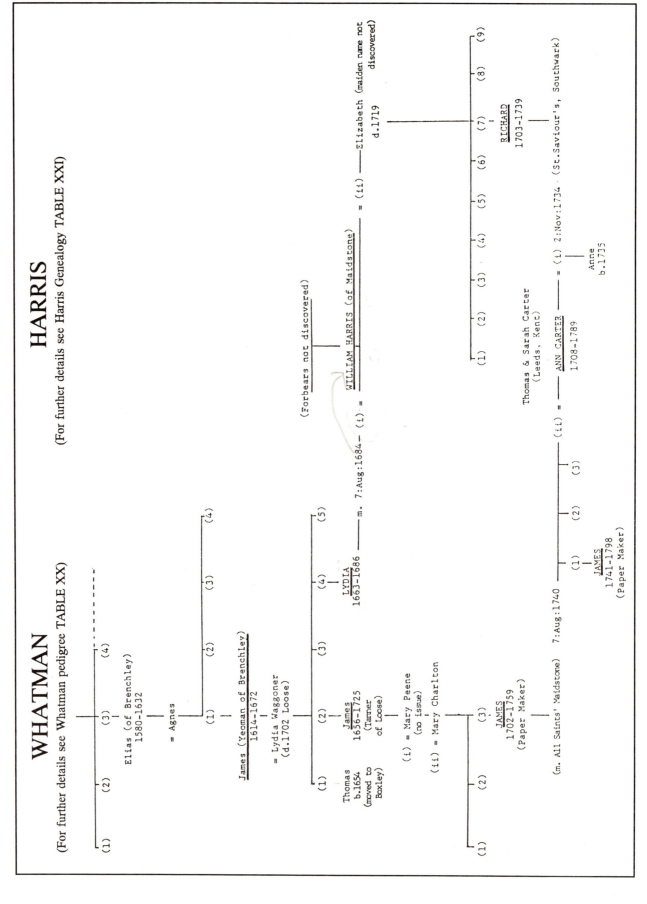

WHATMAN

(For further details see Whatman pedigree TABLE XX)

HARRIS

(For further details see Harris Genealogy TABLE XXI)

Elias (of Brenchley)
1580-1632

= Agnes

James (Yeoman of Brenchley)
1614-1672

= Lydia Waggoner
(d.1702 Loose)

Thomas
b.1654
(moved to
Boxley)

James
1656-1725
(Tanner
of Loose)

(i) = Mary Peene
(no issue)

(ii) = Mary Charlton

LYDIA
1663-1686 —————— m. 7:Aug:1684 — (i) =

(Forbears not discovered)

WILLIAM HARRIS (of Maidstone)
    = (ii) — Elizabeth (maiden name not
              d.1719   discovered)

RICHARD
1703-1739

JAMES
1702-1759
(Paper Maker)

(i) ——— ANN CARTER ——— = (i) 2:Nov:1734 - (St.Saviour's, Southwark)
(ii) = 1708-1789

Thomas & Sarah Carter
(Leeds, Kent)

Anne
b.1735

(m. All Saints' Maidstone)   7:Aug:1740

JAMES
1741-1798
(Paper Maker)

TABLE XIX – THE WHATMAN / HARRIS RELATIONSHIP

The connection between the Whatman and Harris families is a factor of very great significance in any account devoted to the origins and development of the elder papermaking James Whatman. It is clear from the Will of James Whatman, the first as a Tanner of Loose (1656–1725), that a William Harris was his brother-in-law, but the precise relationship between the two families has not been established previously; neither has it been shown conclusively that the William Harris in question was the paper maker at Gurney's mill, Loose (1688/9–1727). The diagram above defines this relationship; and details found in Chap. III and the Harris genealogy (this App.) show that this William Harris was the paper maker of Loose.

The diagram also shows (a) that there was no blood relationship between the Harris and Whatman families; and (b) a second tie between the two families through marriage to Ann Carter of Leeds (this App.).

Neither the birthplace nor the burial of William Harris have been discovered. The first reference we have to him is to be found in the Brenchley Parish Register (KAO P 45 1/2) on the 7:Aug:1684, the date of his marriage to Lydia Whatman, where he is described as "of Maidstone". Several Harris families were living in Maidstone at the time, but it has not been possible to relate William to any of them. Only one possible candidate has been discovered, a twin to Arthur (d.1668), baptised at All Saints', Maidstone on 10:Mar:1667 s.o. Thomas (a "Maultster") and Clare Harris. (Though several contemporary Harris Wills have been examined, including other maltsters of Maidstone, no Will for Thomas (d.1675) and/or Clare has been found.) This William would only have been aged 17½ to Lydia Whatman's 21 at the time of the marriage in 1684; though possible it seems unlikely that this was the paper maker William. If one assumes that William was about 21 (or over) when he married Lydia (a reasonable assumption considering that he was to operate a paper mill in 1689 and leave off papermaking 38 years later), this would take William's birth back to ca.1663 or before, a time when entries in the All Saints' Maidstone Register were still scarce. So though it is conceivable that William may have come from one of the numerous Harris families in Maidstone, it has not been possible to establish a connection.

During the course of the various searches carried out for this book no William Harris baptisms that would fit in with the other facts known about the paper maker have been discovered elsewhere. The parishes include Loose (searched 1650–1665) and all the other parishes round Maidstone and where registers start after the bracketed dates there was no evidence of Harris families to suggest a William baptism. Others included Dover, Canterbury, Goudhurst, Eynsford etc. S.K. Keyes ("Dartford, Historical Notes" 1933, 1938) refers to a Thomas Harris in the Overy Ward where the paper mill was sited and to a James, maltster and constable in 1660, but no William baptisms, though like George Gill's it may not have been entered; the same applied to Sutton-at-Hone and Bexley. For a more promising solution see this App. n.20.

TABLE XX – THE WHATMAN PEDIGREE
WHATMAN

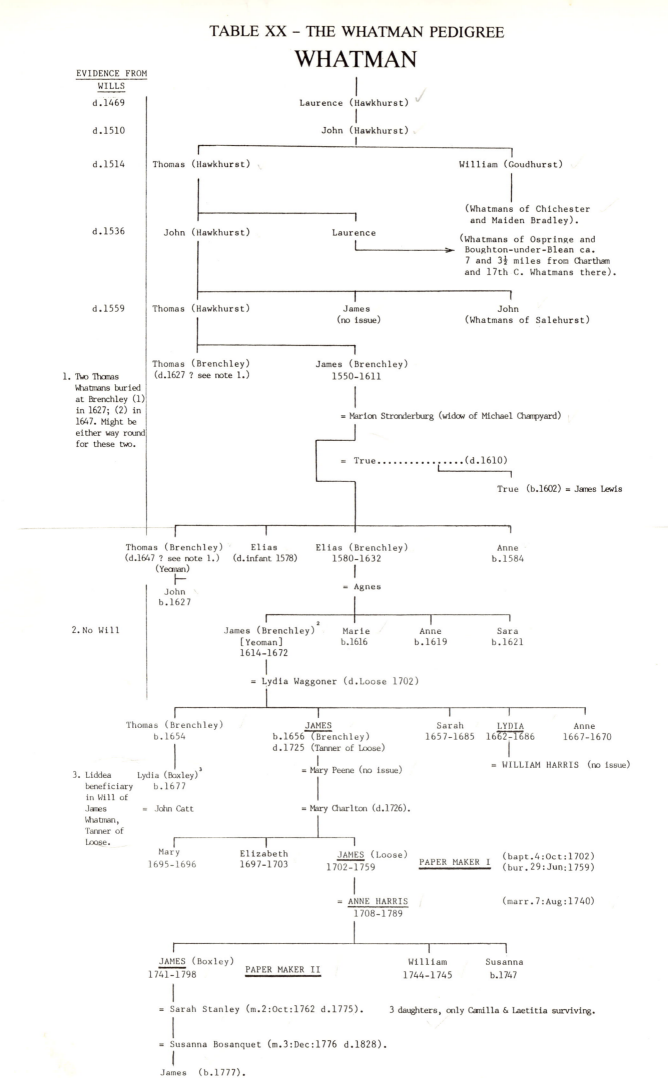

EVIDENCE FROM WILLS

d.1469 — Laurence (Hawkhurst)

d.1510 — John (Hawkhurst)

d.1514 — Thomas (Hawkhurst) — William (Goudhurst)

(Whatmans of Chichester and Maiden Bradley).

(Whatmans of Ospringe and Boughton-under-Blean ca. 7 and 3½ miles from Chartham and 17th C. Whatmans there).

d.1536 — John (Hawkhurst) — Laurence

d.1559 — Thomas (Hawkhurst) — James (no issue) — John (Whatmans of Salehurst)

1. Two Thomas Whatmans buried at Brenchley (1) in 1627; (2) in 1647. Might be either way round for these two.

Thomas (Brenchley) (d.1627 ? see note 1.) — James (Brenchley) 1550-1611

= Marion Stronderburg (widow of Michael Champyard)

= True...............(d.1610)

True (b.1602) = James Lewis

Thomas (Brenchley) (d.1647 ? see note 1.) (Yeoman) — Elias (d.infant 1578) — Elias (Brenchley) 1580-1632 — Anne b.1584

John b.1627

= Agnes

2. No Will

James (Brenchley)[2] [Yeoman] 1614-1672 — Marie b.1616 — Anne b.1619 — Sara b.1621

= Lydia Waggoner (d.Loose 1702)

Thomas (Brenchley) b.1654 — JAMES b.1656 (Brenchley) d.1725 (Tanner of Loose) — Sarah 1657-1685 — LYDIA 1662-1686 — Anne 1667-1670

= WILLIAM HARRIS (no issue)

3. Liddea beneficiary in Will of James Whatman, Tanner of Loose.

Lydia (Boxley)[3] b.1677

= Mary Peene (no issue)

= John Catt

= Mary Charlton (d.1726).

Mary 1695-1696 — Elizabeth 1697-1703 — JAMES (Loose) 1702-1759 — PAPER MAKER I (bapt.4:Oct:1702) (bur.29:Jun:1759)

= ANNE HARRIS 1708-1789 (marr.7:Aug:1740)

JAMES (Boxley) 1741-1798 — PAPER MAKER II — William 1744-1745 — Susanna b.1747

= Sarah Stanley (m.2:Oct:1762 d.1775). 3 daughters, only Camilla & Laetitia surviving.

= Susanna Bosanquet (m.3:Dec:1776 d.1828).

James (b.1777).

THE BACKGROUND TO THE WHATMAN PAPER MAKERS

The pedigree of the papermaking Whatmans is illustrated in TABLE XX. There are, however, several other points to consider about their background other than their line of descent. Is anything known, for instance, about their social standing, the geographical distribution of this family and the reasons for their move to the Maidstone district ? Little has been discovered that would provide an answer to these questions: nevertheless it is possible to make certain observations which can be used to make reasonable assumptions. These are considered below.

1. Social standing and family distribution

We do not know at present what occupation the early Whatmans followed. It is not until the 17th C. that Thomas of Brenchley is described as a Yeoman, though it is self-evident from the Wills of the early Whatmans that they all owned property. The Will of Thomas' younger brother, Elias of Brenchley (1580–1632), shows that he left his house and his land to James; and another house to his daughters on reaching the age of 24. James, his son, left no Will but, since he also has been described as a Yeoman, it is clear that he too had property. It might be reasonable to assume, therefore, that the early Whatmans had followed some sort of agricultural occupation. One might speculate that they had perhaps supplied the local industries with either timber for the iron smelting industry or wool for the weavers; but there is nothing to indicate that they were engaged in either of these industries themselves. The signs are that they belonged to a moderately prosperous middle class and like many families in the past they evidently participated in quite a wide social spectrum.

The Will of the next James[1] (1656–1725), father of the first paper maker Whatman and who became a Tanner of Loose, shows that he too was a man of property. In addition to leaving his house and his tan yards to his widow and, by succession, to his son James the paper maker, he also bequeathed a tenement in Loose to his niece, Lydia; and mention is made of further property in Marden. The impression gained from this is that his son, who inherited all this property in 1726, must have been reasonably well off by the time he came to build his paper mill in Hollingbourne. The ways the fortunes of the family, and their social standing, increased once they had embarked on papermaking careers has been described fully in Thomas Balston's account of their lives[2] and has already been referred to in the main text. The advances made in these spheres, both socially and in their trade, were considerable.

The pedigree shown in TABLE XX indicates that the family were already dispersing in the 16th C., but these off-shoots need not concern us here except to note that the descendants of Laurence Whatman figure in the records of the parishes of Ospringe and Boughton-under-Blean (just West of Canterbury) in the 16th C. and extend to nearby *Chartham* during the 17th C., where the male line appears have to come to an end with the death of Mr. Richard Whatman in 1687 (styled in his Will as Richard Whatman Esq.[3]). The main point to note (one not covered in this pedigree) is that during the 17th C. the Hawkhurst Whatmans had become widely dispersed throughout the High Weald of Kent as well as migrating westward

[1] KAO U 289 T/56.
[2] Balston, T. Bib.1.
[3] PRC 11/51 fo.186.

and northwards to the Maidstone area.[4] It is perhaps worth noting the presence of many of these families in those parishes closest to the paper mills at Goudhurst. Whether this has any particular relevance or not, the point will be appreciated when seen in the context of the account given of these paper mills in App.lV.

2. *The Northward Migration of the Paper Makers' Ancestors*

The pedigree also indicates a general migration northwards for many members of the Hawkhurst family beginning in the 16th C. The reason for this is not known, but it may have been influenced by the decline of the Iron Smelting Industry which was centred on Hawkhurst and which had reached its peak in the Tudor period.[5] Whether the Whatmans had originally been concerned in supplying fuel to the iron smelting industry or not, it is probable that the region further north had become a more prosperous one due to the weaving industry and they may have made the move because of this. However, neither the new foundry at Brenchley nor the Wealden weaving industry thrived for very long after this. The weaving industry, which was centred on Cranbrook, Biddenden,[6] Staplehurst and Tenterden (all a few miles to the East of Brenchley) went into decline in the 17th C.[7] Whatever occupation the Whatmans may have been engaged in (see also below), there must have been a general fall in the level of prosperity in rural districts surrounding these areas; and, at the same time, opportunities for the young and ambitious would have become progressively more scarce.

All this is supposition, but it is paralleled by the movement of the paper makers from this region to the Maidstone district towards the end of the 17th C. and by the end of this century the Brenchley Whatmans had made this move too and concentrated in the Maidstone area. The first to make the move was the eldest son, Thomas, before 1677. James followed and first appears as a resident in Loose in 1688.[8] The fact that he was later described as a "tanner" prompts the further suggestion that though the Whatmans of Brenchley, and perhaps earlier than this, may not themselves have been engaged in the business of tanning, they may have been partly dependent for a living on supplying oak bark to the tanners and that as the Weald became progressively denuded of its oak forests they had been forced to move to new areas. They may thus have had a close connection with tanners such as Richard Peene of Loose whose daughter, Mary, James was to marry. Unfortunately the date of this marriage is not known, but it is quite possible that it took place some years before James moved from

---

[4]   Apart from the Whatmans of Brenchley there was a large family of Whatmans at *Goudhurst* e.g. Richard bapt. there in 1570; Robert 1572, Marie 1575; Robert s.o. Antonie 1573. A James Whatman m. Ann Young there in 1649; two James were buried there in 1660 and 1680; and in 1675 the Ratebook shows that James was assessed for several properties in Goudhurst. A James Whatman m. Elizabeth Tourt in Horsmonden in 1645; the Marden register shows that in 1666 Thomas s.o. William and Mary Whatman was baptised there. Further afield in 1636 Sir Edward Gylbourne Bart. m. Elizabeth Whatman at Shoreham; Mary (widow) was married again at Burham (N.W. of Maidstone) in 1642; and throughout the 17th C. there were Whatman families at *Boxley* (N.E. of Maidstone) and at least two families of Whatman appear in the All SS'. Maidstone Registers by the end of the 17th C. The search has by no means been exhaustive (no fewer than 34 Wills of Kentish Whatmans were proved between 1459 and 1691) but indicates that the distribution of the Whatman family had become widespread by the 17th C. and was not just confined to Brenchley.

[5]   See Chap.II p.68.

[6]   It may be of interest to note (though whether any connection exists with the family and how far back it might date are matters not known at present) that there is a property just to the south of Biddenden (now vineyards) call "Little Whatmans" (MR TQ 850362).

[7]   See Chap.II.p.67.

[8]   The Loose RB. KAO P 233 5/2.

Brenchley, by which time he would have been 32.[9] An alternative to this is that James may quite simply have been apprenticed to Richard Peene some 20 years earlier and married his daughter later, but there is no evidence of him having practised as a tanner at Brenchley during the intervening period.

3. *The Move to the Maidstone District*

James the Yeoman of Brenchley (d. 1672) left no Will so one presumes that Lydia, his widow, became life tenant of the property. Thomas, the eldest son, had moved to Boxley before 1677, where there were better prospects no doubt for himself, possibly encouraged by one of the Whatman families already established in the Maidstone area. Lydia may have been supported by her second son, James, who would have been 21 by then and, if a bona fide tanner, would have completed his apprenticeship. Several different factors must have contributed to their subsequent move north to Loose in 1688. By 1685 James, the son, had married again and Lydia, his mother, would have been ca.55; her three daughters were all dead including Lydia (bur. Brenchley 1686) who had married William Harris, so this, if for no other reason, would have been an opportune time to make a move. Whereas Thomas, at Boxley, may have persuaded them, a more likely reason would have been the need for someone to manage the late Richard Peene's tannery[10] and where, close by, William Harris was converting Gurney's mill to papermaking, 1687–89. This proximity of the Harris family and paper mill almost certainly led the next James Whatman (b. 1702) into taking an interest in papermaking [Map 3, see Vol.1], ultimately embarking on a career in it himself, thus continuing the northward migration which ended in Boxley.

Other near contemporary branches of the Whatman Family

There has been a tendency in past accounts of their papermaking activities to treat the paper maker Whatmans and their ancestors in isolation. This is understandable if one has, for example, a very concise objective; but since so little is known about the elder Whatman and the sources that may have influenced him in his investment decision and his technical achievements, it may be helpful (if not in the present context but to the future workers in this field), if a few notes are recorded here on other branches of the family that lived in or near the papermaking areas with which we are concerned.

The Whatman Families of the High Weald of Kent

It has already been noted that there were Whatman families in the High Weald during the (late) 16th and 17th Cs., some of them described as "Yeoman" or "Husbandman". No evidence has been found to suggest that they were in any way connected with papermaking activity in that area or, indeed, with any of the other trades for which this region was noted. Since their presence there seems to have been confined largely to these centuries and therefore more than half a century at minimum before the elder Whatman showed any tangible interest in papermaking, they cannot

---

[9]   Richard Peene, Tanner of Loose, appears to have died ca.1684 (KAO P 233 5/2). Regrettably Baptist Records do not extend back far enough to determine whether this Richard was related to Simon Pine of Tovil (a close neighbour) see Chap.VII note 225.

[10]   It may also be noted that to accommodate the baptism of James' eldest daughter, Mary, in 1685, his first wife must have died within 5 years of the move to her father's tan yards in Loose. For further information on James' 2 marriages see Bib.1 p.1 n.2.

be seen as having any influence at all on his choice of career though it is possible that his family or he himself may have benefited from them financially.

In Brenchley itself there were Whatmans other than those shown in the pedigree. The parish register[11] during much of the 17th C. is almost illegible in places and this coupled with the fact that the spelling of the name Whatman tended to be variable makes things difficult. Among the doubtful entries a Robert Whatman (bur.1620); a Robert (bapt.1621); and an Edward m. to a Jane Tilson (1620). Two more positive ones are Jane w.o. Abraham Whatman (bur.1633) and the widow of Thomas (bur.1651). Whether other Whatman events went unrecorded during the Commonwealth is not known. It will be remembered that in one case at least the baptism of four of James' and Lydia's children was delayed until 1663, when a multiple baptism took place.[12]

Boxley (KAO P 40 1/1)

There were Whatmans in the parish of Boxley for the greater part of the 17th C. A John and Joyse Whatman were living there by 1608 and had 6 children baptised there between 1608–1622. Among the latter was an Abraham (bapt.1620) but he would have been too young to have been the Abraham referred to above at Brenchley. It is not known whether this John was connected in any way with the Hawkhurst Whatmans.

By 1622 another Whatman family had appeared on the scene as evidenced by the baptism of Ann (Anna) d.o. James (Jacobi) Whatman. Other Boxley entries include the burials of George (1669) and John (1674). More significant perhaps in the present context is the baptism of *Lydia* d.o. Thomas and Bridget Whatman in 1677, which almost certainly must be the family of Thomas s.o. James of Brenchley. (There is no record of a Thomas of that age having been buried in Brenchley; in addition to which James of Brenchley, and subsequently tanner of Loose, left his niece, Lydia, a tenement in Wells Street, Loose, and Thomas was his only surviving brother or sister by whom he could have had a niece).[13] It is likely that the Boxley Whatmans were related to Thomas from Brenchley; this would help explain his move there rather than postulating a venture into the unknown.

Regarding the Boxley Whatmans see also under Maidstone below.

Maidstone (PRs held by Incumbent)

Towards the end of the 17th C. two Whatman families make their appearance in the All Saints' Maidstone parish registers. It is possible that both of these families were the same as the Boxley ones for that period. Between 1679–84 4 children of William and Anne Whatman and in 1703 Mary d.o. William and Martha Whatman were baptised in All SS. William was buried there in 1727 and Martha in 1740, Anne having died in 1684 (bur.All SS. w.o. William). A Thomas and Elizabeth Watman (sic) make their appearance in the All SS. PRs from 1685 with the baptism of two (possibly three) of their children from 1685 onwards. (In both cases the spelling of the name varies, sometimes Whatman, at others Watman, Wattman). The majority of these children died in infancy.

The last Whatman entries in the Boxley Registers for the 17th C. are:
  (i) 1677 Lydia d.o. Thomas & Bridget Whatman (bapt.)
  (ii) 1678 William Whatman m. Ann Miller

To date the burial of Bridget has not been found. If in fact she had died and Thomas had re–married, then this could be the Thomas, husband of Elizabeth, who turns up in Maidstone. Likewise the two William and Ann(e)s could be identical and the William s.o. William baptised 1622 (and grandson of John & Joyse Whatman of Boxley).

Only because of the PR entries has it been deduced that for most of the 17th C. these Whatmans were domiciled in Boxley (Thomas from Brenchley from ca. 1677). But where they lived later or whether they were assessed for property anywhere is an unsolved mystery. The Boxley RB only starts in 1691 with no reference to them after this date nor in the Maidstone RBs. for 1685, 1687, 1692, 1697 and 1702.[14] In the course of searching the RBs of many

---

[11] Brenchley Registers KAO 45 1/1; and after 1655 1/2. There are no Whatman entries after Lydia's burial (1686 corr:date) up to 1691.

[12] See Preface to App.I & II note 8.

[13] Lydia's marriage to ––Catt has not been found; not in the Boxley Register (searched up to 1706).

[14] KAO P 241 11/2, 11/3 and 11/4.

other parishes round Maidstone the name Whatman has not been seen either except for the paper maker's family in Loose.

Ospringe, Boughton–under–Blean and Chartham

The Whatman pedigree shows that in the 16th C. Laurence Whatman founded a branch of the family at Ospringe and Boughton–under–Blean, two villages that are ca.7 and 3.5 miles respectively from Chartham (and also relatively close to Canterbury); Whatmans appear at Chartham early in the 17th C. (The early history of the Whatmans in this area has not been examined in detail). The earliest entry at Boughton–under–Blean is the baptism of Arthur s.o. James Whatman in 1588. Arthur may have been the father of Richard baptised (s.o. Arthur) in 1622 at Chartham. An Arthur and his wife, Martha were buried at Ospringe in 1673 and 1669 respectively, Arthur being styled as Esquire; but whether he was Arthur from Chartham or not is not known. An Arthur Whatman continued to have a family in Chartham up to 1631. In the present context, however, the interest is in the Whatmans who continued to live in Chartham after this date and they include Richard (bur.1687, Mr. Richard); Ann wife of Richard (bur.1677); a Mr. Arthur Whatman (bur.1691/2), the last of the Whatmans found in the parish register there.[15] A Mary Whatman was also married at Chartham to Peter Mount of St Peter's, Canterbury, in 1683. They had a child baptised in Chartham later.

Interest in these Whatmans was aroused originally by the discovery that a Richard Whatman Esq. had left property in Chartham which is referred to in an inventory (PRC 11/51 fo.186) made in 1687 on his death. The question arose as to whether there were any other Whatmans living in Chartham ca.1732/3 when Peter Archer jnr. would have been converting the mill there to papermaking, a noteworthy event in that Peter Archer had installed 2 Hollander beating engines in his new–built paper mill only a month or two ahead of the elder Whatman at Old Mill, Hollingbourne. Installations of this kind were still rare at that time; and although alternative explanations can be put forward, the circumstances surrounding Peter Archer's move suggest that the fulfilment of some technological advance may have been the dominant motive for this.[16]

In the event there appear to have been no Whatmans living in Chartham at the time, though there may still have been relationships through marriage. However, it is unlikely with a gap of six generations that there would still have been contact between the two branches of the family so that the chances of any communication of the papermaking development at Chartham through this source would have been minimal, if at all. Archer only arrived in Chartham in 1732 and his plans to move from Sittingbourne could only have reached Whatman and Harris, if in fact they ever did, by other means, such as, for instance, if there had been any communication between Archer's in–laws in Canterbury and the Deans at Cobtree mill, if related.[17]

Sources of information on the Whatman family

The information given in this section on the Whatman family has been obtained partly from the Kent Archives Office; the All Saints', Maidstone Parish Registers; the Archives of the Cathedral, City and Diocesan Record Office, Canterbury; and from Mr. Hugh Balston (a descendant of the Whatmans) who has kindly let me see a notebook (20th C.) compiled by Margaret E. Malden and given to Mary E.C. Dugdale, which contains the most complete pedigree of the Whatman family seen yet (although in places it is not entirely correct).[18]

[15] No Whatmans were found in the Boughton–under–Blean Registers between 1720–1740 either.

[16] See Chap.IV Introductory Section; and pp.143,145 & n.35.

[17] App.IV note 64.

[18] For the benefit of those who may be interested in the history of the later Whatmans or in Vinters (the house acquired by the younger James Whatman in 1783 (for details see Bib.1 73) and which lay just to the East of Maidstone), the notebook referred to above contains several early photographs of the house and sketch–maps of its layout. Elsewhere there are a number of watercolours illustrating the changes made to the house in the 18/19th Cs. (private collection). House demolished 20th C. after World War II.

There is also a diagram and description of the twin tombstones in Loose Churchyard (restored by James Whatman in 1871) with inscriptions relating to James Whatman, Tanner (bur.1725); his wife Mary; her sister Elizabeth Castreet w.o. James Castreet; and two James Castreets.

TABLE XXI – SUGGESTED PEDIGREE
FOR THE PAPERMAKING DESCENDANTS OF WILLIAM HARRIS

HARRIS

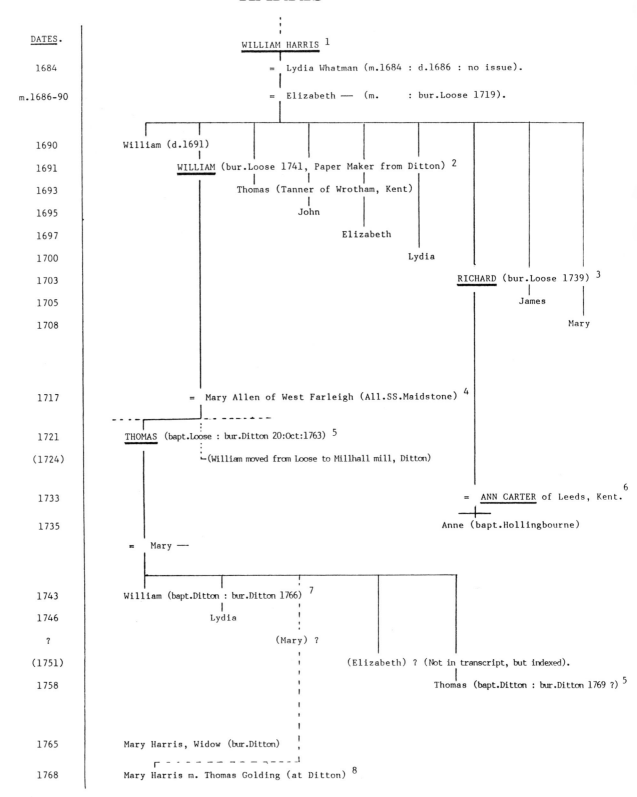

DATES.	
	WILLIAM HARRIS [1]
1684	= Lydia Whatman (m.1684 : d.1686 : no issue).
m.1686-90	= Elizabeth —— (m. : bur.Loose 1719).
1690	William (d.1691)
1691	WILLIAM (bur.Loose 1741, Paper Maker from Ditton) [2]
1693	Thomas (Tanner of Wrotham, Kent)
1695	John
1697	Elizabeth
1700	Lydia
1703	RICHARD (bur.Loose 1739) [3]
1705	James
1708	Mary
1717	= Mary Allen of West Farleigh (All.SS.Maidstone) [4]
1721	THOMAS (bapt.Loose : bur.Ditton 20:Oct:1763) [5]
(1724)	└(William moved from Loose to Millhall mill, Ditton)
1733	= ANN CARTER of Leeds, Kent. [6]
1735	Anne (bapt.Hollingbourne)
	= Mary ——
1743	William (bapt.Ditton : bur.Ditton 1766) [7]
1746	Lydia
?	(Mary) ?
(1751)	(Elizabeth) ? (Not in transcript, but indexed).
1758	Thomas (bapt.Ditton : bur.Ditton 1769 ?) [5]
1765	Mary Harris, Widow (bur.Ditton)
1768	Mary Harris m. Thomas Golding (at Ditton) [8]

[1] Antecedents, birthplace and burial place unknown. See TABLE XIX.

[2] Loose PR. KAO P 233 1/2.

[3] Loose PR. KAO P 233 1/2. Bur. 26:Nov:1739. See also main text; and Bib.1 Chap. III.

[4] This is an assumption; for discussion see p.46.

[5] Again, this is an assumption; for discussion see p.46. The word "Snr." was added to the PR at some time after the burial entry for 1766. Presumably this must refer to the Thomas bapt.1721. The reason for allocating the burial date of 1769 to the Thomas bapt. 1758 is given on p.47.

[6] See this Appendix.

[7] It is not known whether this William ever operated the paper mill. See Note 8 below.

[8] It is known that Thomas Golding took over Millhall Paper Mill, Ditton in 1776 (and the 1816 EGL shows him still as the occupant), but the origin of his wife, Mary (Harris), is uncertain (see text).

THE HARRIS FAMILY PAPER MAKERS

Unlike the Whatmans, the Harris population in the Maidstone district during the 17th and 18th Cs. was so numerous that constructing a comprehensive genealogy is a near impossibility and quite outside the scope of this work. The origins of William Harris, Ancient Paper Maker and paper maker of Gurney's mill in Loose, are not known; but he was succeeded by two papermaking sons, William and Richard, and one grandson, Thomas. Beyond that it is not known whether his great–grandson, William (bapt.1743) ever succeeded his father at Millhall mill, Ditton; or whether his great–grand–daughter (?) Mary, was the Mary Harris who married Thomas Golding who took over at Ditton in 1776. All of these figure in Table XXI. There were, however, in addition to these, four other Harrises who were connected in one way or another with papermaking in that area; but it is not known whether they were related in any way to the Ancient Paper Maker. The discussion that follows is complicated and, unless the reader wishes to follow the reasoning behind the identifications suggested, he or she may wish to omit these sections.

The Harris Population of the Maidstone District

The Harrises seem to have been concentrated mainly in the parishes of Maidstone, where Harrises had featured in the activities of the town from the 16th C. onwards,[19] and Loose, with isolated families in other villages such as East Farleigh. Even so the concentration in Maidstone and Loose was formidable. For instance, between 1654-1742 at least 14 families of Harris were producing offspring in Maidstone alone and these included two sets of William and Elizabeth.

In Loose there were 4 families between 1690-1772 (with another 3 following these) and one branch which moved to Ditton, producing 2 sets of Thomas and Mary. As well as these there were a number of isolated Harrises, connected with papermaking activities, who cannot be allocated to any of these families other than by guesswork. Keeping Table XXI in mind, these include a Richard Harris, "a Paperman" buried in Loose in 1703;[20] an Elizabeth Harris who married George Hunt in Loose (1706) and described as "both of Loose"; 2 William Harris, "Paper Mold Makers", buried in Loose in 1757 and 1765; and in addition to these we have to remember that James Whatman II's mould maker in the 1780's was Thomas Harris, whose domicile has not been located yet. As to whether the latter was connected with any of these is discussed below.

With Maidstone on one side and East Farleigh on the other (where Arthur Harris, the vicar, had a large family between 1687-1700), it can be seen that the problem of constructing a tree will have been a complicated one.

The identification of William Harris snr. and his papermaking descendants

Through the Will of James Whatman (1656-1725), drawn up in 1721, we know that William Harris and his daughter, "Liddea" Harris and his son, Thomas, were all beneficiaries. Two points to note are (a) William Harris must have been alive in 1721; and (b) the tree shows that among his family were Thomas and Lydia. It is possible that Lydia was singled out because Lydia Whatman had been the original link between the families (a name that was to persist for several generations to come); and that Thomas had benefited because by 1721 he must have been well established

---

[19] L. Monckton, K.S. Martin & F. Wallis "Maidstone Records" (1926)

[20] Since there had been no previous paper mill in Loose to Gurney's, one might speculate that this Richard Harris, Paperman, was William's father who had come from outside the Maidstone area unless he had been a local, unrecorded paper mill employee (not seen in any of the PRs). It has already been noted (Chap.II n.21) that William's origin has not been traced and the notion that this Richard (paperman) was his father would do much to account for his status as Ancient Paper Maker which might derive from Richard having been a Master at possibly Eynsford (1650's); or another Richard at nearby Sutton–at–Hone (3 miles further north) in 1654; or even a Richard at Dartford (a further 2.5 miles north) who had 3 sons (1667, 1668 and 1678) but no William whose estimated birthdate is 1663 or earlier. (re Ancient Paper Maker status see discussion in App.IV under "The Russells").

as a "Tanner of Wrotham".[21] The Loose Ratebook shows that a William Harris occupied Gurney's paper mill from 1689–1727;[22] and the only William Harris (snr.) recorded in the Loose Registers for this period[23] is the one shown in Table XXI together with his (second) wife and family. All this information fits together, and is further supported by the fact that the two families, Whatman and Harris, lived in close proximity to one another (see Map 3 Vol.I) in a manner which points to William Harris, the paper maker of Loose and the brother–in–law of James Whatman (from Brenchley) a tanner of Loose, as being one and the same person; it would be very singular, if this had not been the case. To clinch the matter even further, William snr.'s seventh child was Richard, born and buried in Loose, who was closely linked with the elder James Whatman (paper maker) and who was a paper maker himself.[24] If this identification is accepted (as surely it must be), then the following proposals can be made.

William Harris jnr. (1691–1741)

A William Harris became the paper maker at Millhall mill, Ditton, in 1724. At one time it was thought that this might have been the Ancient Paper Maker; but had this been so he would have been ca.60 years old, would have married for a third time and produced offspring who ultimately took over this mill.[25] This idea is a highly improbable one particularly in view of the fact that a much more satisfactory alternative exists.

The more credible alternative is to propose that the paper maker of Millhall mill was William jnr. In Table XXI it has been suggested that William jnr. married Mary Allen of West Farleigh at All Saints', Maidstone in 1717. There is no other basis for selecting Mary Allen than the fact that this is the only William Harris/Mary marriage found that would fit in with the age when one might have expected William to marry; and Mary was the name of Thomas' mother given at his baptism in Loose in 1721.[26]

The fact that William and Mary had a son, Thomas, baptised in Loose in 1721 lends further support to the view that this was William snr.'s son since no other Williams figure in the Loose register and the date of birth would fit in with his age, date of marriage all before their removal to Ditton. Additional support comes from the fact that after his death in 1741, Thomas eventually succeeded at Millhall mill. In 1741 Thomas would have been under 20 and too young to have taken over direct control of the mill. Indeed one finds that the actual succession is Mrs. Harris from 1741–44 and Thomas from 1744 when he would have been 23. (For more on Thomas see below).

Finally, it is proposed that the entry in the Loose Register that reads "William Harris, Paper Maker from Ditton", buried on the 17:Apr:1741, must refer to William jnr. (William snr. would have been at least 77 by that time and is unlikely to have been described as a paper maker of Ditton at that age). The Mary Harris, widow, buried in Ditton (1765) could have been William's widow, although another Mary Harris, widow, was buried in Loose in 1770 and on balance it is thought that the latter may have been William jnr.'s widow, seeing that her husband had wished to be buried in Loose; the former could well have been her son Thomas' widow (see below).

It may be noted from Table XXI that only one child of William jnr. and Mary has been traced, namely Thomas. A William and Frances Harris were producing a family in Loose during the 1740's (see family tree below) and one wonders whether this William might not have been Thomas' brother, perhaps baptised elsewhere.

Thomas Harris (1721–1763) and subsequent succession at Millhall mill, Ditton

Thomas Harris took over Millhall mill in 1744 and at some point before this he married Mary (maiden name not discovered) and their first child, William, was baptised in 1743. It is at this point that more difficult problems arise

[21] The conveyance of The Square and Turkey mill to Richard Harris in 1738 took the form of a conveyance to James Whatman of Loose, tanner (and later the paper maker), and Thomas Harris of Wrotham, tanner (and presumably Richard's elder brother). For details see Bib.1 8.
[22] Spain, R.J. Bib.10 55. (This has been incorrectly identified by Shorter as Little Ivy Mill. For correction see this Appendix note 81).
[23] Loose PR. KAO P 233 1/2.
[24] See this Appendix, Table XIX.
[25] Fuller, M.J. Bib.13 78ff. : and the Ditton "Church Book" KAO P 118/5.
[26] The All SS.Maidstone PRs show that there were other roughly contemporary William Harrises producing families, but their wives in each case were named Elizabeth and were probably descendants of Maidstone families (for details see working papers).

(a) in discovering just how many children Thomas and Mary produced; and (b) just when Thomas, their father, was buried although it seems almost certain to have been 1763. It would be the simplest course to take the burials first.

The Ditton Register[27] records 3 Thomas Harris burials, 1763, 1769 and 1798; and a William Harris in 1766. These have been allocated as follows: Thomas, the father, (less than a month after insuring the stock and utensils in his paper mill)[28] in 1763 and the reason for this is that the word "Snr." was added to the burial register after this entry had been made; Thomas, the son, in 1769 and the reason for allocating this date to him is solely due to the fact that he does not appear as successor at the mill after his father and brother, William, had both died; whereas had he been the Thomas who lived till 1798, one would have expected him to do so, since the next known occupant is not identified until 1776 when Thomas Golding insured the utensils and stock in his paper mill.[29] The William Harris who died in 1766 is assumed here as being Thomas' eldest son; there are no other Williams in the Ditton register who could satisfy this entry. Who then occupied the mill in the intervening years? The answer to this is not known, but some suggestions are made below, when we consider Thomas' children.

As Table XXI shows, in 1768 a Mary Harris married Thomas Golding who is known to have been the occupant of the mill 8 years later. We learn two things from this (a) that Thomas Golding was there 8 years before he took out his policy; and (b) the parish register shows that they were producing a family in Ditton soon after their marriage indicating that they were a young couple and that Mary could not have been Thomas snr.'s widow. On the face of it, it looks as if the Mary Harris, whom we are concerned with here, was related and, indeed, could have been a daughter of Thomas snr.'s not baptised in the parish. Table XXI shows that there is not much of a gap between William and Lydia, but that there is a considerable gap (5 years or even 8) between Lydia and the next child. Elizabeth is somewhat doubtful; her baptism is indexed, but does not appear in the transcript; and Thomas, for some curious reason, was baptised privately on 26:Dec:1758 and admitted into the Church on 12:Jan:1759. There is a possibility then that the Mary who married Thomas Golding may have been a daughter. One might speculate that after Thomas snr.'s death Mary, his widow, had continued to run the mill, perhaps assisted for a time by William and later by Mary until such time as she married Thomas Golding.[30]

William and Frances Harris of Loose : was this a family of Mould Makers?

Parallel to Thomas and Mary of Ditton a William and Frances Harris of Loose were producing a family there between 1737–1752.[31] It is not known from whom this William had descended; but, as mentioned above, only one child of William (1691–1741) and Mary has been traced, namely Thomas of Ditton. It is possible that William and Mary may have had an elder son, perhaps, baptised elsewhere and brother to Thomas of Ditton (one would normally expect an eldest son to have been named William). The tree overleaf illustrates William and Frances' family; and all the baptisms shown are recorded in the Loose Parish Register.[32]

The occupation of William (the husband of Frances) is not known; however, two William Harris mould makers were buried in Loose, 1757 and 1765. It is possible that one of these "mold makers" may have been this William; and, further, his son, Thomas, can be seen as a suitable candidate for James Whatman II's mould maker of the 1780's.

[27] Ditton PR KAO TR 1732/19
[28] Shorter A. Bib.12 185 SFIP 202650 23:Sept:1763.
[29] Ibid. SFIP 373642 20:Sept:1776.
[30] That Mary was a daughter and that Mary, Thomas snr.'s widow, participated in keeping the mill going is the simplest of several alternatives. There were, for instance, two Mary Harrises in Loose (dtrs. of William & Frances and Charles & Elizabeth bapt. in 1745 and 1744 respectively); and a Mary Harris bapt in West Malling in 1746. On the other hand it was customary to give children their parents' Christian names, so one might expect a "Mary" to have been in Thomas snr.'s family. Moreover a birthdate between 1746–51 would suit a marriage date of 1768 better than a Mary bapt. in the 1740's.
[31] A Charles & Elizabeth Harris also make a single appearance in the Loose PR with the baptism of a daughter, Mary, in 1744.
[32] KAO P 233 1/2.

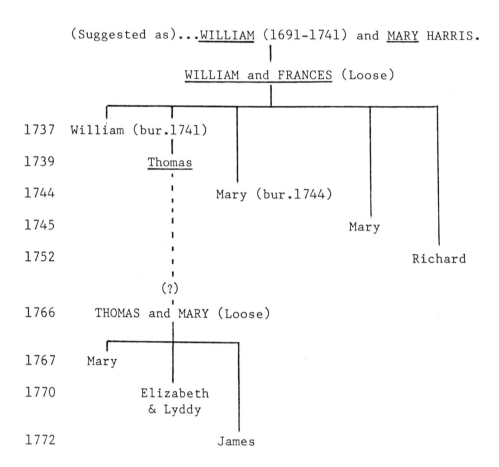

(Suggested as)...<u>WILLIAM</u> (1691-1741) and <u>MARY</u> HARRIS.

<u>WILLIAM</u> and <u>FRANCES</u> (Loose)

1737	William (bur.1741)
1739	Thomas
1744	Mary (bur.1744)
1745	Mary
1752	Richard
	(?)
1766	THOMAS and MARY (Loose)
1767	Mary
1770	Elizabeth & Lyddy
1772	James

In addition to suggesting that William, husband of Frances, may have been the grandson of William, the Ancient Paper Maker because the dates and place of birth of his family would fit in with this, there is to say the least of it very tenuous support given to this idea by the fact that one of Thomas' children, a twin, was named "Lyddy". The name Lydia seems to have been popular in the Harris family, possibly in memory of Lydia Whatman or, going further back, Lydia Waggoner; they may have been legendary figures to the Harris family.

Christopher and Mary Harris of Boxley

During the 1720-1730's a Mr. Harris (and after 1729 a widow Harris; entries disappear by Dec.1736) is recorded in the Boxley Ratebook as an Outdweller. It is not thought that he was any relation of Richard Harris, son of the Ancient Paper Maker, who came to Boxley (Turkey mill) at the end of 1736. The Mr. Harris referred to was undoubtedly the youngest son of Walter and Mary Harris of Maidstone, baptised in 1687.[33] The Will of Walter Harris names Christopher (1716);[34] and Christopher (described as a Maltster) left his estate to his wife, Mary, and he died in 1729 (a date that agrees with the cessation of his assessment in Boxley);[35] and Mary's Will was proved in 1736.[36] None of the beneficiaries mentioned in these Wills show any relationship with the Harris of Loose, a conclusion based on present findings.

[33] All Saints', Maidstone Parish Register.
[34] KAO PRC 17 Book 59 155.
[35] KAO PRC 17 Book 60 444.
[36] KAO PRC 17 Book 61 871.

Summary

In spite of the fact that the name "Harris" occurs very frequently in the Maidstone District, the suggestions made above regarding the descendants of William Harris, the Ancient Paper Maker, are all reasonably credible. One can say that the identity of William Harris, himself, has been established with certainty; his connection with the Whatman family known; and once this has been achieved one can be certain also that two of his sons were paper makers, one at Millhall mill, Ditton, and the other the paper maker with whom James Whatman was closely associated; these were William (1691–1741) and Richard (1703–1739). The succession at Ditton by the Ancient Paper Maker's grandson, Thomas (1721–1763), has the same degree of certainty. But one is on less sure ground after this with the proposal that Thomas' family including his widow and a suggested daughter Mary bridged the gap to Thomas Golding. Again, the relationship and occupation of William and Frances and their son, Thomas, is based on a plausible but less certain proposal. Finally, mention has been made at various points in the text and this Appendix that possibly the Richard Harris, paperman (d.1703 Loose) might have been the Ancient Paper Maker's father, having perhaps been a paper maker at Eynsford or Dartford in the 1650's/60's.

ANN CARTER of LEEDS VILLAGE, KENT

The marriage entry (see below) for Richard Harris and Ann Carter shows that she came from the village of Leeds. The search was, therefore, extended to the Leeds parish register and the 2 following entries were found:-[37]

| 1708 | May 30 | Anne d.o. | Thomas & Sarah Carter | (Bapt). |
| (1721 | Dec.24 | Anne d.o. | John & Anne Carter) | (Bapt). |

The second Anne would have been only 13 in 1734 and it is clear that she could not have been Richard's bride; but she may have been a grand-daughter or niece of Thomas. There were very few Carters in Leeds.

Richard and Anne were married at St. Saviour's, Southwark and the entry in the Register reads as follows:-[38]

1734   (Nov.2)   Richard Harris of Hollingbourne in Kent m. Ann Carter of Leeds in Kent.

It was, apparently, quite common for country people of quite modest origins to marry in London and not in their home parish.

Regarding Ann's background, the Leeds Burial Register provides the following information:[39]

1702	(Jan.19)	Elizabeth d.o. Thomas Carter
1737	(Dec.9)	Thomas Carter (presumably the father).
1744	(Dec.16)	Sarah Carter, Widow (presumably the mother).

Between 1688 and 1744 there were only 8 entries of the name Carter under Burials in Leeds indicating that they were not very numerous in the parish. Among the Leeds Parish Documents (Bills and Vouchers 1721-1747) a bill (d.o.d. 30:Sept:1725) was found "....for timber bought of Thomas Carter for ye use of Leeds Church..."[40] Thomas Carter may thus have been a builder or timber merchant; a more searching examination of Leeds documents might uncover his occupation more precisely.

---

[37] Leeds PR KAO P 222 1/7
[38] St Saviour's, Southwark Marriage Register (Greater London Record Office) P 92 / SAV / 3039.
[39] Leeds Burial Register KAO P 222 1/6.
[40] Leeds Parish Church Documents (Bills and Vouchers 1721-1747. P 222 5/8).
     The paper used for this Bill contained one of two double line "W" Countermarks found on
     their own (or in the case of the other one, see Chap.III note 18, in the same sheet as the Arms
     of Amsterdam). The linkage of the two "V"s is distinctive. (For double line "W" countermarks
     see notes in App.V Table XXVI and pp.278-279).

The events of Ann Carter's life can be summarised as follows:-

30:May:1708	Bapt. Leeds, Kent.
2:Nov:1734	Marr. Richard Harris (Southwark).
7:0ct:1735	Bapt. Anne (dtr.), Hollingbourne.[41]
26:Nov:1739	Bur. Richard Harris (Loose).
7:Aug:1740	Marr. James Whatman I (All SS.Maidstone).
25:Aug:1741	Bapt. James Whatman II (Boxley).
1744	Bapt. William (bur.Loose 1745).
24:Aug:1747	Bapt. Susanna Elizabeth (Boxley).
29:Jun:1759	Bur. James Whatman I (Loose).
1759–1762	Interregnum at Turkey Mill.[42]
1762	Retired to Devonshire with her dtr., Ann (Harris).
1789	Died at Exeter.

Thomas Balston pointed out that it had proved impossible for him to find any direct evidence of the elder Whatman's education or character. The reader will be well aware by this time that despite the mass of genealogical data that has been accumulated in this work on various paper makers there are no records to tell us what sort of lives they lived or what they were like. From this aspect then we really know no more about the elder Whatman's character than Thomas Balston did, although from what has emerged in this book one can derive a certain impression of this. He was obviously a very thoughtful, inventive and methodical person and must also have possessed a good measure of drive in his personality to have achieved what he did. But can one penetrate beyond this impersonal assessment of him ? At present, only in our imagination.

He was the only surviving child of the James from Brenchley and with a father approaching 50 he would have been a lonely boy had he not formed a close friendship with Richard Harris at an early age, a bond that clearly strengthened with time and lasted up to Richard's death in 1739. At some point before November 1734, when James would have been 32 and Richard 31, Ann Carter came into the picture, an event that invites speculation.

Leeds village is some 4 miles from Loose and one can only surmise that they all met whilst the young men were reconnoitring a suitable site for their projected mill, but how this all came about we shall never know. One gets the impression, however, that Ann must have been something of an attraction, somebody with a warm and positive personality. It is possible that both Richard and James fell for her with Richard winning her hand. But it evidently did nothing to disrupt their friendship: That this is not just romantic phantasy is supported by their later relationship.

After Richard's death James who was by then 38, and one would think a confirmed bachelor, could either have looked after Ann's interests dispassionately or just married her to give her,

---

[41] KAO P 187 1/1.

[42] Between 1759–1762 Ann was nominally in charge of Turkey Mill, a period for which there ia at least one Ann Whatman countermark (see App.V Fig.15). For evidence that she probably had some assistance in these duties from her son see Bib.1 18.

his lifelong friend's widow, some sort of protection. But the facts are that within 5 months of Richard's death they were engaged and 3 months later Ann had married James. Even up to this point one could still view all this as a business–like arrangement; Ann was assessed as the occupant and would obviously have needed a new paper maker to run the mill, James being an unmistakable candidate for this task; the marriage settlement makes this quite plain. But was this in fact the sole reason ? Within 3 months of her marriage Ann had conceived her son, James, the next paper maker to be; and followed, later, by two more, William and Susanna, the latter when she was 39.

There was a human side then, albeit only a glimpse of it, to the otherwise insubstantial image we have of the man, though perhaps not such a passionate paper maker as John Larking proved himself to be some 50 years later and at an almost identical age (see App.IV note 121).

As to Ann, she survived her husband by 30 years. She was obviously mistress of the business during the interregnum as her personal countermark demonstrates; and presumably she moved to Devonshire because she preferred the company of her daughter to the changing social régime at Turkey Court following her son's marriage in 1762.

QUELCH FAMILY and JOHN TERRY : and CLEMENT TAYLOR and Family

Although the Quelch connection with papermaking at Dartford in Kent may date back to 1673 (this being the first documented date referring to Edward Quelch, one of the Ancient Paper Makers who signed the Petition in 1690), in this book interest is primarily centred on William Quelch snr. (1679–1775); his son, William jnr. (1704–1777); Quelch snr.'s sons–in–law, John Terry (1711 ?–1775); Clement Taylor snr. (1718–1776) and his four papermaking sons; and more indirectly John Russel's family. It is also appropriate that a cousin of Clement snr.'s be added to this list (see below).

They are of twofold interest in that (a) they had wide ranging papermaking activities in Kent during the 18th C. (see TABLE XXII and Chap.VII under "The Quelch Empire") occupying between them the three mills in Dartford; converting Basted mill in the Nepicar Ward of Wrotham parish; occupying Gurney's mill in Loose on the departure of William Harris; Whatman's Hollingbourne Old Mill; and, by extension to the Clement Taylor family, Basted mill, Poll mill, Upper Tovil mill; Gurney's and Old Mill, Hollingbourne, for a second time; and later associations with members of John Russel's family in the Maidstone district; and, finally, Clement Taylor papermaking activities in Ireland; to this one might add John Taylor of Wrotham's lease of Carshalton mill, Surrey, between 1744–75. And (b) because of their contacts with other noteworthy Kentish paper makers of the period e.g. Richard Archer (Dartford No.2);[43] William Buttanshaw (Hampton's mill, West Peckham);[44] the Finchs (mainly William, Quelch jnr.'s godson) at Dartford No.3;[45] and Thomas Budgen, John Terry's son–in–law (Dartford No.2).[46] They can be seen, therefore, to represent an important (though perhaps not very inventive) body of paper makers in the Kent papermaking scene of the 18th C. There are, however, a lot of gaps and other uncertainties in the genealogical data relating to the Quelch and Terry families which make it difficult to provide anything much more than a skeleton tree to indicate their connections.

The Quelch Family

Dunkin[47] believed the Quelchs to have had a German ancestry and that they had come over with Spilman, a suggestion that has been adopted by some later authors. But dictionaries of names e.g. Bardsley's[48] gives Quelch as "Welsh or Welshman" and this is supported by the facts (a) the name is found in England long before Spilman's time; and (b) there were concentrations of them in Berkshire and Oxfordshire in the 16th and 17th Cs.[49] Though it is an unusual name, in these searches the experience has been that the name is not often found where one might expect it to be e.g there is only one mention in the Wrotham PRs (and

---

[43] See App.IV under Archer and Chap.VII p.286.

[44] William Quelch snr. took on William Buttanshaw, bapt. Wrotham 1715, as an apprentice at Basted mill in 1723 (Shorter, A. Bib.12 196). For further information on William Buttanshaw, his family, and the paper mill at West Peckham see Chap.VII. pp.293–295.

[45] There was a William, James and John Finch all paper makers in Dartford. Shorter has suggested that William worked the mill, Dartford No.3, for John Finch (Bib.12 182). Dunkin, J. Bib.46 gives the relationship of William to Quelch jnr.

[46] Dunkin, J. Bib.46.

[47] Ibid.

[48] Bardsley, C.W. "Dictionary of English and Welsh Surnames" (Frowde, London 1901).

[49] Shorter, A. Bib.12 44 Note 13.

TABLE XXII

DATES	DARTFORD No.1	DARTFORD Nos. 2 & 3	BASTED MILL, WROTHAM	GURNEY'S MILL, LOOSE	OLD MILL, HOLLINGBOURNE
1588-1626	JOHN SPILMAN (Proprietor)				
1626-1641	John Spilman jnr. (Proprietor)				
1651	William Blackwell (first appearance in RB, Shorter believed as occupant of mill)				
1673		(Edward Quelch m. Anne Wood)			
1679		William Wood assessed for mill			
1687				Conversion from fulling to paper mill starts	
1689	WILLIAM BLACKWELL (Ancient P'Mkr.)	[EDWARD QUELCH] (Ancient Paper Maker) late Mr.Wood and Paper Mills			
1690				WILLIAM HARRIS (Ancient Paper Maker)	
1692		COMPANY OF WHITE PAPER MAKERS 1			
1693-1697					
(1698-1704)	(No RB assessments for Blackwell)				
1698		George Hagar 2			
1698-1700		JOHN QUELCH			
1701-1740		Richard Archer (proprietor : later occupier).			
1702-1717	WILLIAM QUELCH Snr. (Assessed for £12 instead of £24 as Blackwell had) been				
1716			WILLIAM QUELCH Snr. 1st.assessment 3		
1717	late William Quelch				
1723			Wm.Quelch takes on Wm.Buttanshaw Appr.		
1724	End of Dartford No.1 Paper Mill				
1727				last Wm.Harris Snr. entry	
1728				WILLIAM QUELCH (could be Snr. or Jnr. see genealogical data. In fact both were there.)	
1733		William Quelch insured the mill in the tenure of Richard Archer. 4			James Whatman built mill and installed Richard Harris as paper maker (1733)
1736					(Richard Harris working Turkey Mill) (1736)
1739		Wm.Quelch insured 2 P'Mills in occup: of Self and John Terry. 5 (Wm.Snr.)		(Wm.Quelch of Loose party to insurance of Old Mill, Hollingbourne) Snr.	Quelch & Terry insured mill; Terry in occupation (see App.I.). (1739-41)
1741				Wm.Quelch insured house, mill & stock at Loose (SFIP 87455 20:Apr:1741) Jnr.	John Sanders, occupant. (1741-5)
1742			Clement Taylor snr. insured mill in own occupation (SFIP 93923 26:Jan)		
1743				Wm.Quelch takes Appr. Abrh'm Musgrove. Wife, Ann, bur.Loose 9:Feb:1745)	
1746				Wm.Quelch jnr. of Dartford insured mill in tenure of Stephen Scott. 6 Scott replaced Quelch in RB	James Whatman I (1746-8)
1747					
1749				Abraham Musgrove replaced Scott	Henry French Snr., occupant. (1749-60)
1751				Abraham Hillier replaced Musgrove	
1759-1760				Henry French snr.	
1760-1770					Henry French Jnr., occupant. (1760-70)
1762			Clement Taylor took on Apprentice 7		
1766-1767		JOHN TERRY d.1766. Widow Eleanor insured (No.2) mill in 1767			
1770-1774		WILLIAM QUELCH Snr. died.			James Whatman II (1770-4)
1775			Clement Taylor snr.died John took over (see Note 8)	Henry French snr. died : Thomas French had been in charge since ca.1771.	Clement Taylor jnr., occupant. (1774-93)
1776		WILLIAM QUELCH Jnr. died.			(Henry French jnr. bur.Hollingbourne) (1776)
1777					
1778		Thomas Budgen insured Dartford Mills 9			

No Dartford RB between 1725-1753

TABLE XXII – THE PAPERMAKING ACTIVITIES OF THE QUELCH, TERRY AND CLEMENT TAYLOR FAMILIES

1 The suggestion was made in Chap.IV pp.140–143 & App.IV pp.142–143 that Peter Archer snr. may have managed Dartford No.2 mill for the COMPANY during this period.

2 As almost certain this George Hagar is one and the same as the Hagar, a dyer by trade, one time opponent of the COMPANY and former proprietor of Eynsham mill. It was proposed in Chap.IV pp.144 + n.32 that his arrival in Dartford was not fortuitous.

3 Wrotham Overseers' Accounts (KAO P 406 11/17). The assessment in 1716 was for £10 and remains at this level until 1720, when the piece terminates. (The 1706–1720 document is described as "unfit for production" but was seen nevertheless). The next piece was also "unfit" and is too fragmented to be able to confirm that the assessment was still £10 in 1723, when we know William Quelch was there at the mill. The evidence given here points (a) to a new assessment of £10 in the Nepicar Ward where the paper mill was established; and (b) it was made against William Quelch, known to have been a paper maker and who continued to be one.

4 SFIP 72367 (corr:date 6:Jan:1737). Quelch must have bought the mill from Richard Archer snr. at the end of 1736.

5 Quelch divided Dartford No.2 mill into two, No.2 and No.3 (1741/2), with Terry occupying No.2 (see Text for discussion).

6 SFIP 104723 (27:Mar:1746). This is the first positive mention of William Quelch jnr.

7 William Appleton. This may have been the action of Clement Taylor jnr. as CT snr. had moved to Poll mill (next to Turkey mill) in 1761.

8 In 1776 John succeeded at Basted; James, another son, at Poll mill; and Clement jnr. at Upper Tovil mill.

9 Initially, it appears that Thomas Budgen, John Terry's son–in–law, took over both mills. Later, one of the Finch family occupied No. 3.

** Much of this Table has been compiled from information obtained from Shorter, A. Bib.12 supplemented and corrected from other sources including the author's own searches.

DARTFORD (Cumberland Transcript KAO TR 1303/2).

DATES	ENTRY	PERSONS CONCERNED	COMMENTS
1673	marr.	Edward Quelch & Anne Wood	Possibly William Wood's sister.
1675	bapt.	Thomas s.o. Edward Quelch	Wife's name not given.
1676	marr.	Edward Quelch & Ellin (sic) Outram	
?	(b.)	(John)	See text for this proposal.
(?1679)	(b.)	(William snr.)	See text for this proposal.
1682 (corr:date)	bapt.	Elizabeth d.o. Edward & Ellenor Quelch	
1684	bapt.	Edward s.o. Edward & Ellenor Quelch	There are 3 Edwards to explain:- 1686 bur. at neighbouring Sutton-at-Hone. [1] 1717 bur. Dartford (see below). 1712 a third Edward producing a family; could be the Edward bapt.1684 (see below).
[1702]	marr.	[Wm.Quelch of Dartford & Eliz.Terry of Wilmington]	[Wilmington PRs]. First reference to WILLIAM SNR.
1702	bapt.	Elizabeth d.o. William & Elizabeth Quelch	
1704 (24:Sept).	bapt.	William s.o. William & Elizabeth Quelch	This is WILLIAM JNR.
1711	bapt.	Eleanor d.o. William & Elizabeth Quelch	marr. JOHN TERRY William Quelch snr.'s journeyman; occupied Hollingbourne Old Mill 1739/41; daughter, Sarah, bapt.Hollingbourne (KAO P 187 1/1) 22:Feb:1741 corr:date. Returned to Dartford No.2 mill 1741. [2]
		(William snr. & Elizabeth Quelch disappear from Dartford PRs until the 1740's and after. They must have had Sarah, Clement Taylor's future wife, sometime between 1711-ca.1720. Where ?).	
1717	bur.	Edward Quelch	This is the second Edward burial (see above)
1712-1720	bapt.	Edward & Elizabeth Quelch 4 children bapt., the last, Edward, bapt. and bur. 1720.	This is the 3rd. Edward (see above).
1720	marr.	Thomas Quelch & Elizabeth Callow	re Elizabeth Callow see next entry below.
1722-1729	bapt.	Thomas & Alice Quelch have 6 children bapt., the first, Edward, bapt.1722.	It is not known who this Thomas was; it could have been Thomas bapt.1675. No explanation is offered for the Elizabeth/Alice problem.

LOOSE (KAO P 233 1/2)

DATES	ENTRY	PERSONS CONCERNED	COMMENTS
1729 (22:Nov).	bapt.	Eleanor d.o. William & Eleanor Quelch	On present data this could be William Snr. or Jnr. The entry below shows that 2 William Quelch families were in Loose then. (See also burials in Dartford, below). Daughter, Eleanor, buried same year.
1732	bur.	Elizabeth d.o. William & Ann Quelch	was this Jnr. ? Ann bur. Loose 9:Feb:1745 (see Note 5).
1737 (29:Mar). (corr:date)	bapt.	John s.o. William & Eleanor Quelch	John was buried the same year.

UNKNOWN

DATES	ENTRY	PERSONS CONCERNED	COMMENTS
ca.1740	marr.	Clement Taylor Snr. & Sarah d.o. Wm.Quelch Snr.	Clement Taylor was born in 1718 [3] (possibly in Ightham). All his children bapt. in Wrotham. Buried Ightham 1776. Sarah must have been born ca.1720 or earlier; bapt. not in Wrotham PRs. Had William Snr. remarried ? [4]

DARTFORD (Cumberland Transcript KAO TR 1303/2).

DATES	ENTRY	PERSONS CONCERNED	COMMENTS
1741	bur.	Elizabeth Quelch	Might have been any one of 4.
1762	bur.	Hannah w.o. William Quelch	Was this Wm.Jnr.'s (or Snr.'s) second wife ? [5]
1771	bur.	Elizabeth w.o. William Quelch	Was this William Snr.'s Elizabeth aged ca. 90 ? [5]
1775	bur.	William Quelch	Dunkin makes this entry refer to William Snr.
1777	bur.	Mr.William Quelch	This must be William jnr. ?

1. Sutton–at–Hone PRs. start 1607. This is the only Quelch entry there (bapt:marr:bur). No Quelch entries at Darenth (PR starts 1678).
2. John Terry died in 1766 according to Dunkin, J. Bib.46. Probably related to Elizabeth Terry of Wilmington, Wm. Quelch snr.'s wife.
3. Inf. Bridge, W.E. "Some Historical Notes, Basted Mill" (1948 : pamphlet).
4. No Quelch entries seen in either the Plaxtol or Ightham PRs; and only 1 not relevant one in the Wrotham PR.
5. The fact that the Quelchs left Loose ca. 1746 and that *William jnr. of Dartford* insured Gurney's for that year (see Table XXII) may point to William jnr. moving because of his wife's death; in other words Ann would thus have been William jnr.'s first wife. But who then was the "Eleanor" w.o. William in Loose ? It is also of interest to note that although there are Quelch entries in the Loose PRs, the name Quelch is entered as O.D. in both Loose and E. Farleigh RBs. up to 1740.

this is not relevant) and none in neighbouring Ightham and Plaxtol registers;[50] and, further back in time the difficulties increase. It is regrettable, but a necessary limitation, that the search for William Quelch snr.'s origin, and that of John Quelch and his subsequent movements, has not been extended to other counties or to London and its immediate environs, where the name is sometimes found at this time. Because of their papermaking interests the Kent Quelchs' family origins were probably Oxfordshire or, possibly, Berkshire (see below).

The marriage of Edward Quelch to Anne Wood (1673) is the first mention of the name found so far in Kent (see Table XXIII). As it seems likely that this Edward was the Ancient Paper Maker, it has been suggested by more than one author that he was related to another contemporary Quelch paper maker, Thomas at Wolvercote mill, Oxfordshire and first mentioned in the Wolvercote PRs in 1657.[51]

Table XXIII illustrates some of the problems surrounding the early members of the family in Kent. For example, was Edward the sole progenitor or were there other Quelch families? The baptisms of John and William snr. have not been found, but have been estimated from William's age at death. The dates of their births, arrived at in this manner, could be fitted conveniently under Edward and Ellenor's marriage (supported by the fact that one of William's daughters was named Eleanor). Again, there is no evidence that Edward Quelch, the Ancient Paper Maker, was ever assessed for a paper mill. Finally, one has to ask the question was Edward, husband to Anne, the same Edward as the husband of Ellin ? There were more Edward burials in this period than can be accounted for, if they were one and the same person. These subjects are discussed below.

The evidence presented by Shorter suggests that William Wood, trade unknown, may have built Dartford No.2 mill in 1679; he was assessed for this property for that year; in 1691 the entry becomes "Mrs. Wood and Paper Mill"; and in 1692 "late Mr. Wood and Paper Mills".[52] William Wood was obviously assessed for the paper mill for the whole of this period. Anne Wood, Edward Quelch's wife, may have been William's sister or cousin; there were at least three Wood families living in Dartford at that time.[53] The inference from this is that Edward must have worked as paper maker for William Wood, the proprietor, and that for some reason the assessments were made against the latter.[54] The mill had passed out of their hands by 1693 and it was not until 1698-1700 that it was worked again by a Quelch, John, the property being assessed this time in his name. Even if the opportunity had arisen John, if born ca.1677,[55] would have been too young to have occupied the mill before 1698, when he would have been 21.[56] On present evidence then Edward Quelch, husband of Anne,

---

[50] The only Quelch entry in Wrotham PRs. (KAO TR 1303/15) is ca.1725 Newman m. Quelch. No Quelch entries between 1724-41 in Ightham (KAO P 202 1/2) and Plaxtol (KAO P 406 1/1) Registers.

[51] Shorter, A. Bib.12 226. See also Note 2 in TABLE XXII.

[52] Shorter, A. Bib.12 180.

[53] Families produced by John Wood (act. from 1675); Mr. Richard and Margaret Wood (act. from 1679); and William and Frances Wood (act. from 1681).

[54] A parallel might be drawn here with the assessments made against the COMPANY who followed, 1693-97; they were made against the COMPANY and not an individual. For discussion see Chap.IV p.139.

[55] Reference to a John founding a paper mill at Brookland ca.1694, if valid, points to either an earlier birthdate or another John altogether (Chap.IV note 32).

[56] The reappearance of Hagar and a suggested relationship of this event to John Quelch's occupation is discussed in Chap.IV p.144.

was an Ancient Paper Maker working for the Wood family unlike William Blackwell, the other Ancient Paper Maker in Dartford, against whom assessments were made for No.1 mill. John Quelch disappeared from this scene in 1700; the question as to whether he may have been the paper maker who appeared at West mills, Newbury, Berkshire in 1748 is discussed in Chapter VII under St. Mary Cray paper mill.

To return to the Edward Quelchs, if one assumes that Edward husband of Anne and Edward husband of Ellin were two different people, then it is possible to account for the 3 Edward burials. The first Edward through his Wood connection must surely have been the Ancient Paper Maker surviving till 1690 to sign the Petition; from this it follows that he was the Edward buried in 1717. If this was so, the Edward husband of Ellin, must have been the one buried at Sutton–at–Hone in 1686. He certainly appears to have had no children after 1684. John and William snr. could thus have been the sons of either Edward or else have come from another family outside Kent.

When one considers the later Quelchs, the first documentary evidence relating to William snr. was his marriage in the nearby parish of Wilmington to Elizabeth Terry in 1702, possibly an aunt or cousin of John Terry, later her son–in–law.[57] At the same time, according to the Ratebook, he occupied Dartford No.1 paper mill, assessed at half the ratable value for which Blackwell had been assessed, from 1702–1717. The low ratable value coupled with Quelch abandoning it in 1717 in favour of Basted mill indicates that the mill had deteriorated to a point beyond worthwhile repair and, even more to the point, that it had been outmoded by Dartford No.2 mill, probably as a result of improvements made to the latter by the COMPANY.[58] In the event Dartford No.1 mill closed ca.1724.

William snr. died in 1775 aged 96[59] (one of two William Quelch burials at this time TABLE XXIII), making 1679 a reasonable date for his birth. The fact that John and William both started their tenancies at about the same time (q.v.) suggests that the former was the elder, but the identity of their parents remains unsolved. Such signs as there are conflict with one another; Edward snr. would account for their papermaking interests; William snr.'s daughter, Eleanor, points to the other couple as parents; and, finally, William snr. does not appear to have had an Edward in his family, so did he come from another family altogether ?

The last ratebook assessment, in 1717, against William snr. at Dartford ties in with his appearance in Wrotham in 1716, the ratebook evidence indicating a new assessment signifying the conversion of a cornmill to papermaking. Developments stemming from this event have been discussed in the text.[60] William jnr. would only have been 12 at the time of this move.

It is clear from the Table illustrating the Quelch family's papermaking activities, and the chronology of these events, that it was William snr. who was the driving force behind these moves; and with such energy it is small wonder that he lived to the age of 96. Both William snr. and jnr. were together for a time at Gurney's mill in Loose and, although William jnr. and John Terry were contemporaries, it was John who was installed in Old Mill, Hollingbourne,

[57] Wilmington (less than 1 mile south of Dartford) Parish Register KAO P 397 1/1.
[58] See also Chap.IV pp.137,145.
[59] Dunkin, J. Bib.46.
[60] Chap.IV under Quelch Empire, and Chap.VII PART IV.

and one sees William snr.'s hand in this arrangement. The motives for all this activity, and the direction in which it was applied, have been examined in Chapter VII.[61] It is more than obvious that having learnt what they wanted to know at Old Mill, they wanted to apply their new found knowledge as rapidly as possible, moving back to Dartford to convert the surviving mill there into two new mills. At the same time, 1741/2, William snr.'s other son-in-law, Clement Taylor snr., took over Basted mill, probably modernising it at the same time.[62] As explained below, and as one might expect, it was William snr. who made the move to Dartford. It is curious, however, that although he created an Empire and undoubtedly made White paper, none that is identifiable appears to have survived.[63]

Although William jnr. was described as "of Dartford" in the Insurance Policy of 1746, it is probable that he had only just made this move because of the death of his wife, Ann, in Loose in 1745 (see Note 5 to the genealogical chart). The 1746 Policy[64] makes it clear that it was "his house in Loose..... and his paper mill near the house". From this one may infer that it was William snr., aged 62, who made the move to Dartford in 1741 and undertook the conversion of No.2 mill assisted by Terry and the experience the latter had gained at Old Mill.

---

[61] Chap.VII pp.280/282.

[62] Chap.VII pp.283/284.

[63] The only Quelch countermarks known to the author are those of Thomas Quelch (d.o.ds. 1680-1685) cited in Shorter, A. Bib.12 265, 344/45; a "W QUELCH" d.o.d. 1780 Eynsford Land Tax Return KAO Q/RP1/130; and a similar "W QUELCH" illustrated in Gravell, T.L. Bib.68 No.573 (d.o.d.1783 Phila. Pa. U.S.A.). The last two of these obviously belong to William Quelch jnr. and in papers made by him before 1777.

[64] SFIP 104723 (27:Mar:1746) William Quelch junior, paper maker, of Dartford insured his house in Loose in the tenure of Stephen Scott, paper maker, and his paper mill near the house.

JOHN TERRY

In contrast to the name Quelch the name Terry is to be found everywhere at this time and among them numerous John Terrys. According to Dunkin (op.cit.) the John in question married Eleanor, daughter of William Quelch snr. and baptised in Dartford (1711). The place of the marriage has not been found, but it obviously would not have taken place until Eleanor had been at Basted for more than 10 years. Since John is alleged to have been a journeyman of her father's, one might surmise that the John baptised in Wrotham in 1711 (s.o.Thomas, a labourer and Elizabeth) might have been the one in question. On the other hand his mother–in–law (as seen under William Quelch snr.) had been a Terry from Wilmington just outside Dartford. John Terrys of about the right age have also been found in other papermaking centres. In view of the common occurrence of his name, one cannot be certain either of his birthplace or whether he was connected with Mrs. Quelch.

The evidence presented in the previous pages regarding Terry's movements, and the mills he occupied, suggests that he was held in high regard by William Quelch snr. as a paper maker, who gave him a key role as the occupant of Old Mill, Hollingbourne, and then accompanied him to Dartford for the conversions there in 1741/2. Apart from this nothing noteworthy is known about the rest of his papermaking career. Although "IT" countermarks have been found, it is thought that these were more likely to have emanated from one of the Taylors, probably James.[65] He was followed at Dartford No.2 by his son–in–law, Thomas Budgen who made countermarked paper.

CLEMENT TAYLOR snr. and Family

Clement Taylor and his family and their activities have been discussed at some length in Chapter VII.[66] It is only necessary to summarize this here. It includes their occupation of Basted mill, Wrotham, as a family from 1742–1802; their involvement while there in various projects promoted by the Society of Arts in the 1750's; their expansion into the Maidstone district, led initially by Clement snr. in 1761 to Poll mill, Boxley, where he was followed after his death in 1776 by James and George. They in turn had a bitter dispute with the younger Whatman ending in their eviction. In the meantime Clement jnr. had occupied Upper Tovil mill in 1772 and enlarged it considerably; he also ran Whatman's Old Mill, Hollingbourne, from 1774 and to judge from the very significant rise in the insurance premium between then and 1791 enlarged this also. He together with his brother, George,[67] crossed swords with many contemporary paper makers ca.1790 over very dubious claims to a Patent on bleaching. Also, at about this time he was getting into increasing difficulties with his workforce and wages, which may have been a contributory reason for making preparations

---

[65] Gravell, T.L. Bib.68 No.407 illustrates, for example, an "IT" countermark in a document dated 1786. John Terry died in 1766 so it is most unlikely that this mark could possibly have emanated from him. Shorter, A. Bib.12 181 reports that a James Terry shared the insurance of Dartford No.2 and No.3 mills with Thomas Budgen and Finch in 1779; but Budgen was certainly producing paper with his own countermark in it very shortly after. So here again it is most improbable that this mark could have been a Terry one. (For further comments on "IT" marks see Chap. VII note 97.)

[66] See Chap.VII PART V.

[67] Clement Taylor jnr. married a Sarah Rolfe of Bearsted and had a son George baptised in 1762. If he survived, he could possibly have been a candidate for the George in this Patent. Whether he was or not is not known.

to set up a paper mill near Dublin in 1792,[68] where presumably labour was much cheaper than in England. Continuing in Ireland, a Taylor worked 2 paper mills there for James Woodmason, who at one time had been Stationer to both the younger Whatman and the Taylor family; this was ca.1819/20.[69] Back again in Kent Clement jnr. succeeded in being elected as one of the Members of Parliament for Maidstone in 1780. Whether this was a real triumph or not for him, we shall probably never know; Maidstone at this time was renowned for its corrupt politics. During the 1790's Clement jnr. was embarking on partnerships with the Edmeads and Russells, but all this was brought to a sudden end by his bankruptcy in 1797. After Clement jnr.'s demise, only John Taylor was left to continue for a few more years at Basted. The Taylor family certainly partook in no small way in the history of papermaking in Kent.

JOHN RUSSEL; and other possible Taylor connections

John Russel was another of Clement Taylor snr.'s sons-in-law; he married Sarah, where and when has not been discovered. John does not seem to have been involved in papermaking, but he shared the lease of the Newnham Court farm land adjoining the younger Whatman's Vinters Estate with James Taylor, the lessee of Poll mill, and thus entered into the Whatman/Taylor dispute though not in such an aggressive role as the Taylor family. He acted as an intermediary between the two parties. But the reason for including him here is (a) he was probably the son of Edward Russell, occupant of Forstal mill, Boxley from 1733 to after 1763[70] or, just conceivably, connected with the Russells of Shoreham[71] and (b) he produced 3 papermaking sons, Edward jnr., Clement Taylor and Hillyer Russell jnr.[72]

Other possible Taylor Connections

As will have been seen in Chapter VII, the papermaking ramifications of Clement Taylor and his family were very extensive, but it is quite possible that this web embraced a number of other as yet unexplored threads that may have drawn paper makers like John Larking and the Wilmotts, at Shoreham, indirectly into its activities, for instance, a connection with the Russells. For the moment, however, there is no positive evidence of liaisons of this kind for the others. There were other Taylor families in Ightham and probably many other relations living in the Sevenoaks district. In the above context it may be noted that:—

---

[68] Shorter, A. Bib.9 232.
[69] Miss P.M. Pollard, letter 4:Sept:1982.
[70] For further details of these Russells and their partnerships see App.III under Synopsis of History of Forstal mill; Forstal mill notes and queried Countermarks; and extracts from the Boxley Poor Ratebooks and Land Tax Returns.
[71] See Chap.VII note 148.
[72] The Presbyterian Chapel, Earl Street, Maidstone Records show that Edward (bapt.1771) was the second, Clement Taylor (bapt.1774) the fourth and Hillyer (twin to Thomas and bapt.1781) the eighth of John and Sarah Russel's children. That they were connected with the Boxley Russells is indicated by the fact that one of their children was named "Hillyer" and it is further assumed that this John was a son of Edward snr. the paper maker at Forstal from 1733–ca.1780 and perhaps a brother of Hillyer snr., Edward snr.'s successor. Edward jnr. married Eleanor d.o. Clement Taylor jnr. (bapt. Ightham 1774).
For a more general discussion of the Russell genealogy problem see App.IV under Alexander Russell.

(i)   1733 John Taylor (widower) m. Clement Larking at West Peckham.
         There was quite a large contingent of Larkings living in West Peckham, which lay
         in between Ightham, the home of the Taylors, and Tonbridge, where John Larking,
         later with papermaking interests in East Malling, is believed to have come from.

(ii)   1769 Geoffrey Taylor (Sevenoaks)   m. Anne Wilmott, d.o. William Wilmott,
                                                  the paper maker at Shoreham.

         1769 William Edmeads (Wrotham)   m. Sarah Wilmott of Shoreham.

         They produced William and Robert Edmeads who, later, were both paper makers in
         the Maidstone district and formed brief partnerships with Clement Taylor jnr. and the
         Russells.[73]

(iii)   In 1744 John Taylor of Wrotham (Kent) leased the "Paper Mill" at Carshalton, Surrey
         from the Scawen family. It has not been possible to confirm John Taylor's earlier
         presence in Wrotham; but a William and Jane (m.1713) had a son, John, baptised in
         Ightham (near Wrotham) in 1716, a date that would fit in with the John who moved
         to Carshalton. The Wrotham PRs. show that a William Taylor of Ightham and
         Elizabeth had a son baptised there in 1736 suggesting that the family had moved
         there, possibly with Clement Taylor snr. who also had a son, William, baptised there
         in 1741 (just before he took over Basted mill). It seems probable that, like Clement,
         John also learnt his papermaking at Basted, but having no kinship with the Quelchs
         he moved elsewhere, in his case to Carshalton. The fact that he rebuilt the Paper mill
         there two years later strongly suggests yet another extension of the Quelch Empire's
         influence. John Taylor was still there in 1776 working with William Curteis of
         Clapham. In the interval the mill was sub–let to the Herbert family, first William
         (d.1755) and after passing to his widow, Susan, it was leased by Robert Herbert,[74] a
         Paper Maker/Stationer and one of the signatories of the Paper Makers' Petition to the
         Treasury in 1765 (see App.IV p.174 : for John Taylor of Wrotham see Paper Maker
         Index).

---

[73] A "T & E" script countermark d.o.d.1799 found in the Susanna Whatman correspondence (PSM) may represent Taylor
     and Edmeads.
[74] Information on John Taylor was kindly supplied by Prof. Alan Crocker.

INFORMATION ON THE FRENCH FAMILY

The French family made White countermarked paper at Old Mill, Hollingbourne (1749–1770); and, later, at Gurney's mill, Loose (1760–1796). Who were they and where did they come from ?

The only positive information that has emerged from this search has been the identification of Henry French snr. as distinct from his two sons, Henry jnr. and Thomas. Shorter's observations of Leeds Ratebook entries had already hinted that a difference existed between the assessments made against "Mr. French" (1749–1761) and those against "Henry French"; but he did not establish the relationship nor discover which of the Henrys was buried in Loose and which in Hollingbourne. The matter is not of great importance, but it does point to the facts that (a) they were all from one papermaking family; and (b) Henry French snr.'s was the hand that directed operations.

A detail that triggered off this investigation was the use of "Mrs. French" for the Hollingbourne ratebook assessment made against Old Mill in 1749 (see Table App.I); and the fact that Whatman "did demise.....the said premises [Old mill] to Mary French etc....." as given in the Tripartite Indenture of 1762 (KAO U 289 T 34). In view of the fact that Whatman had had a disastrous occupant of the mill earlier in Mr. Sanders and then had the mill on his hands for the next few years, could Mary French have been some relation of his or of his wife? Or was there some other explanation ? For instance, could Henry French have been the Hampshire paper maker, formerly at Hurstbourne Priors and described as "a fugitive from debt" in 1743 ? The searches have not revealed the answers to these questions. One thing is clear, French, or his workforce, knew how to use the New Technology to make White paper.

Identification of members of the French family

Apart from the ratebooks and Whatman's account book[75] which refer solely to money matters, the only sources found to date that have proved helpful in identifying members of the family have been the Hollingbourne and Loose parish registers, Insurance Policies and Wills. Information from the former is set out in the Table of Baptisms overleaf.

The baptism of Hannah (1749) is the first French entry in the Hollingbourne PRs. Out of 11 parishes round Maidstone the name French/ffrench had only been found in 3, Leeds,[76] Maidstone[77] and 17th C. only in Aylesford. Turning to the Table, it can be shown that the children baptised after 1762 (Hollingbourne) were those of Henry French jnr. (no burial entry for Mary; nor a Will).[78] Henry snr.'s Will not only confirms this but solves the question of their burials, Henry snr. in Loose (1775); Henry jnr. in Hollingbourne (1776); and that there was another son, Thomas, bur. Loose (1795).[79]

---

[75] Whatman's Rent Book (KAO U 289 [2001] A1) under *Hollingbourn* has entries for Mr. Henry French only from 1759–1767; in other words not in James Whatman I's time. The Hollingbourne RB assessments continued to be made against Henry snr. up to 1760.

[76] The Leeds PR KAO P 222 1/6 shows that a John French was buried there in May 1747; but that is the only mention.

[77] In the latter part of the 17th C. there were at least 3 families of French or ffrench in Maidstone. A John ffrench s.o. Henry ffrench bapt.Nov.1698 is really the only possible candidate. A John French m. Elizabeth Elmet there also in 1732 (All SS. PRs).

[78] There is a Will of a Mary French of Linton (Cant. Archdeaconry Crt. Book 92. 70 9°); n.a. Mary's Estate went to her Wallis brothers in 1745.

[79] Henry French snr's Will PRO PROB/11/1006 fo.94 made in 1775 confirms that he was the father of Henry in Hollingbourne and Thomas in Loose; it makes provision for his *wife* and children with an instruction that the proceeds of the Sale of the mill on Thomas' death (or in the event of the business failing) was to be divided amongst the children.

TABLE OF FRENCH FAMILY BAPTISMS IN HOLLINGBOURNE AND LOOSE

Date	HOLLINGBOURNE Baptisms KAO P 187 1/1 (1556–1799)			COMMENT	
1749	Hannah	d.o.	Henry & Mary French	This is Henry Snr.; evidence below shows that he was in the middle of having a family.	(1)
1751	Charles	s.o.	Henery (sic) and Mary French	bur.1753	
1753	Charles	s.o.	Henry & Mary French	bur.1754	
1756	Jane	d.o.	do. do.		
				No sign of Mary's death in Burials Register. (3)	
1762	Henry	s.o.	Henry & Elizabeth French	Other evidence confirms that the father here was Henry jnr.	
1764	Charles	s.o.	do. do.		
1767	John	s.o.	do. do.		
1770	George	s.o.	do. do.		
1773	Elizabeth	d.o.	do. do.		
1776	William	s.o.	Henry French (decd) & Elizabeth Widow	Hollingbourne Burials show that on 15:Feb:1776 Henry French, Paper Maker, was buried. Other evidence shows this to be Henry jnr.	

Date	LOOSE Baptisms KAO P 233 1/2			COMMENT	
1763	Sarah	d.o.	Thomas & Sarah French	Other evidence confirms that this was another of Henry Snr.'s sons.	(2)
1766	Thomas	s.o.	do. do.	bur. 1766.	
1767	George	s.o.	do. do.		
1773 ?	Catherine	d.o.	do. do.	bur. 1773.	
				Loose Burials show that on 6:Mar:1775 Mr. Henry French was buried. Tombstone describes him as Paper Maker.	
				Thomas French was buried in Loose in 1795. (The mill continued in Widow French's name for 1 more year).	

NOTES
1. The search for these baptisms started at 1725.   2. The search here started at 1730.
3. This observation is made merely to confirm that the "Elizabeth" who follows was not Henry snr.'s 2nd wife.

From this information and Insurance policies for Gurney's mill, Loose, it is clear that Henry snr. had arrived in Hollingbourne with 2 young sons (and possibly other children) from elsewhere. From the baptism of the earliest of the sons' children it has been reckoned that their birthdates must have lain between 1730–43, probably nearer to 1740.

By 1760 Henry snr. must have had sufficient confidence in Henry jnr.'s ability to leave him in charge of Old Mill and launch out by taking over Gurney's Mill, Loose, which had become vacant on the death of Abraham Hillier and which was reoccupied very promptly.[80] The question as to whether Henry snr. modernised or up–graded this mill has been referred to in Chap.VII under William Quelch snr. If nothing else the Frenchs up–graded the paper to White countermarked. By 1771 the mill was handed over to Thomas and insured in his and another's tenure.[81] Widow French continued to run it for a further year after Thomas' death; thereafter it passed briefly into Taylor's and Edmeads' hands (see Chap.VII p.323, n.254). The subject of Thomas French countermarks is discussed in App.V.

---

[80] SFIP 176721 10:July:1760. On a purely speculative basis, was there a Hillier/French connection? Could either Henry jnr. or Thomas have married a Hillier? Out of interest the name Hillier French was to be found in the Maidstone district in the 20th C..

[81] Henry French (snr.), paper maker of Loose, insured his paper mill etc. and two houses adjoining (probably those built by Quelch) in the parish of East Farleigh in the tenure of Thomas French and William Greenway, paper makers (Shorter Bib.12 193 citing SFIP 293987 31:Jan:1771).
Shorter identifies this mill incorrectly describing it as "a paper mill at Loose (probably Little Ivy Mill)"; Balston T. Bib.1 39 n.1 also gives incorrect identification, Little Ivy Mill was not converted to papermaking until ca.1808. Spain R.J. Bib.10 54, 66 correctly identifies 2 separate mills. Map 3 (Chap.III) shows that Gurney's mill lay partly in East Farleigh and partly in Loose as the Insurance Policy quoted above confirms.

Meanwhile Henry jnr. continued to run Old Mill until the Younger Whatman took it over again in 1770 to carry out experiments on his new Engine. The subject of Henry French countermarks has been discussed under Old Mill in App.I.

The search for the origins of Henry French snr.

The result of this has been negative. Either one must assume that after Sanders' failure Whatman had up–graded the workforce at Old Mill which would have allowed Henry snr. to run the mill at arm's length until he had learnt the New Technology or that French himself had had previous experience of it elsewhere and had been vetted by Whatman to ensure that his mill did not fail again. In either case one has to assume that French was a paper maker.

With such a commonly occurring name nothing more than a limited search has been possible. As note 77 indicates there was a small colony of French/ffrench families in Maidstone and although no Henry (whose birthdate was almost certainly after 1700) has been found there, it is possible to envisage a local Henry French having worked as paper maker in the district graduating to a position of competence at Turkey mill after 1740 under Whatman. Against this no suitable Henry French has been found or the baptisms of his two sons. Moreover none of the contemporary French Wills for Kent indicate any connection with a papermaking district.

Outside Kent the only reference known to the author refers to "Henry French, Paper Maker, late of Down Husband (Hurstborne Priors) in Hampshire as a fugitive from debt and was in King's Bench Prison, Southwark, London 20:Aug:1743".[82] Nothing more is known about him. Since Henry jnr. and Thomas are thought to have been baptised ca.1740, it was thought worthwhile making enquiries in this direction. The Rev. Humphrey Llewelyn of Longparish Rectory (parish amalgamation) kindly informed me (9:Mar:1981) that no references to Henry or Thomas' baptism could be found between 1730–40 and that there were no records in the Hurstborne Priors' Register for the period 1740–44, an inconclusive result.

[82] London Gazette (Shorter, A. Bib.12 168).

CRISPE : SALMON : NEWMAN : and STROUD

This section attempts to identify six paper makers, or people sponsoring papermaking ventures, who appeared on the Hollingbourne scene towards the end of the 18th and at the beginning of the 19th Cs. In each case their identity has never been determined satisfactorily; the appearance of their names in ratebooks, countermarks and other documents has been noted but usually left unintegrated into accounts of the paper industry of the area. In actual fact all six of them were local people who came together in Hollingbourne to form partnerships, or who emigrated from there to break new ground elsewhere. It is important, therefore, in the context of this work to know more about their origins. In the pages that follow the details of identifying each of them will be dealt with separately.

In Appendix I it was noted that at one time a partnership between CRISPE and NEWMAN had existed;[83] at another STROUD and NEWMAN, to be followed by NEWMAN and SON;[84] and, as will be seen, a JOHN CRISPE was involved in financing a papermaking venture for the SALMON BROTHERS (from Hollingbourne) in Yorkshire; and that he had known Richard Barnard, proprietor of Eyehorne paper mill, over a number of years. Finally, Thomas Stroud emerges from a papermaking background at Upper Tovil mill which must have begun in Clement Taylor jnr.'s time, so that Stroud may well have been familiar with papermaking in Hollingbourne and at Old mill in particular. It can be seen from this that they were all in a sense interrelated; they were not a collection of paper makers who had strayed in from outside but represent a concentration of local interest at that particular time in the paper industry. Finally, investigating the history of these activities exposes three instances of the increasing unrest amongst the workforce at the turn of the century.

CRISPE

Crispes were associated with at least two papermaking ventures between ca.1795 and some undetermined date in the first decade of the 19th C. Crispe was a relatively common name in the Maidstone district, but as one of the Crispes involved is clearly designated as coming from Sutton Valence (MR TQ 815492), about 4 miles SW of Hollingbourne, and as the other, that is if it is not the same one, was involved in papermaking in the Hollingbourne/Leeds area, it seemed reasonable to begin a search in this area by examining Crispe families centred on Sutton Valence and Leeds (with closely associated Broomfield), all villages in the same area of Kent as Hollingbourne, a village where Crispes appear to have been absent at this time.

According to an Insurance Policy of 1795[85] John Crispe of Sutton Valence was acting as Mortgagee in respect of Langcliffe mill in the West Riding to Henry and Thomas Salmon. The search made here has shown that Henry and Thomas came from Hollingbourne. The Salmon brothers were evidently working Langcliffe mill before the date of this Policy (implying thereby that the mortgage may have pre-dated it also) because a countermark "H & T SALMON 1794" has been recorded. The partnership between Henry and Thomas was dissolved in 1797, but Henry continued to operate the mill until 1815.[86] An Insurance Policy of 1804 makes no further reference to the mortgagee.

Another, or maybe the same, Crispe was associated with Daniel Newman in the Hollingbourne area from at least 1797, the earliest "CRISPE & NEWMAN" countermark to have been discovered.[87] The association seems to have been a twofold one; first, they appear as partners running Hollingworth and Balston's Old Mill for a time (see

---

[83] See App.I pp.4–5 and Synop. Park Mill.
[84] loc.cit.
[85] Shorter, A. Bib.12 250 SFIP 638601 26:Feb:1795.
[86] For further details of the Salmons see this Appendix p.75.
[87] See App.I Park Mill TABLE XVIII.

Synopsis of the history of Old Mill); and, second, they are entered in the Hollingbourne RB as partners in a Paper Mill from 1800–01/2, identified as Park mill (see Synopsis).

Shorter believed that the latter was a John Crispe of Leeds[88] on the basis that his name, "John Crispe", appears on page 2 of a printed list of Kent and Surrey Paper Makers dated 30:Apr:1801 (although it may be noted that in this list *the name of John Crispe is quite separate and not adjacent to Crispe & Newman* which also appears in it); he supports this proposal by citing the burial of a daughter of John Crispe entered in the Leeds burials for 1804. But it has already been pointed out, in the Synopsis for Old Mill, (and proven beyond doubt below) that this John Crispe had died in 1793.

The question arises, are we dealing here with one Crispe or was another involved ? Disentangling the Crispe genealogy has been a complicated business and, *unless the reader wishes to follow the identification (see below), it may be simpler for him to accept the findings summarised at this point.*

Firstly, the search has been concerned with identifying, out of a potential four, the John Crispe at Sutton Valence. The conclusion reached here is that John Crispe snr., who was buried there in 1811 (est. age 66) and described as a Yeoman, was the mortgagee for the Salmons' venture in 1794. As far back as 1791 he had acted as Land Tax Collector for Sutton Valence in conjunction with Richard Barnard, by then proprietor of Eyehorne mill, Hollingbourne and, possibly because of the success of his Langcliffe investment, he may also have embarked on a partnership with Daniel Newman in Hollingbourne; but it must be said that there is an alternative.

A less likely candidate for Newman's partner would have been Edward Crispe snr. of Leeds. The advantages of this proposal are (a) he would have known Newman, who came from Broomfield and had resided in Leeds for a time; (b) he was the proprietor of a water cornmill in Leeds and could thus be said to have an indirect interest in water paper mills; (c) it would offer an explanation of the double reference to Crispe in the printed list of paper makers; and (d) Thomas Stroud (see later), Newman's next partner, was to marry into the Crispe family and take over some years later much, if not all, of the Crispe agricultural interests in Leeds. One has to remember that there is nothing to indicate that Newman's partner was a "John" Crispe or not. There are other factors that enter this situation. For instance, John and Edward may have been related and both taken part;[89] and then there is the question as to why the partnership with Newman ended in 1802. Was it a case of age, illness or of a decision to recover their money from an industry passing through a difficult phase ?[90] Edward, in point of fact, died in 1803. Though we are nearer now to solving the question of this "Crispe's" identity, on present evidence the result cannot be regarded as conclusive.

---

[88] Shorter, A. Bib.12 195 Col.2.

[89] That there may have been a family link between John Crispe of Sutton Valence and Edward of Leeds is suggested, albeit very remotely, by the presence of a document in the Overseer's accounts for Leeds village dated 1798 and bearing a countermark ".....(T) SALMON" (truncated but with sufficient to indicate a "T" for Thomas and thus made in 1797 or before). There is nothing of particular significance in the contents, just a bill for a prayerbook and lettering for Churchwardens' names etc; but one of the overseers at the time was Edward Crispe and one wonders how a paper made in Yorkshire should find its way into Leeds parish documents unless there had been a connection between the two Crispes.

[90] Since John Crispe evidently had the money and since Crispe appears to have been the senior partner in Newman's case, it is probable that the man concerned would have been middle–aged rather than young. In 1800, when operations started at Park mill, Edward would have been 57; John Crispe of Sutton Valence an estimated 55; and Daniel Newman estimated as in his early 40's. The partnership ended in 1802 and Edward died in 1803; so it could be argued that this was another point in favour of Edward being Newman's partner. On the other hand, whereas the 1780–1790's may have been a period favourable to investment, at the beginning of the 19th C. the industry was facing a severe shortage of raw materials as well as growing unrest in the workforce, which certainly manifested itself in the Maidstone district; so the investor in this case may have decided that it was time to withdraw his money, coinciding perhaps with John Crispe's cessation as mortgagee to the Salmons; this argument could apply to either Edward or John.

Points to be noted regarding the information contained in the Table below:-

(1) The parishes of Leeds and Broomfield have always been very closely associated with one another; the two churches of St. Nicholas and St. Margaret are only a mile apart and an undefined parish boundary at that time passed slightly to the west of Leeds Castle which lies between them.

(2) All the Crispe burials are entered under Leeds Burials and where a Crispe from Broomfield was buried in Leeds, it is entered as such; thus "John Crispe of Broomfield" was buried in Leeds in 1793.

(3) *Broomfield* (a) in a search started at 1730 no Crispe Baptisms were found before 1762, Ann d.o. John and Mary Crispe.
        (b) in a search started at 1750 no Crispe Burials found in Register.

(4) *Leeds* The search started in 1720

(5) In a column to the left of the dates given in the Table below an "L" signifies an entry in the Leeds Register; a "B" in the Broomfield Register.

	DATE	BAPTISMS			DATE	BURIALS	COMMENT
L	1726	John	s.o.	Thomas & Ann Crispe			
L	1728	William	s.o.	Thomas & Ann			
		Mary	d.o.				
L	1730	Elizabeth	d.o.	Thomas & Ann			
L	1731	Mary	d.o.	Thomas & Ann			
L	1734	Thomas	s.o.	Thomas & Ann			
L	1736	William	s.o.	Thomas & Ann			
L	1738	Ann	d.o.	Thomas & Ann			
L	1740	Jane	d.o.	Thomas & Ann			
L	1743	EDWARD	s.o.	Thomas & Ann			This must be the founder of the generation that followed at Leeds.
					L	1747	William Crispe	
					L	1748	Thomas Crispe (corr:date).	This may have been the h.o. Ann ?
					L	1760	Elizabeth Crispe snr.	
B	1762	Ann	d.o.	John & Mary Crispe			
B	1764	John	s.o.	John & Mary			
B	1766	Mary	d.o.	John & Mary			
					L	1767	Anne Crispe (Widow)	This must be Thomas' widow.
B	1769	Elizabeth	d.o.	John & Mary			
B	1770	Priscilla	d.o.	John & Mary			No further baptisms for JOHN & Mary
L	1772	THOMAS	s.o.	EDWARD & Elizabeth Crispe			s.o. EDWARD bapt. 1743.
					L	1773	Priscilla Crispe (from Broomfield)	There was a Priscilla at Sutton Valence after whom she may have been named ?
L	1776	WILLIAM	s.o.	EDWARD & Elizabeth	L	1776	John Crispe	Since he was not designated as from Broomfield, possibly John bapt.1726 ?
L	1781	Edward	s.o.	EDWARD & Elizabeth			Last bapt. for Edward & Elizabeth.
B	1786	Mary Ann	d.o.	John & Ann Crispe	L	1786	JOHN CRISPE (from Broomfield)	This must have been h.o. Mary.
B	1787	WILLIAM	s.o.	John & Ann			This William m. Elizabeth Stroud in 1812.
B	1789	Elizabeth	d.o.	John & Ann			
B	1790	Ann	d.o.	John & Ann	L	1790	Elizabeth w.o. EDWARD CRISPE	
B	1792	Harriet	d.o.	John & Ann			No more Crispe Baptisms in Broomfield
					L	1793	JOHN CRISPE (from Broomfield)	This must have been the h.o. Ann. See Land Tax evidence; and s.o. John & Mary, bapt.1764.
					L	1799	Hannah w.o. Thomas Crispe.	
					L	1800	Thomas Crispe.	Thomas bapt.1734.
					L	1802	Edward s.o. EDWARD & Elizabeth	"John" is entered but partially deleted.
					L	1803	EDWARD CRISPE.	bapt.1743.
					L	1804	Harriet d.o. John & Anne Crispe.	No "decd." or "widow" used. but John must have been dead and Anne gone from Broomfield.
					L	1807	WILLIAM CRISPE (aged 32)	This was EDWARD's William bapt.1776. His age appears in the Register, 32.
					L	1810	Mary Jane Crispe (from Maidstone)	
					L	1810	Edw.-Thos. (infant) s.o. of Thomas & Lydia.	This may have been Thomas' (bapt. 1772) family; Thomas continued to run Abbey Corn mill up to ca.1809/ 10 and then disappears from Leeds.
					L	1818	Lydia-Louisa (infant)	

The identification will be dealt with in 3 stages. First, the elimination of Dr. Shorter's "John Crispe of Leeds" based on the entry in the Leeds Burial Register for Harriet's burial there in 1804. Second, outlining the genealogy of the Edward Crispes of Leeds because Edward is a possible candidate as partner to Daniel Newman; and also these Crispes were connected with Thomas Stroud, Daniel Newman's next partner. And third, the identification of "John Crispe" of Sutton Valence, mortgagee to the Salmon brothers and also a candidate for the Newman partnership.

The elimination of John Crispe of Leeds

The Table opposite, and the accompanying notes, show that apart from John Crispe baptised in Leeds in 1726, no other Johns were living in the parish. It is believed that this John was buried in Leeds in 1776; there seems to have been no other candidate for this burial.

There were, however, two John Crispe families living in the associated parish of Broomfield. It is not known where the first of these Johns came from, but presumably because they were all buried in Leeds there must have been some connection with the Crispes in Leeds. It is not relevant to the subject in hand. It can be shown that the second John Crispe of Broomfield had died by 1793, evidence based on the Land Tax Returns for Broomfield. Therefore, the burial of "John Crispe from Broomfield" in 1786 must refer to the first John (in any case the second one was just starting a family).

The case against Shorter's John Crispe rests, therefore, on showing that the entry that he cites for Harriet's burial in 1804 is incorrect in so far as the words "deceased" or "widow" were not used. The Land Tax Records for Broomfield, coupled with extracts from the Ratebook, make it quite clear that by 1793/5 the entries are for Widow Crispe and Mrs. Crispe, having been up till these dates in the name of John Crispe. It has already been shown that the elder John Crispe had died in 1786, so this John Crispe must have been his son (bapt.1764). In addition to this, these records show that his widow, Anne, had left Broomfield by 1800. In other words the burial entry for their daughter, Harriet, in 1804 would not have had "from Broomfield" appended to it.

BROOMFIELD LAND TAX   KAO Q/RP1 56 (Starts 1780) compare it with Leeds Land Tax App.I. pp.32–33.

DATE	PROPRIETOR		OCCUPIER	COMMENT
1786	William Wilkins		Mrs.Crispe	Probably the widow of John Crispe d.1786.
1788	do.		John Crispe	Clearly h.o. Ann and now aged 24.
1792	do.	£8	do.	
1794	do.		do.	John Crispe of Broomfield bur.Leeds 1793.
1795	do.		Widow Crispe	This must be Anne (see 1798 below).
1797	Executors of William Wilkins		Mrs.Crispe	
1798	Late Mr.Wilkins		Anne Crispe	
1800	(no mention)		(no mention)	
1801	(no entry in Wilkins place)		(no mention)	
	Richard Day & Christopher Sears	£3	Mr.Crispe	William, John's son would only have been 14. It is possible that Edward snr. of Leeds paid this Tax ? or else the entry is a mistake. See entry for 1802.
1802	do.		Mrs.Crispe	
1803	do.		do.	
	TAX RETURNS FOR 1804–1808 missing			
1809-14	(No mention)		(No mention).	

BROOMFIELD RATEBOOK   KAO P 222b 12/3 (1774–1816) spot checks; see also Leeds RB App.I. pp.32–33.

DATE	ASSESSMENT			
1774	John Crisp (sic).	£16		
1775-80	do.	£16		John Crispe, Overseer.
(1781)	John Owlett	(£84)	For Park Barn Land	Owlett appears in the Leeds Records also.
1785	John Crispe	£16		
1788	John Crispe	£16		
	John Owlett	£84		
1790	do.			D.Newman, Overseer (first & only mention).
1792	John Crispe	£16		
	Crispe & Owlett	£ 6		
	John Owlett	£84		
1793	Mrs. Crispe	£16		
	Owlett (only)	£ 6		Mrs.Crispe evidently did not continue here.
1794-99	Mrs. Crispe	£16		
1800-06	(no Crispe entry)			Appears to have been replaced by John Honey

The Crispes of Leeds village : the EDWARD CRISPES

The Table on p.68 takes the Crispe family in Leeds back as far as 1720, a point where the search was started. The ancestry of these Crispes has not been pursued further, but to judge from the absence of burials of other Crispes than those shown in the Table it is possible that Thomas and Ann Crispe may have come to Leeds from outside. The only point of this observation is that, if one knew for certain where these Crispes had come from then one would be able to establish whether a connection existed between the Leeds and the Sutton Valence Crispes. There are many hints in results obtained from these searches that this was so. On present evidence one can only postulate a relationship such as that illustrated below, which would have made Edward of Leeds and John of Sutton Valence first cousins.

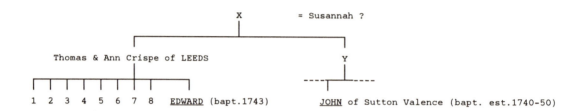

The Ratebook and Land Tax Returns show that during the latter part of the 18th C. and early 19th Edward Crispe and his family were the only Crispes to be assessed in Leeds. It is not known where the first John Crispe of Broomfield came from, but he too may have been a first cousin or, more likely, a nephew of Edward.[91]

In view of the fact that Newman's Crispe partner was active only between ca.1797–1801/2 examination of the Leeds PRs. (see Table) shows that apart from Edward snr. (1743–1803); and his two sons, Thomas (1772–) and William (1776–1807), all the others including a William from Broomfield (bapt.1787 see also below) and Edward's elder brother, Thomas (1734–1800), can be eliminated.

Whereas Thomas and William (25 and 23 respectively in 1797) can be seen as potential partners, but hardly as a senior partner, to Newman (35 at least and with papermaking experience), Edward snr. (at 54) is more likely to have had the means to have backed Newman's papermaking activities.

As regards the termination of this partnership the evidence is equal either way; Edward snr. on the one hand died in 1803; and on the other, presumably because of his age, Thomas and William were assessed for the family's property in Leeds from 1801 (see App.I RB extracts for Leeds). They continued to manage this, as indeed they may have done earlier, (William dying in 1807), until Stroud took much of it over in 1811 at a time when he was engaged in running his paper mill at Aylesford.[92] William Crispe (s.o. John and Ann of Broomfield) inherited the Leeds property on Stroud's death. The fate of Thomas is not known.

On balance, then, if a Leeds candidate is sought, remembering that by 1797 Newman should have been capable of running Old Mill himself, Edward snr. is the most likely Crispe to have been the partner. Moreover, he had probably known Newman for many years, even as far back as 1782.[93]

---

[91]   A check was made to see if there were any other Crispes in the immediate locality. None has been found, for example, in Hollingbourne for that period; none in Langley, a parish that lies between Leeds and Sutton Valence (KAO P 219 1/2) searched Baptisms 1756–1806; and none in 3 parishes lying to the East of Sutton Valence, Chart Sutton (KAO P 83 1/1), East Sutton (KAO P 360 B 1/1) and Ulcombe (KAO P 376 1/2) all searched for periods to cover John & Mary of Broomfield (bapt.1734–41) and John & Mary of Sutton Valence (bapt.1770–75) and no sign of any Crispe families there either.

[92]   See App.III Pratling Street mill.

[93]   See Section on Newmans (this Appendix) for details.

The Sutton Valence Crispes

It is obvious from what has been said in the previous pages (and in App.I) that the families of Newman, Crispe and Stroud were interlocked in one way or another in the Hollingbourne/Leeds area. To this may be added the fact that Daniel Newman also cannot have failed to learn about the John Crispe and Henry and Thomas Salmon enterprise at Langcliffe mill in the West Riding. The Ratebook entries for Leeds (see App.I) show that in 1791 Newman succeeded Robert Salmon (as an Outdweller) in a property that belonged to Old Mill. Robert Salmon would have been 65 at that time and must have leased this house on behalf of his two sons whilst they were serving their apprenticeship or learning about the management of a paper mill under Clement Taylor jnr. at Old Mill. Newman had another contact with Robert Salmon Esq. when Lord Fairfax demised the house, Cotnams, in Eyhorne Street formerly rented by Salmon to Daniel Newman in 1793. Equally John Crispe of Sutton Valence must have had a good idea through both the Salmons and Richard Barnard of the papermaking set–up in Hollingbourne, so it is not unreasonable to assume that it was this Crispe who became involved in a papermaking venture there with Daniel Newman. Can he be identified with any certainty ?

The first mention of Crispe in the Sutton Valence PRs is of a John and Elizabeth who produced a large family between 1766–84 (eldest son William and with no surviving John). Another John, married to Mary, appears in 1803 for the baptism of their daughter, Mary Ann; his origin has not been traced but, to judge from the records, he was probably a young relation. (Another John s.o. William, baptised 1795, is too young to be relevant). Full details of the Sutton Valence Crispe genealogy have been consigned to "working papers". The most important feature to have emerged from this search is that there were two adult Johns in the parish by 1803, if not before this date.

Examination of the LT Returns (see TABLE) shows that (a) the husband of Elizabeth was designated John snr. (1806) and that he had been assessor from 1781–98 partnered by Richard Barnard in this from 1785 soon after the latter had become proprietor of Eyehorne paper mill in Hollingbourne;[94] and (b) certainly by 1807 John jnr., thus described, had become the tenant of property owned by John snr., indicated as (c) in the Table.

There are two anomalous Sutton Valence Crispe burials in the 18th C., William (d.1772) and Susanna (d.1795), which may refer to John snr.'s parents; certainly Susanna's burial is contemporaneous with that of Widow Crispe and it is clear that John snr. inherited her property at that time, the same year as the Langcliffe mortgage. Other evidence deduced from this Table shows that it was John jnr. who died in 1807. Everything points to the latter as being a young man (30–40 years junior to John snr.) and obviously not wealthy and *a fortiori* not a likely candidate as senior partner in the Crispe and Newman papermaking venture at Old and Park mills.[95]

Were there any reasons for prompting John Crispe snr. of Sutton Valence into sponsoring a papermaking venture of any kind? As the Land Tax Extracts show he had a 13–year long association as Richard Barnard's opposite number as an assessor and collector of Sutton Valence. They must have known one another and each other's businesses intimately. Richard Barnard became the proprietor of Eyehorne paper mill in 1782. His prolonged involvement with this mill and papermaking must have excited John Crispe snr.'s interest. But it was not until 1809 that he installed his nephew, Richard jnr., as paper maker there. So for 17 years the mill cannot have been anything more than an investment. This could account for Crispe's funding of another papermaking enterprise, this time in Yorkshire and fired by the success of this, one can imagine him becoming interested in Newman's activities. In the late 1790's Newman may have been offered the lease of Old Mill by Hollingworth, which required more finance than he possessed. It can be seen then that there is a good case for believing that John Crispe snr. of Sutton Valence was involved in both these papermaking enterprises.

---

[94] That this was the same Richard Barnard is shown by the 1790 and 1802 Poll Books where the entry reads "Sutton Valence m. & l. in Hollingbourne" (mill and land).

[95] There are two curious coincidences worth recording here relating to John jnr. and his unknown origin. First, a John and Mary Crispe (1786–92) were exact contemporaries of Daniel and Judith Newman (1782–86) in Broomfield, but it has been shown conclusively that John died in 1793 so cannot be connected directly with John jnr. and Mary of Sutton Valence despite the fact that in each case their first child was named Mary Ann, a gap of 17 years between the two. And, second, John jnr. appears for the first time in the Sutton Valence records in 1803, soon after the Crispe and Newman partnership had come to an end.

(Whilst searching the Sutton Valence PRs. it was noted that between 1772–4 infant mortality was very high, suggesting some epidemic).

EXTRACTS FROM SUTTON VALENCE LAND TAX RETURNS (KAO Q/RP1 374)

Q/RPl/374	Proprietor 1781-85	Occupier	1787-1791	1794	1795-1798		1799-1801		1802-5
(a)	Widow Crispe £5	herself		do.	John Crispe £5	himself	John Crispe £7	himself	Priscilla Crispe £2 herself
(b)	Widow Crispe £2	do.	£1 (1787) £2 (1789-91)	do.	John Crispe £2	himself			do.
(c)	Miss Bouverie £35	John Crispe		do.	Miss Bouverie £35	Jo.Crispe	Philip Pusey £35	Jo.Crispe	do.
(d)	geo.Children £23	John Crispe		do.	geo.Children £22	Jo.Crispe	geo.Children £22	Jo.Crispe	do.
(e)	Jo.Filmer £37	John Crispe		do.	Jo.Filmer £35	Jo.Crispe	Sir Bax.Filmer £35	Wm.Crispe	do.
(f)	Widow Crispe £2	Poulter	Widow Crispe(1787) Miss Crispe (1791)	Priscilla Crispe	Priscilla Crispe £2	Thos.Field	Priscilla Crispe £2	Thos.Field	do.

	1781	1785	1787	1789	1791	1794	1795-1798	1801	1802-5
Assessors	Jo.Crispe	Jo.Crispe	do.	Woollett	Jo.Crispe	do.	do.	Wm.Acton	do.
	Jo.Pope	Rd.Barnard	do.	do.	do.	do.	do.	do.	Richard Barnard gone by 1805
	Thirkey	Rd.Barnard	do.	do.	do.	do.	do.	do.	
Collectors	Bentley	Wm.Crispe	do.	Woollett	Jo.Crispe	do.	do.	Wm.Acton	do.

	1806	1807	1808			1809-10	1811	1814	
(a)	Priscilla Crispe/herself	do.	Priscilla Crispe	£2	herself		(Propr)	Priscilla Crispe	
(b)	Jo. Crispe Snr.	do.	John Crispe	£7	himself		(Propr)	Geo. Crispe	(aged 33 in 1814)
(c)	do. 1805	(occupier) J.Cr.Jnr.	Geo. Children	£2	himself	Geo. Children/Geo.Crispe	(no Geo. Crispe)	Geo. Crispe	
(d)	do. 1805	(occupier) Wm.Cr.	Hon. P. Pelsey	£55	Wm. Crispe		(occup:)	Wm. Crispe	(aged 48 in 1814)
(e)	do. 1805	(occupier) Wm.Cr.	Sir Ed. Filmer	£55	Wm. Crispe		(occup:)	Wm. Crispe	
(f)	do. 1805	(proprietor) Pr. Cr.	Priscilla Crispe	£2	Thos. Fielder		(propr.)	Priscilla Crispe	

	1809-10
Assessors	William Crispe
	Pettit
Collectors	Pettit
	William Crispe

Sutton Valence Burials. (KAO P 360 1/4)	1807	JOHN CRISPE
	1811	JOHN CRISPE YEOMAN

This must be John Crispe jnr.

George Children was the occupant of (c) in 1808 in place of John Crispe jnr. for 1807, while John Crispe snr. (so designated in 1806 but not in 1807) continues to occupy (a) and (b). Added to this the last child of John & Mary to be baptised was in 1806. It is not known where this John Crispe came from.

This must be John Crispe snr.

First, because he was the only other John in Sutton Valence of age; second, because he was succeeded in (a) and (b) by his son, George, (b.1781): (William, with a large family, evidently did not wish to move); and third, he is the most likely one, at an age ~ 66, to have been described as "Yeoman".

In the event the Crispe involvement was relatively short-lived. When Henry Salmon insured Langcliffe mill again in 1804, there appears to be nothing to suggest that John Crispe was mortgagee any longer.[96] Likewise, the Crispe and Newman partnership had also come to an end by 1802. According to Shorter[97] the number of paper mills in England & Wales increased from ca.345 in 1775 to as many as 428 in 1797 (an increase of 83 in 22 years). The expansion in Kent had taken place somewhat earlier, although it still continued there; but the emphasis was now shifting to other areas. Papermaking during the 1780's–1790's was no doubt considered to be a worthwhile investment. By the beginning of the 19th C. the industry was facing not only a shortage of raw materials but there was increasing unrest among the labour force and it is probable that the smaller type of mill would have been the first to suffer. As will be seen when the situation at Ford Mill (Little Chart) is examined later, the results could be disastrous. The Sutton Valence Crispes appear to have been farmers (and possibly millers as well), so Crispe snr. could well have considered 1802 to have been a good time to make his exodus.[98]

---

[96] Shorter, A. Bib.12 250. SFIP 761466 (6:Apr:1804) Insurance taken out by Henry Salmon only.

[97] Shorter, A. Bib.9 76.

[98] It was pointed out on p.67 that in spite of eliminating a number of "John Crispes" as candidates for the partnership "CRISPE & NEWMAN" and in spite of the arguments favouring John Crispe of Sutton Valence, who we know was already involved in one papermaking venture (Langcliffe mill), we still cannot be absolutely certain that the Sutton Valence Crispe was Newman's partner. In a printed document referring to "District Meetings" to be held quarterly during 1803, among the names listed of those appointed to meet on behalf of the Farningham (Kent) district was a "Mr. Crispe"; the others included Charles and Thomas Wilmott (Shoreham & Sundridge), Mr. Staines (Foots Cray), Messrs. Floyd & Fellowes (Eynsford), Mr. Robson (Hawley), Mr. Budgen (Dartford), Messrs. Taplin (probably Tapley & Co. but identity of mill not known, possibly one occupied later by John Hall at St Mary Cray or one at Darenth occupied later by Weatherley & Lane?). Who then was this "Mr. Crispe"? John Crispe jnr. (d.1807) could have been a candidate; John snr. did not die until 1811 but at 58 what interest could he have had in West Kent papermaking? Though it could be argued that John jnr. might not have required financing while working with Newman at Old Mill, the facts remain that (a) Crispe was clearly the senior partner in both the Old Mill *and* Park mill ventures; and (b) nothing is known of any Crispe paper makers.

THE SALMONS OF HOLLINGBOURNE AND LANGCLIFFE MILL, WEST RIDING

Salmon was a well established name in Kent and is to be found, for instance, in the 17th C. All Saints', Maidstone Parish Registers. How long they had lived in Hollingbourne has not been established here, but the Table overleaf shows that Henry and Thomas, the paper makers, were baptised there during the 1760's. Robert Salmon Esq. was assessed for quite a number of properties in Hollingbourne, one of these being a house leased from Old Mill and later occupied by Daniel Newman. In view of the fact that both Henry and Thomas Salmon appear later as persons capable of founding and managing a paper mill in 1794, the indications are that they had undergone training of a suitable nature at Old Mill.

Presumably because the availability of sites for paper mills in the locality was becoming ever more limited and perhaps also because Lord Fairfax had terminated the Salmons' lease of Cotnam House, the house that was later to become associated indirectly with Park mill, the Salmon brothers may have been forced to look for a mill site elsewhere. Why they picked a mill in a remote district of the West Riding for conversion to papermaking is a question for which an answer has not been found yet. One might speculate that perhaps Ann, their mother, had come from that area; or that the North–West might by that time have been regarded as an area of great promise for industrial development; or that perhaps a Hollingbourne paper mill employee had come from the Settle area mentioning the mill at Langcliffe as one suitable for conversion.

A limited examination of Yorkshire records throws no light on the reasons for the Salmons' move to Langcliffe.[99] The Langcliffe Land Tax Returns confirm that Henry Salmon was proprietor and occupier of the property in 1794, 1805, 1810 but not for 1815.[100] They also show that in 1790 Benjamin Farrend Esq. was the proprietor of Langcliffe mill, but do not describe what sort of a mill it was. For confirmation, it was determined that no record of the surname Salmon was found in the indexes to the West Riding Registry of Deeds, 1792–4, which is regarded as a very comprehensive Archive. (The maiden name of Ann Salmon is not known, though Wills might provide a clue).

Among the Ferrand Papers in the Bradford Central Library there are three that refer to Langcliffe mill:–

(i) A copy of a Lease and Release dated 8/9:July:1728, by which a Water Can Mill at Langcliff called Langcliff Mill in the occupation of Mr. Fisher was conveyed from Charles Nawell of the parish of Giggleswick gent and William Carr of Langcliff gent to Benjamin Ferrand of St. lves Esq.

(ii) An Agreement that Mr. Fisher shall continue in occupation.

(iii) A draft Conveyance dated *1792*, by which a Water Can mill in Langcliff called Langcliff *Old* Mill was to be conveyed by Benjamin Ferrand of St. Ives Esq. to William Sutcliff of Settle apothecary (along with Giggleswick and Settle Mills).

There is nothing in any of these documents that indicates that the Water Can mill was a paper mill which, had it been by the 1790's, would have contained a variety of distinctive equipment, but not mentioned in the conveyance. So one might conclude that not until Sutcliff's day at least, or even more likely not until the Salmons' day, was it converted into a paper mill, assuming that it was not a separate entity from from Langcliff Old mill. There was no indication in the Salmons' Insurance Policy that their mill was "newly built". Langcliffe mill remained Salmon property until 1851. John Salmon was in partnership there in 1836 with John Hartley and Robert Rugg; the Ruggs were from Kent (inf. kindly supplied by Mrs. Jean Stirk: extract from "the Ancient Parish of Giggleswick"). The paper mill was still operating in 1957.

---

[99] I am greatly indebted to the County Archivist; the Yorkshire Arch. Soc.; and the Central Library at Bradford for searches made on my behalf.

[100] Gravell T.L., Bib.68 No.619 records an "H. SALMON 1815" countermark.

GENEALOGICAL DATA RELATED TO THE PAPERMAKING SALMONS OF HOLLINGBOURNE

DATE	HOLLINGBOURNE BAPTISMS KAO P 187 1/1 (1156-1799)	COMMENT
1765	HENRY s.o. Robert & Ann Salmon	See below for data on Langcliffe mill, Settle, West Riding.
1767	Ann d.o.	
1769	THOMAS s.o.	See mill data below.
1772	Elizabeth d.o.	
1774	Mary d.o.	
1776	Mary d.o.	

	HOLLINGBOURNE BURIALS KAO P 187 1/6 (1813-1834)	
1815	ROBERT SALMON (aged 89)	

	HOLLINGBOURNE LAND TAX KAO Q/RP1 180	
1815	Robert Salmon still appears in the Return	
1816	do.	
1817	Henry Salmon	
1818	do.	

DATA ON LANGCLIFFE MILL, SETTLE, WEST RIDING, YORKSHIRE[1]

1794	Countermark " H & T SALMON 1794 "	Recorded by Shorter.
1795	The mill was insured by Thomas and Henry Salmon, Paper Makers and Mortgagers; and JOHN CRISPE, Mortgagee, of Sutton Valence, Kent.	SFIP 638601 26:Feb:1795.
1797	The partnership between Henry and Thomas was dissolved, Henry continuing to run the business.	
	"....(T) SALMON " Countermark	Leeds Parish Document (KAO P 222 5/13) d.o.d. 1798.[2] As Thomas was involved the paper was clearly made before 1797.
1804	The mill was insured by Henry Salmon	SFIP 761466 6:Apr:1804.
1805	" H SALMON 1805 " Countermark	J.M.W.Turner's "Sandycombe and Yorkshire" Sketchbooks of 1812 (T.B. CXXVII). [3]
1809	" H SALMON 1809 "	As above (T.B. CXLVI). [3]
1810	Langcliffe Land Tax Return shows Henry Salmon still at the paper mill.	West Yorkshire County Record Office. [4]
1816	Excise General List shows George Woods as the occupant of Langcliffe mill. [5]	EGL 1816 Halifax Collection, Excise No.106.

[1] The data on the mill and Insurance policies taken from Shorter, A. Bib.12 250.
[2] Author's search.
[3] Finberg, A.J. "Inventory of Drawings of the Turner Bequest" (Nat.Gall. Vols I & II 1909).
[4] Letter (7:Jan:1981) Mrs E.K. Berry, County Archivist.
[5] Henry Salmon succeeded his father at Hollingbourne in 1816.

Among the countermarks cited in the Table, two it will be noticed, have been found in two of Turner's sketchbooks, the Sandycombe and Yorkshire Sketchbooks, which Finberg dated 1812 (TB CXXVII); and the Yorkshire 3 Sketchbook, which he dated 1816 (TB CXLVI). Between 1810 and 1824 Turner was a frequent visitor to Farnley Hall where one of his most zealous patrons, Walter Fawkes, lived. Farnley Hall is just over 30 miles SE of Settle, so in these cases it seems quite possible that Turner may have obtained these sketchbooks from a Yorkshire supplier, e.g. say in Harrogate or York.

DANIEL & DENNY NEWMAN, PAPER MAKERS OF HOLLINGBOURNE, KENT

The search has not revealed where the Newman family came from and with a name like "Newman" it would be difficult to be certain that one was identifying the right family. The evidence presented in the Table below suggests that they may have come to the Hollingbourne area from somewhere else. No Newmans were seen in Broomfield prior to the baptism of Maria in 1782; and the first mention of Newman in the Hollingbourne Ratebook appears to be for the year 1785; and though not specifically searched for the name was not noticed in the Leeds Parish Registers, when examined for Crispe baptisms and burials, before Judith's baptism. The Table below illustrates Daniel Newman's appearances in Broomfield, Leeds and Hollingbourne, which may be related to the information given in Appendix I,[101] and the appearance of the John Newmans in Hollingbourne in 1785. This evidence does, however, not answer a number of important questions (see below).

BROOMFIELD

DATE	BAPTISMS & BURIALS KAO P 222b 1/2 (1776–1812)			COMMENT
1782	Maria	d.o.	Daniel & Judith Newman	This is the first mention of Newman discovered so far. Not mentioned in the RB; but was named as Overseer for 1790 Land Tax.
1786	**DENNY**	s.o.		

LEEDS

	BAPTISMS KAO P 222 1/7 (1689–1795)			
1791	Judith	d.o.	Daniel & Judith Newman	In this year *Mr.* Newman appears for the first time as an *Out-dweller* in the Leeds RB. (See Extracts, App.I). In 1795 he is entered as Daniel Newman and continues as such up to 1823.

HOLLINGBOURNE

	BAPTISMS KAO P 187 1/1 (1556–1799)			
1786	Ann	d.o.	John & Sarah Newman	Search started 1772; 1786 first mention of Newman. The RB indicates that John & Sarah Newman had been living in Cotnams between 1785–1790. (For these years it was in Mrs. Newman's name).
1790	William	s.o.	do. do.	
1794	Jemima	d.o.	Daniel & Judith Newman	The first unequivocal evidence of Daniel Newman's presence in Hollingbourne comes from this baptism and the Land Tax entry for the same year, 1794. The 1791 RB entry, Mr. Newman, might indicate his presence there ? On the other hand it is not until 1793 that the entry reads Mr. Newman Late Salmon, the year when Lord Fairfax demised Cotnams to Daniel.

---

[101] See App.I under Park mill for Newman properties; occupation of the mill and countermarks; further refs. Hollingbourne & Leeds RB. and LT. entries.

The Table opposite shows Daniel Newman making his first appearance in Broomfield, where there was no paper mill, in the year 1782 on the occasion of Maria's baptism, indicating that he must have been at least 20 (if not more) by then. In 1791 he was described in the Leeds Ratebook as Mr. Newman. Who was this man and what sort of training had he had that enabled him to manage Old Mill for Clement Taylor jnr., as has been suggested, and later to enter into partnerships in papermaking enterprises.

Added to these unanswered questions there seems to be more than a hint of the existence of some relationship between the Newmans and Lord Fairfax at Leeds Castle. Daniel appears to have been looked upon favourably by the latter. Was he related to him legitimately or even, perhaps, illegitimately ? Lord Fairfax died in 1793,[102] but already in 1792 the Hollingbourne Land Tax Return makes it clear that he was making the house, Cotnams, ready for Daniel; although, from the Ratebook, Newmans, and in particular a Mrs. Newman, may have occupied it since 1785. It must have been sub-let to them by Robert Salmon, if this was the case. In the event Lord Fairfax terminated Salmon's lease and conveyed it to Daniel Newman. Presumably, John Newman in Hollingbourne was a brother of Daniel's. Whether there is any significance in the fact that Daniel first went to live in Broomfield which, geographically, is on the Castle's back doorstep can only be a matter for conjecture. And who was this "Mrs. Newman" ?

In all this one thing seems clear, Daniel Newman needed money to finance his papermaking operations; John (or perhaps it was Edward) Crispe cannot have offered him any papermaking expertise, it can only have been money. Stroud, who succeeded the Crispes as Newman's partner, came on the other hand from a papermaking family. Daniel's son, Denny, made his first appearance in the Ratebook and Land Tax Returns in 1809/10 when he would have been aged 23.

---

[102] The seventh Lord Fairfax bequeathed Leeds Castle to his nephew, the Rev. Denny Martin; and, later, it passed to General Martin and eventually to Fiennes Wykeham, a second cousin.

SEARCH DATA ON STROUD

ALL SAINTS', MAIDSTONE

DATE	PARISH REGISTERS FOR BAPTISMS AND BURIALS	COMMENT
1762	Mary d.o. Thomas & Martha Stroud	Searched 1760-1775.
1767	THOMAS s.o. do. do.	Baptised 7th Feb. The year ties in with his age at death.
1815	THOMAS STROOD (sic) from LEEDS, aged 48.	This information ties in with his Will, which was proved 23:Mar:1815 (Stroud was buried on the 25th Feb).

EXTRACTS FROM THE MAIDSTONE RATEBOOKS [TOVILL (sic) WARD][1]

DATE	SUM	RATEBOOKS KAO P 241 11/13, 11/14	PREMISES	COMMENT
1800 (Oct).	£208 £35 5/-	J RUSE Esq. do. Robert Peters.	House & Mill 14 Cottages	Upper Tovil paper mill. This must be the apprentice that Thomas Stroud took on in 1800, but not at "Cotterams' Mill" as suggested by Shorter (Bib.12 196). In any case this was probably Thomas snr. (see Text).
1801 (Sept)	£208 £65	J RUSE Esq. JOHN PINE		Lower Tovil paper mill.
1802 (Oct).	£208 £6	J RUSE Esq. William Stroud		First appearance, not there in May. Possibly may have been another son ? (not checked). See addition to note 108
1803 (Aug).	£208 £6	J RUSE Esq. William Stroud		Letter received by William Balston from "Ruse's mill" dated 29:Sept:1803 (see Text) signed by Stroud but initial not clear "T" ? or "S" ?
1804 (May)	do.	do.		
1805 (1st.A)	£208	J RUSE Esq. William Stroud, ENGINEER.		This title usually meant Beaterman or Foreman.
1805 (Oct).	£208	J RUSE Esq.		The two TURNERS (Richard & Thomas) joined RUSE in 1805, Countermark " RUSE & TURNERS 1805 ".
	£6	William Stroud & Rangers		Thereafter the entries are for Rangers, Stroud gone.
	£	Robert Vigor / Thomas Stroud		This was presumably Thomas snr. ? Thomas jnr. was assessed for the first time in the Hollingbourne RB in 1805.
1807 (Feb). (May) (Sept)	£208 £6 £	J RUSE Esq. Rangers Thomas Stroud	4 Cottages	

Footnote

[1] The Maidstone Land Tax Returns for TOVIL (College Road & Tovil) KAO Q/RP1 242 were searched for the years 1801 and 1806 (the next surviving Return is for 1817). *No* references to Stroud.

THOMAS STROUD (1767-1815)

Though some data has been collected it has not been an easy task to locate facts about the younger Thomas Stroud. In his Will he described himself in 1813, two years before his death, as a Paper Maker at Pratling Street, Aylesford,[103] and a farmer of Leeds, Kent. Whereas his name is found in the Leeds and Hollingbourne records, he does not appear in the Aylesford Parish Records during his lifetime (nor in those of the surrounding parishes); neither has the place of his marriage been found nor the baptism of his 5 children, nor the parish where his daughter was married (see below).

As the Table opposite shows he was born in Maidstone in 1767 the son of Thomas and Martha Stroud. That this is the Stroud we are concerned with is confirmed (a) by his age at death, recorded in the All Saints' Register; and (b) by his Will (see below).[104] Having identified him as the Thomas Stroud who appeared later as the paper maker in Hollingbourne and the farmer of Leeds, one can now place the reference to a Thomas Stroud taking on an apprentice, Robert Peters, in 1800 more precisely.[105] Shorter suggested that this may have been evidence of Thomas' presence at "Cotteram's mill" otherwise identified as Park Mill in Hollingbourne (Cotteram's being a corruption of Cotnams House where Daniel Newman lived). But the data presented opposite shows that Peters was resident in the Tovil Ward (Maidstone) in 1800 and, moreover, this is where the Stroud family were concentrated.

In 1750 Simon Pine took on, amongst others, Thomas Stroud (the father) as an apprentice at Lower Tovil mill;[106] but it is clear from other evidence that later he worked at Upper Tovil mill, initially under Clement Taylor jnr. and latterly under Joseph Ruse. This is confirmed by a letter written by Thomas Stroud (the father) on behalf of Joseph Ruse to William Balston and dated 29th September 1803[107] informing the latter that Mr. Ruse's men had walked out and "laid the mill still". Thomas, the son, was taken on as an apprentice in 1774.[108] That a further distinction can be made between father and son is to be found in the Maidstone Poll Book for 1807 where Thomas snr. is described as a paper maker of Maidstone and the other as a paper maker of Hollingbourne.[109] Shorter suggested that Thomas snr. may have been the paper maker at Pratling Street but on present evidence this is unlikely; the proposal has been made that Charles Brenchley was the paper maker there from 1804-1812,[110] in addition to which the data presented in the Table opposite shows Thomas Stroud (the father) as the occupant of property in the Tovil Ward of Maidstone from 1805-7. Moreover, his name cannot be found as Master Paper Maker for any of the Maidstone mills at this time, so to support the reference in the Poll Book the assumption has to be made that by this time, if not before this, the management of the more important and larger paper mills was delegated

[103] For his occupation of Pratling Street mill see App.III.

[104] The Will (see below for details) confirms everything we know about Thomas Stroud, the son. The 1790 and 1802 Poll Books show two other Thomas Strouds in East Kent neither of which fit our Thomas in any respect.

[105] Shorter, A. Bib.12 196. In Bib.7 (Apr.) 62 he extends this reference from 1800-1803, but does not specify who the apprentices taken on were.

[106] Shorter, A. Bib.12 190.

[107] Master Paper Makers' Correspondence (PSM) in a folder entitled "Absentees". Joseph Ruse, at this time (1799-1804), was the sole proprietor of Upper Tovil mill.

[108] Shorter, A. Bib.7 (Apr.) 62. ✳ Incorrect.— It was William Stroud son of Thomas Stroud, papermaker.

[109] ibid.

[110] See App.III under Pratling Street mill for this proposal.

to a senior paper maker or "Engineer", the title given to William Stroud in 1805 (see Table), possibly another of Thomas snr.'s sons.

Thomas, the son, may have been employed at Tovil mill for a time and it was, perhaps, here that he gained managerial knowledge. In 1802 the countermark "STROUD & CO 1802" appeared which must indicate that Thomas the son had set up as a paper maker in his own right[111] (seeing that there are "J RUSE" Countermarks from 1799–1804,[112] it is out of the question that Ruse would have allowed him or his father to use their name in paper made at Upper Tovil mill). In any case this was the year that the Crispe and Newman partnership came to an end and, although Stroud's name is not associated with Newman until 1805 (nor is it found elsewhere under other property) in the Hollingbourne RB, the Park Mill entries are for "Newman & Co." for 1803/4. That the younger Stroud was present in Hollingbourne as early as 1803 is confirmed by a manuscript document (in duplicate), enclosed with a printed Resolution dated 19th September 1803, giving a list of employees and absentees at Stroud & Newman's "Cotterams mill".[113] Furthermore, another printed Notice of a meeting held at the George & Vulture, Cornhill, on the 13th June 1803 showed Stroud & Co. represented among other Master Paper Makers. In connection with this it is worth noting that neither Crispe nor Newman's names appear in this list under the "Makers of Kent". Taking all these things into consideration it is safe to say that the countermark referred to above must represent the younger Stroud's partnership with Newman at Park mill.

The next evidence of Thomas' activities in Hollingbourne is to be found in Ratebook entries from 1805 onwards.[114] In 1806 in a letter written by the Hollingworths to William Balston there is a further reference to him. They wrote "we should be glad to know if you have heard of the turn out at Stroud's mill"; this was at a time when William Balston, though now concerned with Springfield mill, must still have had close contact with Hollingbourne affairs. The quotation also shows, as the Countermark showed, that Stroud was the dominant partner though how he came to be in this position is not clear. It seems somewhat doubtful that at this stage he would have brought money to this partnership although it has to be admitted that by 1810 he had acquired very considerable farming liabilities in the parish of Leeds;[115] so money must have come from somewhere, perhaps through his marriage. It may be that because Daniel Newman was an established Hollingbourne figure by then that the Ratebook entries for 1806–12 are Newman & Stroud. In 1812 Thomas Stroud signed himself as Churchwarden for Hollingbourne. This is the last certain reference to him in Hollingbourne. The following year the Park Mill countermark has become "D NEWMAN 1813".

Little has been discovered about his family. According to Shorter the Kentish Chronicle reported the marriage of his daughter on the 24th July 1812.[116] No trace of this marriage has been found in any of the parishes with which Stroud is known to have been associated. This must have been the wedding of his daughter, Elizabeth, to William Crispe of Leeds

[111]   See App.I Park mill Countermark Table.
[112]   "J RUSE" countermarks known to the author are for the years 1799, 1800, 1801 and 1804. It has to be said that none has come to light yet for 1802/3. On the other hand paper makers did not always change the dates on the moulds annually to begin with, that is for the first decade after 1794.
[113]   PSM Master Paper Makers Correspondence.
[114]   See App.I preamble to Park mill (pp.20–23).
[115]   See App.I Leeds RB extracts.
[116]   Shorter, A. Bib.12 196.

(mentioned in his Will); this William, it will be remembered, was the son of John and Ann Crispe and baptised in Broomfield in 1787. Other members of the family will be mentioned when Stroud's Will is examined.

It seems clear from the entries in the Leeds Ratebook that Thomas' farming interests began in 1810 when he was assessed for the very considerable sum of £220–10. The other entries referring to him in the Leeds Ratebook and Land Tax Returns have already been discussed except to point out that after his death the Land Tax for the farm was charged first to Elizabeth, his widow, (1816/1817) and subsequently against his son–in–law, William Crispe.

There is no doubt that the Stroud and Newman partnership had come to an end by 1813, so the next Stroud countermark to be found, "STROUD 1814", clearly refers to a new venture. It is only from his Will, made in 1813, that one hears of him at Pratling Street paper mill in Aylesford. There are several indications that he had become a paper maker on his own. For instance, in the front of Catherine Balston's "Mill Letter Book" (PSM) there is a list of paper makers written to on the 5th May 1813. Separate letters were addressed to Newman and to Stroud & Son (sic).[117] Stroud also figures in a subscription list (PSM) dated 7th September 1813 and apart from the Leeds RB and Land Tax Records and his 1814 countermark, this the last reference to him.

On the 25th February 1815 he was buried at All Saints', Maidstone, aged 48. He had obviously been living in Leeds at the time of his death. His Will[118] confirms the Leeds Ratebook and Land Tax entries that his widow was Elizabeth; it also shows that he had a daughter, Elizabeth; three sons one of whom was another Thomas; and Sarah. It would seem from the tenor of the Will that the three sons and Sarah were all minors at the time of making the Will (1813). The reference to his daughter, Elizabeth, runs as follows "...my daughter, Elizabeth, wife of William Crispe of Leeds aforesaid Farmer..." thus confirming Crispe's involvement.

The Thomas Stroud assessed in the Aylesford RB for 1817 and 1819 may have been the surviving son, who had perhaps reached his majority by then and that the assessment referred to his occupation of Pratling Street;[119] the next known occupant of this mill was Charles Wise in 1821. No other references to Stroud have been found. The most important information contained in his Will relates (a) to his occupation; and (b) the Pratling Street paper mill. The will was made in 1813 and he was described then as "of Pratling Street in the parish of Aylesford in the County of Kent Paper Maker and also of Leeds in the same County ffarmer". In the Will his wife is instructed to carry on the business of Paper Maker and ffarmer; and, later, mention is made of "...the *Leasehold Mill and Premises at Pratling Street* aforesaid and all my farm and premises at Leeds.....Elizabeth sole Executrix".

---

[117] Catherine Balston seems to have been confused here as the entries should surely have been Newman and Son and, separately, Stroud, because there are several other entries in her book between the 19:May:1813 and the 15:July:1814 referring to letters either sent to or received from Newman or Newman & Son, Hollingbourne; there are no further letter entries under Stroud. Dated 17:July:1813 there is a letter in the PSM from Catherine to William Balston, in which she wrote "Mr. Stroud has not called yet, but I have written a note ready saying that to save him the trouble of calling again as you was going from home, you had drawn the bill".

[118] PRO ref. PROB/11/1566 fo.159 Proved 23:Mar:1815.

[119] See App.III note 33.

The name Munn is commonly found in Kent and many examples of this are to be found in the Maidstone district, for instance, in Maidstone itself, Boxley, East Farleigh and even in the parish of Leeds. The object of the search, the results of which are shown in the Table below, have been aimed primarily at identifying the origin of Lewis Munn snr. and determining whether a relationship exists between him and Lewis Munn jnr. There is positive evidence that Lewis snr. was the son of Thomas & Elizabeth of East Greenwich, but whether his grandfather was the Thomas Munn buried at Sutton-at-Hone in 1772 is a matter that has not been pursued here; his Will mentions two sons, William and Thomas, and a grand-daughter, Elizabeth, showing that he was not a young man. Likewise, a secondary interest in a possible relationship existing between the paper maker Munns and the Munn Stationers of Bond Street cannot be positively established on present evidence. That Lewis had a brother James is demonstrated in Thomas' Will,[1] but it has not been possible to establish that he was the father of William & James, though James, their father, like Thomas, was based in Greenwich.

DATE	PARISH	BAPTISMS & BURIALS				COMMENT
	Unknown	Charles	s.o.	Thomas & Elizabeth Munn	(bapt)	Where the Parish is "Unknown" the baptisms are placed in the order in which the names appear in the Father's Will (the order in the Mother's will is slightly different). Another difficulty is that the sexes are separated so one cannot be sure that a daughter may not have been baptised before a son.
	Unknown	William	s.o.		(bapt)	
	Unknown	Elizabeth	d.o.		(bapt)	
	Unknown	James	s.o.		(bapt)	
	Unknown	Jane	d.o.		(bapt)	
	Unknown	Sarah	d.o.		(bapt)	
	Unknown	Martha	d.o.		(bapt)	
1747	East Greenwich	Susannah Sarah	d.o.		(bapt)	[Greenwich Baptisms 1715-1796].
1749	East Greenwich	LEWIS	s.o.		(bapt)	In all the contemporary Munn families encountered in various searches, this is the only mention of a LEWIS.
1751	East Greenwich	Mary Litchfield	d.o.		(bapt)	
1752	East Greenwich	Thomas	s.o.		(bapt)	Thomas is not mentioned in either Will.
						There were a number of other Munn families living in East Greenwich; John & Mary (early 1730's); John (Wheelwright) & Sarah (1760's); William (Gent) & Catherine (1770's), William later described as an attorney.
1770	Sutton-at-Hone	LEWIS MUNN B.B.	s.o.	Sarah Richardson	(bapt)	This must surely have been Lewis snr.'s illegitimate child; in the event it died. In spite of extensive searches no other Lewis Munns have been found anywhere (see Text).
1770	Sutton-at-Hone	LEWIS MUNN RICHARDSON			(bur.)	
1772	Sutton-at-Hone	Thomas Munn			(bur.)	His Will shows that his wife, Lydia; 2 sons, William & Thomas; and a grand-daughter, Elizabeth, survived him. Though a Thomas & Elizabeth were married at Sutton-at-Hone in 1719, they are clearly not the same as the Thomas & Eliz. of East Greenwich as the PRs show (see Text).
1774	Hollingbourne	William	s.o.	LEWIS & Sarah Munn	(bapt)	The marriage of Lewis & Sarah has not been found. A search in the Hollingbourne PRs starting 1755 revealed no earlier Munn entries there. As mentioned above similar searches in the Maidstone area also failed to show a Lewis Munn.
1776	Hollingbourne	Mary	d.o.		(bapt)	
1778	Little Chart	Jane	d.o.		(bapt)	The first Munn baptism at Little Chart. Lewis Munn and Thomas Sweetlove (of Leeds) insured Ford paper mill in Little Chart in 1776 (see Text).
1779	East Greenwich	Thomas Munn			(bur.)	LEWIS Munn snr.'s father, see Will.[1,2] His widow, Elizabeth, died in Bromley, Kent, 1795. see Will.[3]
1781	Little Chart	Elizabeth	d.o.	LEWIS & Sarah Munn	(bapt)	
1783	Little Chart	GEORGE	s.o.		(bapt)	George figures later at Thetford mill, Norfolk and, later, at Solesbridge, Hertfordshire, with Lewis jnr. (see Text).
1784	Little Chart	Sarah	d.o.		(bapt)	
1786	Little Chart	RICHARD	s.o.		(bapt)	Richard figures at Thetford, Kings Lynn and Castle Rising, Norfolk.
1788	Little Chart	Maria	d.o.		(bapt)	
1790	Little Chart	Louisa	d.o.		(bapt)	This is the last of Lewis snr.'s family
1797	Little Chart	Lewis	s.o.	William & Elizabeth Munn	(bapt)	William, the eldest son appears to have remained at Little Chart. (The eldest son by this marriage) ? (See Text).
1798	Little Chart	William	s.o.		(bapt)	
1804	Little Chart	LEWIS MUNN Snr.			(bur.)	See Text.

[1] P.R.O. PROB/11/1049 fo.20 The Will of Thomas Munn, died 1779, in the Parish of East Greenwich, Kent.

[2] A son, James, is mentioned in Thomas' Will (see Note 1). A James George Munn s.o. James & Charlotte Munn, Painter Stainer, was bapt. in East Greenwich 1790. Paul Sandby Munn was also s.o. a James Munn of Thornton Row, Greenwich, a carriage decorator and Landscape painter, born Greenwich 8:Feb:1773 but his baptism has not been found (or in the neighbouring parishes). See Text.

[3] P.R.O. PROB/11/1261/331. The Will of Elizabeth Munn, Widow in the Parish of Bromley, Kent 1795.

LEWIS MUNN Snr. (1749–1804)

Although no evidence has been found to show that Lewis Munn snr. ever worked in either of the Hollingbourne paper mills, the fact that he spent two years there (1774–76) and then set up a paper mill (1776) with Thomas Sweetlove of Leeds (in all probability a millwright) in Little Chart ca. 8.5 miles to the S.E. of Hollingbourne on the upper reaches of the Great Stour (MR TQ 942460) points to his having either worked or studied in Old Mill or Eyehorne. The problem as to which of these mills he was trained in has been discussed in App.I (p.19 q.v.) The view was held there that he might have trained earlier at Dartford and that at either William Quelch or John Terry's suggestion he moved to Hollingbourne where Clement Taylor jnr. was about to lease Old Mill, possibly with Whatman's new Beater installed there.

The partnership with Sweetlove only lasted 4 years but Munn carried on operating the newly established Ford mill until his death twenty–four years later in 1804. The first countermarks to appear from this mill suggest that Munn, only 27 at the time, was the senior partner in this venture.

The first task in Munn's case has been to determine his origin which almost certainly was East Greenwich, bapt.1749 (see Table opposite, and below). Having established this Munn is of particular interest on four separate accounts:–

(1)   Was he related to the Munn Stationers of Bond Street and their brother, Paul Sandby Munn, the artist?

A reasonable case can be made out that this was so.

(2)   Where did he train as a paper maker? Why did he move from the London area to a remote village in central Kent?

The suggested answer to these two questions is probably at Dartford under John Terry; Munn was in that area at a time when he would have been learning his trade. And Hollingbourne was a place known to John Terry and was clearly an important nursery for prospective White paper makers, an occupation which would be in keeping with family Stationer connections.

(3)   Was the Lewis Munn, paper maker, in Hertfordshire who appeared there at the end of the 18th C. a relation of his? And coupled with this question does this explain the extension of the Munn family's papermaking interests to Hertfordshire and Norfolk?

Although documentary evidence for this has not been discovered yet, other evidence points to Munn snr. as the father. For instance, like Lewis jnr., George and Richard transferred their papermaking activities from Kent to Thetford and Castle Rising mills in Norfolk and later to Solesbridge mill, Hertfordshire, where George took over from Lewis jnr., a strong indication that they were his brothers.

(4)   Finally, does the practice of using "KENT" in his later countermarks suggest any
       common element between the Munns and other contemporary paper makers who
       made use of a county in their marks?

       Certainly a case can be made out for this, but not proved. One has "R WILLIAMS
       KENT" and "JOHN HOWARD SURRY" and in both cases it is possible to point to
       a very tenuous link with Mill End Mill, Hertfordshire.

Evidence to support the answers to the questions posed above is set out in some detail below
for the benefit of those readers who may wish to follow these subjects further.

Establishing the Identity of Lewis Munn snr.

Though scattered entries of Munn were found in the parish records of Maidstone and one or two outlying villages,
no Lewis Munns were found; in fact Lewis snr., Lewis jnr., of Hertfordshire and Lewis snr.'s grandson are the only
ones to be found anywhere to date. No Munns were found in Hollingbourne prior to William's baptism there in 1774.
On this basis and more extended searches, the conclusion was reached that Lewis snr. did not come from the
Maidstone papermaking district and in view of the apparent scarcity of Lewis Munns the Greenwich born one has
been provisionally accepted as the future paper maker.

The question as to why this Munn became interested in papermaking depends to some extent on whether a
relationship between Lewis, Paul Sandby Munn and the Stationers, William and James can be established and
whether, if this is the case, Lewis' father was in any way connected with paper.

The Relationship between Lewis and James Munn

Although it is more than probable that Paul was Lewis' nephew it has not been possible to prove a relationship
between these two. Though not shown in the East Greenwich registers the wills of Thomas and Elizabeth Munn
provide conclusive proof that Lewis had a brother James (both wills show distinctive combinations of names like
"Mary Litchfield" and "Susannah Sarah" both found in the parish registers).

The only member of James' and Charlotte's family (Paul's parents) found in the register is a James George, baptised
in 1790, 17 years after Paul's birth.[120] But neither the baptism of Paul nor those of his brothers, William and James,
have been found despite searches made in the neighbouring parishes of Charlton[121] and Woolwich[122]. It is alleged
that Paul, if not the other two also, was born in East Greenwich and to account for this, the suggestion is made here
that they may have been baptised privately at the Chapel of the Royal Military Academy at Woolwich.[123]

There seems to be a reasonably good case for assuming here that Lewis snr. and the other Munns were related; James,
Lewis' brother, a painter stainer;[124] Paul a watercolour artist; William and James jnr. stationers, all connected in one
way or another with "paper". Unfortunately, no evidence has come to light to show what occupation Lewis' father,

---

[120] That we are considering the same James here is shown by the fact that not only was James' wife named Charlotte, as
       was Paul's mother, but James is described in the parish register as a "painter stainer".
[121] Records of the Society of Genealogists.
[122] Greater London Record Office P 87/MRY 6–7 & R 301.
[123] Paul Sandby Munn was the godson and, later, pupil of Paul Sandby, the distinguished 18th C. Artist (1730–1809) held
       in very high esteem in his age and who is regarded as the artist who raised Watercolour to the status of an Art in its own
       right and who, among many other distinguished activities, was appointed Drawing Master at the Royal Military Academy
       at Woolwich. Paul Sandby on his journeys from his house in Tyburn, and after 1772, 4 St George's Row (overlooking
       Hyde Park), to Woolwich, used to stay with James Munn in Greenwich and because of this close association may have
       arranged for some of James' children to have been baptised in the Military Academy's Chapel. (see Note 124 below).
[124] Martin Hardie (Watercolour Painting in Britain Vol.II 137 [Batsford 1967]) suggests that James Munn, Paul's father, may
       also have been a Drawing Master, at Greenwich.

Thomas, might have followed, but from the above one might speculate that he too had some connection with paper that might have prompted Lewis to pursue papermaking as a career.

The distribution of Munn countermarks (see Table) shows that his paper was used widely, even it seems, found in American documents, but no examples have come to the author's notice as yet of his paper being used by artists[125] nor that they were handled by William and James, the Stationers. Nevertheless the subject of a connection between the paper maker and Stationers has been raised in the eventuality of the discovery of documents demonstrating this. The Bond Street premises were designated at the beginning of the 19th C. Stationers and Printsellers.

William and James Munn conducted their business in part of the premises at No.107, while their brother, Paul Sandby Munn, the Artist had his studio in another part. Paul was a notable minor artist of the period (one of the original members of Girtin's Drawing Society, an associate of the Old Watercolour Society and an Exhibitor at the Royal Academy); he is of some interest also because of his association with John Sell Cotman, one of the greatest of the English Watercolourists, during an early and formative period of his life. Very few connections are known, indeed any that were productive, between the papermaking industry and leading artists of this period, so that any indication of a link, however tenuous, is worth noting. Cotman became a member of the Drawing Society in 1801 and possibly made a tour of Wales in the company of Paul Sandby Munn in 1802; certainly for part of his first tour to Yorkshire in 1803, but of importance in this context he lived and worked at No.107 Bond Street between 1802-1804. There is no documentary evidence to show that Cotman, or indeed other members of the Drawing Society, bought paper from the Munn Stationers, but it was the practice of many artists of that time (and Cotman was no exception to this) to produce their finished works in their studios.

Lewis Munn snr.'s move to Hollingbourne

If one accepts that Lewis had decided to follow a papermaking career, was it a matter of chance that he ended up in Hollingbourne or were there other reasons that led him there? If Lewis' father was a Stationer or even a printer then he must have had contacts with local paper mills of which there were an ever-increasing number in the not too distant Darenth and Cray valleys just at a time when Lewis might have been indentured to some trade; the 1750's-60's saw no fewer than 5 new mills coming into operation there over and above the established ones.[126]

It is not feasible to invoke here any possibility of Paul Sandby influencing the decision of Lewis to make this move though his connection with his brother James might conceivably have aroused Lewis' interest in White paper at a later date. Sandby's connection, for instance, with the younger Whatman was not until much later, 1794 (see PLATE 15) though he was clearly aware of the pre-eminence of Whatman's papers long before this as indeed his brother, Thomas, both of whose output in watercolour had been very considerable. Indeed he had been present at a dinner (1786), with other celebrities, given in honour of Alderman Boydell, an occasion at which the decision was made to print a National Edition of Shakespeare, which in the event was printed on paper specially made by Whatman.[127] All this would have been too late to have influenced Lewis and other reasons for his move must be sought.

The genealogical Table shows that a Lewis Munn had an illegitimate son baptised at Sutton-at-Hone in 1770; the Greenwich Lewis would have been 21 then. Is there any reason to suppose that these two were one and the same? First, there was quite a considerable colony of Munns living at Sutton-at-Hone including a Thomas and Elizabeth married in 1719; but no direct connection can be established between these two and Lewis' parents; the Wills

---

[125] The author's own searches have covered several thousand watercolours and drawings by British Artists of the period and though there are many instances of Maidstone countermarks, none of Munn's have been seen; but it has to be admitted that no more than a handful of Paul Sandby Munn's drawings have been examined.
That there was an interest in the quality of "Drawing" paper, as such, at the beginning of the 19th C. can be seen in William Balston's correspondence with his Stationers. There are many references to the quality of these and of Plate Papers. The following extract may be cited as an instance of this, Key of Dalton & Keys writing to William Balston on 18:Feb:1809 (PSM) refers to "that particular smooth feel we are of the opinion extremely well adapted for Drawing paper".

[126] James Cripps of Cray Valley mill had been a witness to a Luke Munn's wedding in 1755 (see Chap.VII note 177).

[127] Balston, T. Bib.1 108.

EXTRACTS FROM LITTLE CHART, KENT, LAND TAX RETURNS (KAO Q/RP1/76)

DATE	PROPRIETOR	TAX	OCCUPIER	COMMENT
1780	Henry Darrell	£9	Lewis Munn	This property must have been Ford Mill.
1790	do.	£9	do.	
1800	do.	£9	do.	
	Lewis Munn	£2	Himself	
(1802)				Lewis Munn took an apprentice, William Stephens (Shorter, A. Bib.12 196).
1803	Henry Darrell	£1 [1]	Lewis Munn	The paper makers walked out. [2]
	Lewis Munn	8/-	Himself	
1804	Henry Darrell	£1	Lewis Munn	
	Lewis Munn	8/-	Himself	Lewis Munn bur. Little Chart 4:Sept:1804. (The 1804 Tax Return was countermarked "L MUNN 1803").
1805	Henry Darrell	£1	Sary Munn	
	Sary Munn	8/-	Herself	
1806	Henry Darrell	£1-16	Widow Munn	
	Lewis Munn	8/-	Lewis Munn	Does this refer to Lewis jnr. ? or is it a mistake ?
1807-1809	Henry Darrell	£1-16	Widow Munn	
	Widow Munn	8/-	Herself	In 1808 and 1809 Widow Munn is also assessed twice for other occupants. (see also under 1810).
1810	Henry Darrell	£1-16	Widow Munn	
	Widow Munn	8/-	Herself	One of the occupants of the other properties was George Munn (aged 27 now). (The countermark of the Return is now "L MUNN KENT 1808"). [3]
1811	Henry Darrell	14/-	Widow Munn	(Countermark "L MUNN KENT 1810").
	Widow Munn	£1-16	Herself	There is a curious unexplained reversal in the assessments; possibly should have been 16/- : see 1815.
1814	(as before)		(as before)	George Munn also there. George Munn attended a meeting of the Master Paper Makers of the County of Kent at the Star Inn, Maidstone, 9:Apr:1812.
1815	Henry Darrell	14/-	Edward Paine [4]	(Countermark "JOHN DICKINSON & CO 1810).
	Sarah Munn	16/-	Herself	
1816	Henry Darrell	14/-	Edward Paine	(Countermark "J DICKINSON & CO 1814").
	Sarah Munn	16/-	Herself	

[1] The Tax shown here does seem to be exceptionally low for a Paper Mill. It rises slightly again in 1806 and then falls after 1810 to 14/- and yet from the countermarks the mill was still continuing to operate. One wonders whether the dramatic fall in 1803 can be attributed to the "walk out" and whether after this date the mill never worked at full capacity (see Note 2).

[2] In 1803 Lewis Munn's Vatman, Coucher, Layer and Engineer all walked out and left the mill in July (PSM List of absentees 29:Sept:1803). With his vat crew and head beaterman gone Munn's mill would have been completely paralysed.

[3] The introduction of "KENT" into the countermark appears in 1798 (the earliest found), probably introduced to distinguish Lewis snr.'s paper from Lewis jnr.'s, which must have been just on the point of making its first appearance (see next Table). It was still in use 6 years after Lewis snr.'s death.

[4] Edward Paine succeeded Thomas West at Conyer mill ca. 1809 (West had occupied this mill at Hothfield, a couple or so miles East of Little Chart, from 1740-1809). The 1816 EGL shows Paine occupying both Ford and Conyer mills. The subject of the two countermarks "EDWARD PAINE, LITTLE CHART, 1810" and "EP 1811" has already been discussed in Chapter VII under Conyer mill (see p.326).

demonstrate this.[128] It can only be surmised that there was some family link to account for Lewis' presence there. Second, he appears to have married the illegitimate Lewis' mother, Sarah.

A more plausible reason for Lewis' presence in the area would be that he was learning his papermaking there at one of the many new paper mills. If he had served an apprenticeship there, it is likely that he would have started when he was about 12, that is to say during the 1760's. It will be remembered that at this time both of the William Quelchs and John Terry (d.1766) would still have been operating at Dartford No.3 and No.2 paper mills. All three of them would have known the Maidstone district and Old Mill, Hollingbourne in particular; and further, in 1774, when Lewis first appeared in Hollingbourne, Clement Taylor jnr. (whose father was William Quelch jnr.'s brother-in-law) was just about to take over Old Mill, Hollingbourne, (TABLE XXII) from Whatman who had been experimenting there with his new Engine only two years earlier. Moreover if Munn had set his sights on building a paper mill of his own, Quelch and Terry could well have pointed out that there was greater scope there than in a crowded district like Dartford.

It is possible, therefore, to present a credible reason for Lewis' move from Greenwich to the Dartford area and thence to Hollingbourne; but there are unanswered questions eg. no record of Lewis' apprenticeship; his training may have been the result of a private arrangement with one or other of the Dartford mills and extended later to Hollingbourne? To undertake the conversion and equipping of Ford mill he may have had private means that enabled him to by-pass the customary route to Master Paper Maker.

There is still the unresolved question as to which of the two mills there he might have gone to to gain this experience. The pros and cons of this have already been discussed in App.I (see Eyehorne mill/Robert Williams). On balance the Old Mill solution is favoured, but the notion of a relationship with Williams cannot be discounted easily and further details in support of this are given at the end of this Appendix. The main interest has been to discover that he spent 2 years in Hollingbourne evidently learning before starting a mill of his own.

Lewis Munn at Hollingbourne and Little Chart

No trace of Lewis' marriage to Sarah has been found, but it is quite in order to suppose that he did the right thing and married Sarah Richardson, perhaps in the Dartford area. If all these assumptions are correct, Lewis would have been 25 when he arrived in Hollingbourne, plenty of time for him to have had another Lewis born to him between 1770-4, although this baptism has not been found.[129] All the evidence points to his being the elder brother of George and Richard (see below) who, like Lewis jnr., were to continue their father's occupation of making paper at a later date north of the Thames. The genealogical Table indicates that Lewis and Sarah had 2 children baptised in Hollingbourne and the rest of the family at Little Chart.

In 1776 Thomas Sweetlove of Leeds village[130] and Lewis Munn insured the utensils and stock in their paper mill in Little Chart.[131] Shorter cites a second policy dated 26:Mar:1779 where Lewis only insured a paper mill and a corn mill in Little Chart being described as a paper maker and a miller;[132] this implies that either Ford mill served two functions or that another mill was involved (there was a corn mill further upstream). The Land Tax Returns (see

---

[128] The Will of Thomas Munn of Sutton-at-Hone in the County of Kent (d.1772) P.R.O. PROB/11/977 refers to his wife, Lydia; 2 sons, William & Thomas; and a grand-daughter, Elizabeth. The PR of Sutton-at-Hone KAO TR 2322/69 (1607-1812) shows that Thomas Munn of Darenth m. Elizabeth Dirling of Sutton-at-Hone in 1719. There is no evidence of them having a family in Darenth and the earliest reference to children of theirs in Sutton-at-Hone is not until 1743, a daughter Anne followed by a son, Thomas in 1746. The latter is probably the Thomas & Ann who started a family in 1773. An Elizabeth died in 1769 and it is possible that Thomas snr. re-married Lydia, in which case the Will mentioned above clearly refers to a different family to the Thomas of East Greenwich. The only alternative is that Thomas & Lydia belonged to a different family to the Thomas & Elizabeth of Sutton-at-Hone and could in theory have been Lewis' grandfather or great-uncle. But there were really so many Munns about that it would be very difficult to prove the point. (For other Munn Wills see Working Papers).

[129] No trace of a Lewis Munn baptism found in Hertfordshire records between 1766-1797. To judge from his first appearance at Mill End Mill ca.1798/9 a birthdate between 1770-4 would be consistent with assuming the role of a Master Paper Maker there.

[130] Was this the same or the father of the early 19th C. Thomas Sweetlove a Millwright of Maidstone ?

[131] Shorter, A. Bib.12 196 SFIP 375175 (18:Oct:1776). It is thought that Ford mill was a 1-vat mill.

[132] Ibid. SFIP 411778.(26:Mar:1779).

MUNN COUNTERMARKS

D.O.D.	COUNTERMARK	PROVENANCE
1777 1778	" M S " " M S "	These two Munn & Sweetlove countermarks were found in Leeds Parish Documents, Bills and Vouchers covering the years 1771-1779. One of these bundles was marked "Sweetlove". Thomas Sweetlove, Munn's partner came from Leeds). [1]
1784	" L M " (English Arms in other half)	Gravell, T.L. Bib.16 5:Jul:80 and 4:Dec:81; found in Latimer Papers University of Delaware, U.S.A. (Shorter offers no alternatives to Lewis Munn).
1792	" L MUNN "	Whatman & Balston Archives (illustrated Shorter, A. Bib.12 328 No.112).
1796	" L MUNN 1794 "	Sutton Valence Land Tax for 1796 KAO Q/RP1/374.
1798	" L MUNN 1797 "	do. do. for 1798 do.
1799	" M " (script) below Posthorn/Shield	Gravell, T.L. Bib.68 No.494 (unattributed but compare countermark for 1821 below). Philadelphia document.
1800,1802	" L MUNN 1798 " 　 KENT	Sutton Valence Land Tax for 1800, 1802 KAO Q/RP1/374. (For further comment on use of " KENT " see pp.82, 85).
-	" L MUNN 1801 "	PSM. Sheet of paper in Springfield Collection.
1804	" L MUNN 1803 "	Little Chart Land Tax for 1804 KAO Q/RP1/76.
1811	" (M) 1805 "	PSM. Letter from Edmund Vallance (brother) to Catherine Balston (sister). Since the Vallance family came from Hertfordshire, this letter (posted in London) could have been on paper made by Lewis Munn jnr. at Mill End mill.
1810,1811	" L MUNN 1808, 1810 " 　 KENT	Little Chart Land Tax for 1810, 1811 KAO Q/RP1/76. [2]
1816	" ((M)) 2 concentric circles round M.	Harrietsham Document. KAO P 173 5/2 Pt.I. Countermark attributed here to Munn.
-	" ℳ 1821 "	PSM. Sheet of paper in the Springfield collection with "Munn" written on it in a contemporary hand. This could be a countermark emanating from Lewis jnr. at Mill End mill or from George and Richard Munn at Thetford; more likely to have been Lewis jnr.
After 1824	" G R MUNN 1824 "	PSM. Master Paper Makers correspondence. Document "Rules and Articles to be observed by Journeymen Paper Makers".

[1]　The Leeds Documents referred to here are KAO 222 5/8 and 5/11. It should be noted that the countermark found in bundle 5/8 is in the wrong bundle datewise; it should be 5/11.

[2]　The Countermarks that follow these in the Little Chart Land Tax Returns are the "JOHN DICKINSON & CO 1810" and "J DICKINSON & CO 1814" ones, see previous Table.

As one might expect the Land Tax Returns for the adjoining parish of HOTHFIELD (KAO Q RP1 187) are studded with MUNN Countermarks. The only one that is not in the above list is a double line block letter "M" in the Return for 1781. (Other marks noted were "S.LAY" in the 1787 Return; "PAINE & SONS" in 1792; "A STACE" in 1804; "C BRENCHLEY 1804" in 1805; and "T STAINS 1803" in 1809. Not all of the Returns were examined).

TABLE) taking the events occurring in the paper mill into account suggest that either Munn had ceased to operate the corn mill at the time of the walk–out or that the corn mill was an altogether separate enterprise leased to another. It may be noted that the proprietor of the mill site was Henry Darrell, but with no known connection with the paper trade; and that after 1779 Lewis snr. was on his own. To account for the drop in the Land Tax after 1800 one might speculate that land formerly associated with the mill had been transferred to another part of Darrell's estate (or sold off). The returns show a Lewis Munn entry for 1806, 2 years after Lewis snr. had died; this entry almost certainly refers to Lewis snr.'s dwelling house (unless a mistake was made) temporarily in the occupation of Lewis jnr. and later reverting to his mother. Other points to note in this Table are the paralysing effect the walk–out must have had on the paper mill in 1803; but that up till 1814 it appears to have continued to operate as a Munn paper mill, possibly under the management of George but in the tenure of Lewis jnr. and hence the continued use of the "L MUNN KENT" countermark. It has also been assumed here that the Edward Paine who took over the mill in 1815 (see Note 4 below Table on p.88) was the same Edward who had bought the leasehold of Chartham mill in 1792, later partnered by the Pike brothers in 1796, Paine selling out his share to them in 1799.[133] Quite why he decided to enter the paper trade again is not clear unless it was a son of his.

Lewis jnr., George and Richard Munn

No documentary evidence has come to light yet to show that Lewis jnr. was Lewis snr.'s son. There are, however, a number of facts that indicate that, if not his son, he was at least a close relation. In support of this the following points may be made:–

(1)   There was time enough before Lewis snr. arrived in Hollingbourne for a Lewis jnr. to have been born to him.

(2)   There is a mention of a Lewis Munn in the Little Chart Land Tax Return for 1806 two years after Lewis snr. had died (see Land Tax Table).

(3)   LEWIS MUNN KENT countermarks continue for at least 6 years after Lewis snr.'s death. Had the mill been taken over by George Munn, who was evidently still there since he appears in a list of Master Paper Makers attending a meeting in Maidstone in 1812 (see Land Tax Table), the countermark could have been altered to "G MUNN" as it was to be later when George and Richard had moved to Thetford.

(4)   Catherine Balston in her Mill Letter Book (PSM) records a number of letters sent to or received from what appears to be "Munn L" during 1813–1814.[134]

(5)   Finally, there is the fact that a George Munn took over Solesbridge paper mill, Hertfordshire, from Lewis jnr. in 1842. It might be argued that this may have been another George, but it seems an extraordinary coincidence that the partnership of George and Richard at Thetford should have broken up just at this time (see below).

(6)   Lewis jnr. appeared at Mill End mill, Hertfordshire, sometime between 1796–1799, probably 1798 since it is for this year that there is the first evidence of LEWIS MUNN KENT countermarks. He continued to work this mill until 1832. In the meantime, however, he took over Solesbridge mill (not far from Mill End mill) from Joseph Huffam in 1816/17[135] and worked it until 1842 when George Munn took it over (EGL Dec.1842) and ran it for another 5 years, the mill being taken over by Mr. Bell. The fact that it passed out of the Munn family's hands after 30 years suggests that George Munn was some other relation to Lewis jnr. than a son e.g. a brother from Little Chart and Thetford.

---

[133] See Chap VII Table XIV.

[134] Letter to *Munn L*   26:Mar:1813   Letter from *Munn L*   2:Apr:1813

Letter to *Munn L*   4:Apr:1813   Letter from *Munn L*   5:Jan:1814

Letter to *Munn L*   16:Jan:1814

There was also a letter from "Munn, Little Chart" dated 31:May:1813.

[135] Finerty, E.T. Bib.69 (May) 422.

George (bapt.1783) was a Master Paper Maker so it is more or less certain that the George and Richard (bapt.1786) Munn, who are shown to be at Thetford mill in Norfolk in the 1816 Excise General List, were sons of Lewis snr. (and not some other Munns). This would agree with the fact that Ford mill at Little Chart had by then (1815) passed to Edward Paine. It is also supported by the fact that a sheet of paper belonging to "Munn" and countermarked "M 1821" has been preserved in the Springfield collection, indicating that William Balston was evidently still interested in these Munns, whether it was Lewis jnr., George or Richard. That George & Richard were still at Thetford is shown by a "G R MUNN 1824" countermark, but it is believed that the partnership may have ended before 1845, when Richard Munn & Co. were known to be at Bishop's mill, Thetford.[136] At this time Richard was also operating mills at Kings Lynn and Castle Rising in Norfolk.[137]

---

[136] Stoker, David Bib.70 247/8. Richard was still there in 1858. Mackay & Watson took it over in 1865.
[137] Ibid.

ROBERT WILLIAMS

Reference has already been made to the unusual concurrence of the paper makers' names, Robert Williams and Lewis Munn, at Hollingbourne and in the Rickmansworth area in the 1770's and again at the turn of the century (see App.I). The question was asked whether the two Robert Williams were related. The evidence from the Rickmansworth area shows that they were certainly not father and son as a result of his marriage there. However, we do not know where the Rickmansworth Robert had been prior to 1756, the year of his marriage to Elizabeth Bowers. There is evidence to suggest that he may have been in his 40's by this time (see below); if this was true then he could have had an earlier marriage, which has not been found, and a son, Robert, by it. Though it might have been theoretically possible for Robert Williams to have run a mill in Rickmansworth and in Hollingbourne at the same time, the fact is that they were two different people.

The Hollingbourne Robert Williams

A search of the parish registers[138] showed the following:-
   (a)   1763   Mary d.o. Robert & Mary Williams (bapt. 30:Nov:1763).
          This is the first evidence of Robert in the parish; the search was started in 1757.
   (b)   His wife's name was Mary.
   (c)   He had another 6 children after Mary, the last being baptised in 1777.
   (d)   1803   Robert Williams died.

The Rickmansworth Robert Williams

The Rickmansworth and Hertfordshire Records[139] show:-
   (a)   1756   Robert Williams m. Elizabeth Bowers (26:July:1756).
          This is the first mention of him the Rickmansworth PR (D/P 85 1/3 searched 1723–1766).
   (b)   His wife's name was Elizabeth.
   (c)   He had 3 daughters the last of which was baptised 3:July:1763, Elizabeth. No further Williams' baptisms recorded after this.
          The Hertfordshire Bishops' Transcripts do not show the baptism of any Robert Williams in any other relevant parish.
   (d)   1780   Robert Williams died.

From the two sets of results set out above it will be seen that the two Roberts had wives with different names and the baptism of Mary (Hollingbourne) is only separated from Elizabeth (Rickmansworth) by 4.5 months. The two Roberts were clearly different people; and further, the evidence shows that the Rickmansworth Robert could not have produced by *this marriage* a son, Robert, to fit in with the activities of the Hollingbourne one.

Further information on the Rickmansworth Robert Williams

Hertfordshire Records show that there was a Robert Williams, *Paper Maker*, of Croxley Hamlet, Rickmansworth, whose name does not appear in the Militia List after 1758. Two facts may be deduced from this information; (a) it has been suggested that this may indicate that the paper maker was over 45 by this date, no longer liable for service; and (b) he is described as a paper maker by this date. If he was in fact over 45 then, he would have been aged ca.67 at the time of his death. This point is made because it does allow the possibility of Robert having had an earlier marriage than the one shown above, in 1756. The second point raises the question as to where Robert was working as a paper maker and whether this might not also indicate that he had come to Hertfordshire from somewhere else.

---

[138] KAO P/187 1/1 et al.
[139] The search (Nov.1981) was kindly undertaken by Mrs. Catchpole of the Hertfordshire County Record Office.

In 1758 Robert is described as a Paper Maker of Croxley Hamlet, about 1 mile NE of Rickmansworth; and from his marriage that he was there in 1756, which means that he had almost certainly been made free at the *latest* a year or two before this. The first evidence of him occupying a paper mill is not until he insured Mill End mill, about 1 mile SW of Rickmansworth, in 1773;[140] this is not to say that he might not have been working there long before this, but Finerty[141] does not mention him at all as being at Mill End mill (even after 1773), which became a paper mill in 1755 in the tenure of either George Andrews or William Sands (one of these paper makers was at Mill End and the other at Scots Bridge, nearer to Croxley Hamlet). They both took apprentices in 1755, this being the first documentary evidence of the existence of these mills. From what has been said here, even though Williams may have worked at one or other of these mills, he must have already been a fully-fledged paper maker at the time they are believed to have become operational. It is reasonable to assume then he had been working as a paper maker (aged perhaps at least 40 in 1755) elsewhere, possibly in the not far distant papermaking centres in Buckinghamshire or even Middlesex.

Robert Williams was evidently considered a noteworthy paper maker as he was interviewed and selected by the Court of the Clothworkers Company to be the occupant of Hamper mill (not very far from Mill End), which they had acquired in the 1770's and partially modernised. The conditions of the Lease, which was for 42 years, stipulated that Williams spent £700 completing the modernisation within 18 months.[142] After William's death in Jan.1780 the mill passed to the Stationer, William Lepard; and Mill End mill to Joseph Dell, Mark and Richard Howard,[143] thence to Mary Dell and Mark Howard in 1786, Mary Dell in 1796 and Munn and Simmons probably by 1798. It is noteworthy that 1798 is the first date for an "L MUNN KENT" Countermark introduced to distinguish the Ford mill papers from the Mill End mill papers; it is noteworthy because at about this time Robert Williams at Eyehorne mill in Hollingbourne also introduced "R WILLIAMS KENT" countermarks, a practice not followed by his successors.[144] (Further comment on the use of this type of countermark see below).

The Williams / Munn Relationship ?

Cumulatively these coincidences indicate the possibility, but no hard evidence to confirm this, that the Hollingbourne Williams was related to the Rickmansworth one. If this was so, the latter would have had intelligence of the industry's great expansion in Hertfordshire at that time. Finerty proposes 8 new paper mills between 1753–1768 and a further 6 between 1769–1788. Lewis Munn snr., who must have known Williams, could have acted on this with an eye to Lewis jnr.'s future. There must have been better prospects for the latter north of the Thames than in a backwater in south-eastern Kent. The Munns may also have had relations in Hertfordshire that encouraged this move.[145]

In the preceding pages attention has been drawn to what may be no more than a coincidence, namely, that both the Kent Williams and Munns had used "KENT" in their countermarks presumably to distinguish their papers from those made at Mill End Mill; this was certainly the case with the Munn countermarks. There is, however, a further twist to this practice which again may be no more than a coincidence, albeit a very striking one. At Mill End Mill (i) a Mark and Richard Howard together with Joseph and, later, Mary Dell were successors to the Rickmansworth Williams; and (ii) these in turn were succeeded by Munn briefly associated with a Simmons. The latter has not been identified.[146]

---

[140]   Shorter, A. Bib.12 178 SFIP 333134 (23:Nov:1773)   Robert Williams, paper maker, of Rickmansworth insured his utensils and Stock in his paper mill called Mill End Mill.

[141]   Finerty, E.T. Bib.69 (Apr) 314.

[142]   Ibid. (June) 510.

[143]   Shorter, A.Bib.12 178 SFIP 438860 (29:Jan:1781).   The mill is described here as a paper mill and a corn mill.

[144]   See Appendix I Eyehorne mill Table of Countermarks. The Hollingbourne Williams died in 1803, so that this "R WILLIAMS KENT" countermark must have been roughly contemporary with Lewis Munn snr.'s first "KENT" countermark.

[145]   Wills (Soc.Gen.refs.) Thomas Munn of Hertfordshire 1787 Apr.178: ditto 1790 Aug.389., neither of these searched; but the Rickmansworth PRs 1766–1797 produced no Lewis Munn (jnr.) bapts. nor a search of the Mormon microfiche of Hertfordshire bapt., marr. and bur. (search carried out 25:May:1982 Herts.R.O.).

[146]   Shorter, A. Bib.12 179. (The reference to "Simmonds" in PSM is only found for the Meeting of 21:Jan:1799. The spelling "Simmons" is used for Meetings held on 12th and 26th June 1799 PSM).

But coincidence or not a large family of Simmons paper makers had operated Sickle mill, which Shorter locates at Haslemere, Surrey, from 1735.[147] A John Howard leased this mill from the Simmons between 1802–11 and used the countermark "JOHN HOWARD SURRY", a unique use of the county in Surrey paper marks.[148] It has not been possible as yet to show any relationship between the Surrey Simmons and Hertfordshire Simmons. But Crocker has pointed out that the Surrey John Howard came from Woburn and a connection existed between his son and James Simmons III as late as 1840.[149] What has not been established is a relationship between Mark and Richard Howard and the Woburn family. It might legitimately be asked, what prompted the use of "SURRY" ?

[147] Inf. Simmons and Sickle Mill, Crocker, A. Yearbook Int. Assoc. Paper Historians 1988 7–9. Shorter, A. Bib.12 240 noted the burial of a Thomas Howard, paper man in 1740 in Haslemere.

[148] Crocker, A. ibid (illustrated).

[149] Crocker, A. Letter 12:Nov:89.

APPENDIX III

The Sandling Paper Mills

Preface to Appendix III

The paper mills in the Maidstone district feature prominently in various parts of this work, especially in Chapters II, IV and VII. In Appendix I detailed coverage is given to the paper mills in the Hollingbourne area (and in Appendix II the paper makers that occupied these and related mills). The reason for this was due to incomplete or inaccurate accounts of them in the literature. Other paper mills on the Rivers Len and Loose have been adequately covered by Mr. R.J. Spain,[1] occasionally supplemented in the text with additional data where needed; these accounts cover Turkey and Poll mills, both in the parish of Boxley. This leaves the paper mills to the North-West of Maidstone, three paper mills in the Sandling Ward of Boxley, one in Aylesford; and five paper mills on the opposite bank of the Medway (see Map 5) Millhall mill, Ditton, and three paper mills in East Malling all on the East Malling stream; and Snodland on the Leybourne stream. The last five of these mills are partially covered in work either published,[2] or in progress, by Mr. M.J. Fuller: and, once again, supplemented in the text and in App.IV with additional data where needed. Out of all these mills the ones inadequately covered in the literature are the three mills in the Sandling Ward, Boxley, and Pratling Street mill, Aylesford.[3] The purpose of this Appendix is to provide a revised account of these four paper mills only two of which are of any significance to the subject of this book, Forstal and Cobtree mills, the other two featuring only in the 19th C. but have been included to complete the picture.

Forstal paper mill, already discussed at some length in the main body of the work,[4] was quite possibly the oldest paper mill in the Maidstone district; on present evidence the earliest documented but it is a matter of speculation as to whether its history extends much further back than this together with its near neighbour, Millhall mill, Ditton and, if so, which preceded which.

One might repeat the opening remarks of the Preface to Appendices I & II only with much greater emphasis in this case. Glib references are made to the Sandling mills in the literature and although one can trace various water-mill sites from early editions of Ordnance survey maps, how does one identify them with the paper mills that have been named here ? The first task in this appendix is to determine the location of these 4 mills.

LOCATION

Shorter (1957) originally distinguished two mills in the Sandling Ward of Boxley (both of which he called "Sandling") active towards the end of the 17th C. and at the beginning of the 18th. Later (1960) he extended this to four, Forstal, Cobtree, Pratling Street and Sandling, the last two operating only in the 19th C. He does not, however, locate them precisely. There is no difficulty about Pratling Street because this is marked as a paper mill in early Ordnance Survey maps (see MAP 10 at end of book); the others are more difficult to pin-point. The quotation he cites from Hasted

---

[1] Spain, R.J. Bib.8, 10 and 11. The mills are also covered by Shorter, A. Bib.7 and 12; and Balston, T. Bib.1 Turkey and Poll mills together with an unsatisfactory account of Old Mill, Hollingbourne and Upper mill, Loose.

[2] Fuller, M.J. Bib.13.

[3] Shorter, A. Bib.7.

[4] See Chapter II pp.75–78; Chap.IV p.151; and Chap.VII p.311.

is not helpful, namely, that there were two paper mills on the west side of Boxley parish near Aylesford "on the *rivulet* which takes its rise under the chalk hills", (1782). In fact Forstal and Cobtree, the mills referred to here, were situated on two different streams, one that rises just north of Cossington, which in the 19th C. did have two mills on it, Pratling Street (described as a "New Mill" in the 1816 Excise General List) and Forstal; and the other, the Sandling stream, which had several different kinds of watermill operating along it.

Forstal and Cobtree mills ceased working ca.1840. Neither of them appear in Andrew, Drury and Herbert's map of the district (1769); and in the 1864/4 6 inch O.S. map (Map 10) the site of Forstal has been erased by a new Brick and Tile works; and in Sandling itself there is no indication of a paper mill there either. But the 1844 Boxley Tithe Map (*Western* Section)[5] shows quite clearly what must have been the Mill Pond and Buildings that had once been Forstal paper mill near the junction of the Cossington stream and the Medway. This is confirmed by knowing (i) that one of Shorter's "Sandling" mills was described as "Forestall",[6] and that the succession of earlier occupants can be linked to this name; and (ii) the description of the property found in the Deed of 1671 (referred to later) tallies with the Forstal site shown in the 1864/5 O.S. map. The 1844 Boxley Tithe map also refers in the apportionments (e.g. item 716) to Paper Mill Buildings which indicate that the mill nearest the junction of the Sandling Stream with the Medway must have been Cobtree mill.[7]

The 1864/5 O.S. maps show only two other mill sites in the Sandling area. One of these is marked "Boxley Mill (corn)", which according to Spain[8] had been established as a corn mill over a very long period; the other is shown as a mill for "sifting seeds". The problem is to accommodate the fourth mill described as "Sandling" (Excise No.513) and Dr. Charles' Fulling mill;[9] the former mill was active at intervals between 1819–1839 and the latter related to a papermaking felt business already in decline, but not sold off until 1840, having been assessed for a fulling mill from 1800 onwards. Inspection of both the O.S. and Tithe maps shows that there are not enough mill sites to go round unless we make some assumptions. The apportionments for the Tithe map were not examined for these mills since by 1844 both businesses had come to an end; but all likely ponds or collections of buildings along the streams, such as along the lower stretches of the Boarley rivulet, were scrutinized carefully and no signs were noticed of any sites or likely buildings that would have accommodated either a paper or a fulling mill. Provisionally, therefore, it has been concluded that the mill for "sifting seeds" may originally have been the Charles' fulling mill and that he allowed this to be converted to papermaking sometime before 1819.

The only mill site not obviously identifiable in the 1864/5 O.S. Map 10 is Forstal. It is proposed here that this was:–

Latitude	Longitude	Mill
0° 29' 41"	51° 18' 5"	Forstal mill

In terms of modern Map References the location of the mills concerned are:–

	MR	Identity	Excise No.
TQ	740589	Forstal mill	319
	742593	Pratling Street	318
	750582	Cobtree mill	299
	752583	Sandling mill (or the Charles' fulling mill)	513 & 329
	(715589)	(Millhall mill)	(317) (App.IV further details)

---

5   Cathedral, City and Diocesan Record Office, Canterbury.
6   Shorter, A. Bib.7 citing Pigot's 1832 Directory.
7   Shorter's description (Bib.7) is ambiguous. The pond, he mentions, lying to the East of Cobtree manor has no outflow and the contours to the East of it rise and not fall.
8   R.J. Spain, private communication.
9   See this Appendix under "William Charles and Family"; and App. V The Provenance of the Hand Felt.

The Forstal site lies just on the East side of the Aylesford/Boxley parish boundary and thus appears among Boxley Poor Rate assessments rather than those of Aylesford. Its position, as will be seen below, is important in determining the identity of the first paper maker formerly assigned to it.

The same boundary appears to run directly through the middle of the Pratling Street paper mill, which does not figure in the Boxley records, only fitfully in the Aylesford ones and not at all under Burham.

THE OCCUPANTS

The section that follows is arranged in the form of four synopses of the history of each of the mills, corrected and brought up to date where necessary; and to avoid overloading them with detail supplementary notes precede each account, with detailed RB and LT extracts placed later in the appendix.

It is clear that Dr. Shorter (and Thomas Balston) missed the early years of the Boxley Poor Ratebook, 1691–1711, which appear for some reason at the back of this piece (KAO P 40 12/1). The discovery of this throws a little more light, albeit not nearly enough, on the early occupants of Forstal mill and the succession there. This new information supersedes some of the suggestions made by Shorter e.g. that the presence of Peter Musgrave, Thomas Radford and Peter Archer in Boxley from 1678 onwards may be indicative of a "nursery" there for Kent paper makers.[10] Whereas the expanding industry in the Maidstone district as a whole undoubtedly served as a training ground for 18th C. paper makers, it has been shown in Chapter II and App.IV that in the more restricted confines of Boxley these ideas needed reviewing, Peter Archer, for instance, arrived there as a Master Paper Maker, Musgrove very probably a trained paper maker and a certain query hangs over Radford's activities.[11] Coupled with the uncertainty surrounding the early history of both Forstal and Millhall mills (see Chap.II pp.75–78) a thorough revision of the Succession at the mills in this area was obviously required (for further details of the early occupants see App.IV).

FORSTAL PAPER MILL

It is not known when this mill acquired the name "Forstal". The 1816 EGL merely describes it as a "mill"; by 1832 it was known as "Forestall" and, as will be seen below, we can confirm that it was located in the area now known as Forstal (see Map 10). However, frequent reference has been made in this book to Thomas Taylor and Thomas Willard as the first of its paper maker occupants to have been discovered. How do we know this ?

Several general clues provide an answer to this. First, one can follow the succession back from 1839 to 1691 continuously using the surviving Boxley Poor Ratebooks. These show that up till 1717 there were only 2 paper mills in Boxley, Turkey mill in the Weavering Ward and "a mill" in the Sandling Ward. It has been demonstrated that mill sites on the Sandling stream were very limited and, in fact, that there was nowhere else where a mill could have housed the paper makers then active in the Sandling Ward before 1691 other than on the Cossington Stream, namely, near Forstal, as well as those whose succession can be traced with certainty after 1691.

But there is more positive evidence than this e.g. Thomas Willard's connection with the mill at Forstal and by extension Taylor's also. The Duke family of Cossington (see pedigree overleaf) among other property owned a paper mill in the district and a Deed of 1671[12] describes the sale of this establishment by George Duke of Cossington, in the parish of Aylesford, Kent Esq., to Edward Duke, gentleman of the same place; it included the *paper–*

10 Shorter, A. Bib.12 57.
11 For further details of these 3 see App.IV. under Peter Archer snr.; Peter Musgrove snr.; and Thomas Radford.
12 The text of this Deed is to be found in Notes and Queries, F. William Cock Vol. CLVIII.1930.368. (A copy may be seen in the County Library, Canterbury Division, Kent.)

EXTRACT FROM A MSS PEDIGREE OF THE DUKE FAMILY OF POWER HAYES

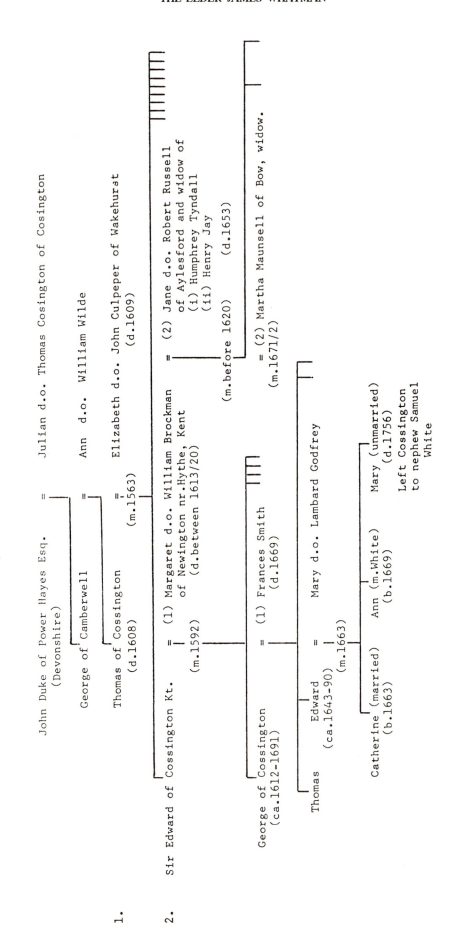

1.

John Duke of Power Hayes Esq.   =   Julian d.o. Thomas Cosington of Cosington
(Devonshire)

George of Camberwell   =   Ann d.o. William Wilde

Thomas of Cossington   =   Elizabeth d.o. John Culpeper of Wakehurst
(d.1608)   (d.1609)
                          (m.1563)

2.

Sir Edward of Cossington Kt.   =   (1) Margaret d.o. William Brockman   =   (2) Jane d.o. Robert Russell
(ca.1612-1691)   of Newington nr.Hythe, Kent   of Aylesford and widow of
                          (m.1592)   (d.between 1613/20)   (i) Humphrey Tyndall
                                                                                         (ii) Henry Jay
                                                                              (m.before 1620)   (d.1653)

George of Cossington   =   (1) Frances Smith   =   (2) Martha Maunsell of Bow, widow.
(ca.1612-1691)   (d.1669)   (m.1671/2)

Thomas   Edward   =   Mary d.o. Lambard Godfrey
               (ca.1643-90)
                          (m.1663)

Catherine (married)   Ann (m.White)   Mary (unmarried)
(b.1663)   (b.1669)   (d.1756)
                                                    Left Cossington
                                                    to nephew Samuel
                                                    White

1   The Will of Thomas Duke Court of Rochester KAO DR₆/PW21 : Elizabeth Duke Court of Rochester KAO DR₆/PW21.
    The Will of Sir Edward Duke Court of Rochester KAO DR₆/PW_WR20.

2   The Will of Sir Edward Duke PCC P.R.O. PROB. 11-183 : Jane Duke P.R.O. PROB 11-227 : William Brockman not found : Robert Russell of Ailsford P.R.O. (1626)
    PROB. 11-149.*

mill and its fittings, 9 acres of land and all water rights etc. belonging to the premises. These premises were in the parishes of Boxley, *Burham* and Aylesford. The tenant was Thomas Willard and the Deed was signed in 1671. This is the most specific description of the location to be found in the Cossington Deeds.

There are two points to note about this Deed. First, examination of Map 10 shows that the site we have assigned to the paper mill lay just within the parish boundary of Boxley and also immediately to the West is a small pocket (15 acres) of land in the parishes of Larkfield and *Burham*; in addition, to the West of this strip (indeed just over 200 yards from the site) one finds oneself in the parish of Aylesford. The description of the premises in the Deed certainly fits the distribution of the land surrounding the paper mill site as shown in the O.S. map. Second, it is clear from the wording of the Deed that Willard must have been at the mill prior to 1671. As tenant, Thomas was "made attornment" (i.e. making legal acknowledgement of the new landlord) by giving "sixpense to Edward Duke, the purchaser". This document then identifies tenant with site precisely. The questions that follow are who was there before him? and who came after ?

There are two further Cossington Deeds of an earlier date[13] where Taylor is mentioned as the occupant of this paper mill. None of the Deeds refer to the occupants' trade; but, since it can be shown that Willard was a paper maker (see App.IV) and all the Deeds are similar in character, it is legitimate to assume that Taylor was one also. Who was Taylor ? and can one go back further than the 1665 Deed ?

Taylor's origin has not been discovered nor any idea of the length of his tenure. There is a possibility that he might have been a local from Aylesford, in which case this presupposes the earlier existence of a paper mill in the area; or another possibility is that he might have been one of two Thomas Taylors born in Goudhurst (1612 and 1640). If this was the case, then since he preceded Willard at Forstal, the 1612 Thomas Taylour seems the more likely one (See App.IV for Taylor).

With regard to the second question no earlier Deeds than those already discussed have been discovered relating to Forstal mill. Resort has therefore been made to the Duke family Wills (see Pedigree opposite). Thomas was the first member of the Duke family to hold the Manor of Cossington (Hasted 2nd Ed. IV.435). Neither his Will nor that of his widow, Elizabeth, (1608, 1609) mention any "mills". However, in his son's Will, Sir Edward (d.1640), George, the heir, was left "....all other messuages, *mills*, edifices, buildings, gardens, orchards lying in the parishes of Aylesford, Boxley, *Burham*, Newington, Halstow, Romney Marsh etc. .....".

---

[13] KAO U 1823/24 T3 part I. The 1665 Deed was an Indenture relating to George Duke of Cossington, County of Kent, Esquire and Frances, his wife of the first part; Lambard Godfrey of Grays Inn, County of Middlesex, Esquire of the second part; the heir apparent Edward Duke and Mary his wife of the third part; and Edward Stott of Lyminge and Richard Duke of Maidstone of the fourth part etc. It included the following:
"and one other Messuage or Tenement with a Paper Mill, mill stream and mill pond, workhouse, outhouses, buildings and appurtenances thereto adjoining and belonging and garden and orchard and one piece of land containing by estimation nine acres more or less lying and being together in Boxley aforesaid and now in the occupation of the said Thomas Taylor". Thomas Taylor is also mentioned in the occupation of other property as well.
A similar Deed of 1668/9 also refers to Taylor and the paper mill; neither is as specific as the 1671 Deed in specifying the parishes; but the reference to "nine acres" identifies it as the Forstal site..

What these "mills" were we do not know; but the conjunction of a mill and the special configuration of the three parishes Aylesford, Boxley and *Burham* is surely a significant pointer to the earlier existence of Forstal i.e. prior to 1640: and if purpose built, the paper mill. It is clear that Edward must either have built these "mills" or else come by them through marriage (1592 and ca.1620). One might envisage a windmill, for instance, coming from the Brockmans on the edge of Romney Marsh. But, if it was Forstal, then a local source; maybe through his marriage to Jane (Russell of Aylesford). Her father's Will (1626) makes no mention of a mill, but it is possible that it had already come to Edward Duke through a marriage settlement. Since Robert Russell's pedigree could not be traced the subject has not been pursued; one can only suggest that the history of Forstal paper mill may have gone back to a date before 1640.

Nothing is known for certain as to who followed Willard and when. It is not until one reaches the first surviving Boxley RB. (1691) that one finds Peter Archer installed there. Working backwards Archer arrived in the parish towards the end of 1683 as a Master Paper Maker and is unlikely, therefore, to have worked under another paper maker e.g. Gill at Turkey mill or Willard, if still at Forstal. Willard would have been 50 at this time and it has been suggested that he may have been an invalid (still had 11 years to live) or an unsatisfactory tenant. If the former suggestion appeals then one could accommodate Peter Musgrove snr. as an assistant to an ailing Willard in 1678.[14] The mill may, of course, have been dormant, but this seems unlikely to judge from the number of paper makers in Boxley or in that area at that time.[15]

The rest of the succession at Forstal is relatively straightforward and will be found in the Synopsis that follows, though it must be said that it has not been discovered where the Hillyers and the Russells came from.[16]

Finally, it is worth noting that at some point in the history of this mill the ownership passed out of the Duke family's hands either directly or indirectly into the Gill family's, remaining in the latter's hands from 1716–1826.

[14] This proposal and Peter Musgrove snr.'s subsequent activities are discussed more fully in App.IV, see Musgrove.

[15] There was Thomas Newman, James Smith and by 1680 others like Manktellow who could have worked Forstal mill had it been idle then.

[16] The problems surrounding John Hillyer's origin is discussed in App.IV q.v. In the case of the Russells who succeeded him the question of their origin and probable relationship with Alexander, the Ancient Paper Maker, is also examined in App.IV under Russell.

FORSTAL PAPER MILL, BOXLEY   (Excise No.319)

This mill was formerly referred to as Sandling Mill (No.19, Kent) in Dr. Shorter's "English Paper Mills" (Bib.12) but later corrected (Bib.7). A synopsis of the history of this mill is set out in the Table below with the sources of information indicated in the right hand column. A note on the possible countermark relating to this mill follows; and, later, extracts from the Boxley Poor Ratebooks and Land Tax Returns.

Dates			References
Early	Unknown	Suggestion has been made that this mill could have been founded late 16th C. by Netherlandish Refugees.	Chap.II p.77 + notes
Before 1600	Unknown	Mention of "mills" in Sir Edward Duke's Will (1640) together with reference to certain parishes can be interpreted as an indication of the paper mill's existence before 1640.	App.III
1665	*Thomas Taylor*	Origin unknown; could be either Aylesford or Goudhurst. Still there in 1668/9.	App.IV and KAO U 1823/ T3
1668/9	Thomas Willard ?	He left Goudhurst sometime after 1666 and probably arrived in Boxley at about this time.	App.IV, p.132
1671	*Thomas Willard*	Definitely occupant of the mill; source implies that he had arrived before this date.	Cock, F.W. Notes & Queries (1930)
(1678)	Peter Musgrove	Suggested that he may have assisted Willard temporarily.	App.IV
1683–93 (1691)	*Peter Archer*	Left Goudhurst towards the end of 1683; first documentary reference to him in Boxley is 1:Jan:1684. First RB entry is in 1691 (see extracts later in this Appendix).	App.IV KAO P 40 1/2
1693–1724	*Goodman Hillary* alias *John Hillyer*	(Name spelt variously). Origin unknown, but suggested as Surrey or West of London. Married Constant Russell in Boxley (1686). William Gill snr. insured mill in his tenure 21:Nov:1716 SFIP 7477.	KAO P 40 12/1 KAO P 40 1/2 App.IV Guildhall Library MSS. Dept.
1725–33	*William Russell*	Origin unknown; suggested as son of William, Ancient Paper Maker, Tredway mill, Buckinghamshire.	See App.IV for discussion
1733–62 ?	*Edward Russell*	The Boxley Ratebook covering 1763–81 is missing and by 1781 the occupant was probably Hillier Russell snr.	See RB Extracts, this App.
1781–90	*Hillier Russell* Mrs. Russell	There is some conflict of evidence between RB and Land Tax entries for this period. Hillier Russell, paper maker of Aylesford, d. 1690 (Kentish Gazette). Mr. Gill Proprietor.	App.III Shorter Bib.12 186. App.III
1790–1800	*Mrs. Russell*	Boxley RB shows Mrs. Russell from 1790; LT not till 1792. ca. 1800/01 a Mr. Russell reappears; this might have been Edward jnr. By 1802 Hillier Russell jnr. came of age and from then on he was the occupant.	For these Russells see App.II, p.60
1802–11	*Hillier Russell*	(see above).	

Dates			References
1812–21 (1816–20)	*Clement Taylor Russell* or Occupier. *William Trindale*	C.T. Russell was made bankrupt in 1816, but his name continues to appear as Occupant in LT Returns as late as 1825.[1] By 1814 Mrs. Gill had become Proprietress. 1816 Excise General List shows him at a "Mill", No.319. For the years 1817–20 the Boxley RB (Sandling Ward) shows entries against "C.T. Russell or Occupier"; in 1820 this was deleted and replaced by "Trindale".	Shorter Bib.7 1960 (Apr.) 62 1816 E.G.L. KAO P 40 12/7
1821–22	*John Pine*	Boxley RB has C.T. Russell & Pine. The Sept. 1821 EGL has John Pine; and the Boxley LT for 1822 has John Pine.	KAO P 40 12/7 KAO Q/RP1/42
(1823) 1824	(C.T. Russell) *Charles Wise*	The Boxley LT shows C.T. Russell for 1823/5; but Excise Letters show Charles Wise here in 1824. E.G.L. May 1824.	KAO Q/RP1/42 Shorter Bib.7
1826–27	(unoccupied) ?	Boxley Land Return for 1826 has "late C.T. Russell, Paper mill and Land". Last entry for Mrs Gill as Proprietress.	KAO Q/RP1/42
1828–38	*George Fowle*	The Boxley Land Tax Return for 1828 has the entry "George Fowle / Self; Paper Mill and Land" and likewise for 1830; but in each case the assessment had fallen from £21 to £17. The reason for this is not known; either the mill may not have been working for this period or, possibly, Land had been transferred and grouped under other property for which Fowle was assessed e.g. Cobtree manor (?).	KAO Q/RP1/ 42
(1832)		The EGL for Nov. 1832 showed Fowle at Mill No.319; and Pigot's Directory for 1832–4 described him as a Maker of White paper at *Forestall Mill,* Aylesford.[2]	Shorter Bib.7 1960 (Apr) 62
1838–39	*John Lavender*	EGL for Sept. 1838 and Pigot's Directory both show Lavender at Mill No.319 and Forestall respectively. Shorter (loc.cit.) stated that no later references to papermaking at this mill have been found. For a Note on a possible countermark relating to this mill see next page.	Shorter ibid.

[1] In spite of his being made bankrupt in 1816 C.T. Russell still continued to figure in the records and it is not until the 1826 Land Tax Return (KAO Q/RP1/42) that the entry reads "late C.T. Russell, *Paper Mill* and Land." It is possible, as will be seen under Cobtree mill, that occupants like Russell might have sub-let these mills to other paper makers e.g. Trindale and, later, John Pine. Quite how Russell managed to maintain this position is not clear. Certainly this practice of sub-letting seemed to be much more common at this time than formerly.

John Pine (see Chap.VII Table XII; and RB extracts this Appendix) may also have had some business association with William Thomas of Cobtree mill from ca. 1788. This became a firm partnership at some point early in the 19th C. temporarily, it seems, dissolved in 1814 when Pine's Lower Tovil mill was burnt down (1814), but reinstated at Basted mill after 1814 (see Chap.VII p.285). Pine's appearance at Forstal in 1821–22 is more likely to have been due to the installation of a paper machine at Lower Tovil in 1821 rather than to any of the effects of the earlier fire i.e. as a means of maintaining production. Thus from 1816–28 Forstal seems to have been a convenient alternative production facility for paper makers needing this rather than the established papermaking concern it had been in the past.

[2] This is the first reference to this mill known to the author under the name by which it is known today.

FORSTAL MILL COUNTERMARKS ?

In 1832 George Fowle was described as a "maker of White paper at Forestall". In 1782 Hasted stated that Forstal and Cobtree mills were used for the making of "an inferior kind of merchandise". To judge from the Gill family's very long association with this mill (over 100 years) and to account for Thomas Willard's move from Goudhurst to Forstal ca.1670, and for the paper makers who followed him, it would appear that this paper mill had been a White mill for most, if not all, of its *known* history. However, it may only have produced inferior grades of White that never merited the use of a countermark, hence the seeming absence of marks that can be associated with this mill in the course of its long history.

Some of its occupants are known to have produced White countermarked paper, but in these cases, e.g John Pine and Charles Wise, they were also operating other mills simultaneously with their tenure of Forstal mill; and the only Lavender countermarks so far discovered must be attributed to papers made at Pratling Street.

There is one possible, though somewhat unlikely, candidate for a countermark belonging to this mill:–

"RUSSELL 1798" (in a drawing attributed to James Wyatt, RIBA Drawing Coll.E6/31[3])

This mark is unusual in that it is not "RUSSELL & CO". As discovered in Chapter VII and elsewhere the Russell involvement in papermaking in the Maidstone district is very complicated (summarised for this period in the Note below).[17] It is possible that this mark might have originated from Russell & Co. at Upper Tovil mill or Mrs.(?)Russell at Forstal; and because C.T. Russell was in at least 2 partnerships at this time one cannot attribute the mark to any mill with certainty.

A number of "HR" marks have also been discovered in late 18th C. documents, some almost certainly attributable to Hendrik Raket (Holland); others, but much less likely, to Henry Revell of Egham's Green Mills (Buckinghamshire) in that they pre–date Hillyer Russell's papermaking activities. But there is one, in a surround that differs from the others, which might be a Forstal countermark.[18]

---

[17] Numerous RUSSELL & CO countermarks are to be found between 1797–1810; as well as RUSSELL & EDMEADS. Shorter, A. Bib.7&12 and Spain, R.J. Bib.10&11 whose accounts partially complement one another describe Russell activities, but the former fails to distinguish Gurney's from the two Ivy mills, important in the context of Russell & Edmeads.

Spain, (Bib.11 175) shows that Upper Tovil mill was auctioned April 1798 (after Clement Taylor's bankruptcy, 1797) and that the Loose RB entry for 1799 (first half) showed Russell & Co. as occupants; in the second half it shows "Joseph Ruse for House, Land & Mill late Russell & *Edmeads*". (If not a mistake, it is not clear who this Edmeads might have been, perhaps William at Gurney's ? see below).

A "RUSSELL & CO 1797" countermark complicates this situation (Bib.12 191) Shorter suggesting that Edward and Clement Taylor Russell were the occupants of U. Tovil mill, supporting this assertion with evidence that they took an apprentice in 1797. The Maidstone RB showed them as the occupants in 1798. It is claimed that these facts would account for RUSSELL & CO 1797–99 countermarks. (C.T. Russell appears at Forstal mill much later, but there is no certain evidence of any partnership between these Russells and Mrs. Russell). Shorter (Bib.7 54) also records that Messrs. Russell were at Otham mill (1810–12) which would explain a RUSSELL & CO 1810 countermark (see Chap.VII TABLE XII note 18).

As mentioned, there was a Russell & Edmeads partnership at Gurney's mill (ca.1797–1802) reported by Spain (Bib.10 56). Sometimes the RB entries read "Russell & Co."; but this was clearly a different partnership to the Upper Tovil one, where there may not have been any Edmeads (see above). An R & E 1801 mark (see Chap.VII note 254) appears in a letter written by W. Thomas from Sandling (i.e. Cobtree mill) to William Balston in Oct.1803. Though this letter is likely to have been written on paper made at Gurney's mill, one might ask whether Russell control extended to Cobtree at this time and that Thomas was in fact using paper made in his mill, Cobtree, under licence as it were ? (For further comments on these Russell partnerships see this App., Table of Extracts for 19th C. from Boxley Ratebooks, note 1). Whereas RUSSELL & CO countermarks are common, the countermark cited in the text above is the only one known to the author.

[18] An "HR" mark enclosed in a circle is found in the 1786 Sundridge (Kent) Land Tax Return (KAO Q/RP1/370). The Boxley Land Tax shows that Hillyer Russell snr. was the occupant of Forstal paper mill from 1780 and this "HR" mark might have belonged to him; a possible alternative might be Henry Revell (as suggested in text) but no marks of his are known to the author.

(continued...)

COBTREE PAPER MILL, BOXLEY

There have been several references to this mill in the main body of the text of this book, especially in Chapter II where it is shown in relation to other early Kent paper mills and, in particular, its position in the second phase of Maidstone's development as a papermaking district;[19] in Chapter IV where mention is made of the Dean family, one of the main occupants and their interest in Upper mill Loose;[20] Cobtree mill itself;[21] and William Gill's interest in this mill;[22] and, finally, consideration of this mill and Upper mill after 1740.[23] From what sources has this information been derived?

Shorter,[24] who added nothing new to the early history of this mill in his later account, assumed that the paper mill may have existed as early as 1700, basing this notion on a David Dean baptismal entry (Boxley PRs) of 1700 in which he is described as a "paperman"; he added that this conclusion was by no means certain. In fact there is nothing to indicate the existence of a third paper mill in Boxley before May 1717 in the ratebook, when we find the entry "£4 Gill for ye new paper mill" sandwiched between two other assessments, one for Jobe Potter above and one for John Springate below. In the November 1716 assessments there is nothing between these two names; whereas after Gill's appearance the mill continues to lie between them into David Dean's time. David Dean's first assessment (Oct.1718) was for £30, a sum considerably higher than that currently made against Turkey and Forstal mills (£16 each). Dean's ratable value continued at this figure until May 1719, dropping to £20 in October and again to £14 in 1720. It is not clear just why these initial assessments were so high but they may represent arrears arising from the low charges made during Gill's first 18 months; or they may have included other buildings which Dean got rid of when the mill became operational.

Apart from the fact that the paper mill appears as a new entry in the ratebook i.e. not apparently replacing another named occupant of, say, a fulling mill, there is nothing else to suggest that Cobtree was "new-built". Indeed its very site suggests that another kind of watermill may have existed there at an earlier date and had perhaps fallen into disuse. Nor has the name of the owner of the site and the paper mill in its early days been discovered; there is no evidence to suggest that Gill had freehold of the land as was the case with Forstal, although by 1780 one finds that a member of the Gill family owned both Forstal and Cobtree mills[25]. David Deane snr.'s Will (made 1732), the first occupant of Cobtree after William Gill snr., indicates that the latter had originally held the leasehold (at least this is the

[18](...continued)

A number of "HR" marks without a surround and accompanied by some sort of cypher or device in the centre of the sheet are to be found in the Hawkhurst (Kent) Land Tax Returns for 1780–83 inclusive (KAO Q/RP1/168). The lettering, however, closely resembles 3 HR marks described and illustrated in Gravell, T.L. Bib.68 Nos.346–8 found in American documents dating from 1738–81 a range quite outside Henry Revell's activities and much more probably belonging to Hendrik Raket. The Hawkhurst papers could well have been imported Dutch ones still.

[19] Chap.II Table IV and pp.86–87.
[20] Chap.IV note 62.
[21] Chap.IV pp.153/154.
[22] Chap.IV p.164; also mention in Chap.III p.121.
[23] Chap.VII Cobtree p.311. and Upper Mill, Loose p.313.
[24] Shorter, A. Bib.7 (April) 62 & Bib.12 187.
[25] See extracts from Boxley RB.s and Land Tax Returns later in this Appendix.

interpretation placed on this).[26] In the case of Forstal mill it is clear that James Brooke acquired this mill and the freehold after William snr.'s bankruptcy in 1731, doubtless with the object of passing it on to William snr.'s prospective heir, at that time "Brooke" (d.1744) but in the event to William jnr., bequeathing it in 1750.[27] In the case of Cobtree there does seem to have been a parallel rescue mounted by another Stationer, Edward Rowe, near East India House, Leadenhall Street, who insured a paper mill "known by the name of Sandlin' mill" near Aylesford in the tenure of David Dean.[28] This was six months after William snr.'s final bankruptcy. It is thought that later the leasehold of this mill must have been acquired by Brooke, who left the lease of "a papermill near Maidstone, Kent" to his nephews (sic) Thomas Wright and William Gill, probably Cobtree (?).

[26] PRC 17/96 f.364. ...to my son Thomas Dean "that paper mill late Mr. GILLS in the parish of Boxley which I now hold by lease now in the possession of Edward ROWE, Stationer in the City of London".

[27] See App. IV p.172.

[28] Shorter, A. Bib.12 187 citing SFIP 54351 (5:June:1731). One might speculate that with Turkey and Forstal mills on his hands Brooke may have felt that Cobtree was more than he could afford and so let it go to Rowe.

COBTREE PAPER MILL, BOXLEY   (Excise No.299)

The synopsis of the history of this mill is set out in the Table below with the sources for this information indicated in the right hand column. A note on countermarks originating from this mill follows; and later in this Appendix there are extracts from the Boxley Poor Ratebooks and Land Tax Returns.

Date			References
May 1717	*William Gill*	"£4 Gill for ye new paper mill", the first evidence of a third paper mill in Boxley.	KAO P 40 12/1
1718–32	*David Dean*	Oct. 1718 RB entry shows "£30 Mr. Deann, for the paper mill". This high assessment probably reflects (a) that Cobtree was now fully operational as a paper mill; and (b) some back payments due for the previous 18 months while the mill was being set up. This must have been David Dean snr. who first made his appearance as "paperman" in 1700.	KAO P 40 12/1 Chap.IV n.62
(1731)		"Sandlin' Mill" insured by Edward Rowe, Stationer, in the tenure of David Dean. (This was ca.6 months after William Gill snr.'s bankruptcy in January 1731).	SFIP 54351 (5:June:1731).
1733–50	*Thomas Dean*	Son of David Dean (bapt. not found) see Will. There are only 2 Dean entries in the Boxley PR's for this period, John (1697) and David jnr. (1700) and neither of them were "baptised". They may thus have belonged to some Nonconformist persuasion and had later members of their family registered elsewhere. (There was also a Joseph who took over Upper mill, Loose in 1733).	PRC 17/96 f.364
1750–80	*Mrs. Elizabeth Deane* \ *Buried 1771*	The Boxley RBs up to 1763 show Elizabeth Deane as Occupant. The RB from 1763–80 is missing, but the first entry in the next one is "Thomas, late Dean"; this was David Dean (see below). So at some unknown date David succeeded Elizabeth.	KAO P 40 12/2,3
1780	*David Dean* \ *Buried 1781*	The Boxley Land Tax Return for this year shows Mr. Gill as Proprietor and David Dean the occupant of this mill.	KAO Q/RP1 42
1781– 1808	*William Thomas*	The Boxley RB entries start with Mr. Thomas, later William and by 1793 this becomes Mrs. Thomas. But the Land Tax Returns continue to show William Thomas up to 1808. A possible explanation for this could be that William Thomas was working with John Pine at Lower Tovil mill from 1793 (?); certainly, Pine and Thomas were in partnership there between 1810–1815.[1]	KAO P 40 12/4–6 KAO Q/RP1 42 Shorter, A. Bib. 7 (July) 58.

---

[1] See Chap.VII Table XII. It is quite possible that John Pine was stretched late in the 18th C. in trying to cope with the family's interests in their 3 paper mills and felt the need for help at Lower Tovil mill which may initially have been a loose association with William Thomas that blossomed into a partnership by 1810. Although this partnership was dissolved in 1815, according to Bridge, W.E. "Basted Mill, some historical notes" (Pamphlet, 1948), Thomas Wildes who had bought Basted in 1815 leased it to John Pine and William Thomas of Tovil for 14 years; certainly William Thomas is shown there in the 1816 E.G.L.

Date			References
(1809– 17)		According to the Boxley Ratebook Mrs. (Fanny) Thomas was assessed for Cobtree mill from 1809–12 inclusive. In 1813 the entry is Stacey Wise late Thomas, Wise remaining there till replaced by Charles Wise in 1817. The Land Tax Returns present a similar picture. But, as seems to have been the case with other mills at that time, Cobtree appears to have been sub–let to other Occupants, the only one of which one can be absolutely certain being Florence Young shown as Occupant in the 1816 EGL.	KAO P 40 12/6, 7 KAO Q/RP1 42 1816 EGL
		However, other Youngs, to judge from countermarks in local documents, may also have been occupants between 1809–16; but it is not known who these Youngs were or where they came from.[2]	
		There is, for example, a "GREEN & YOUNG 1809" countermark in a sheet of William Balston's accounts for 1817. Similarly, there is a "JOHN YOUNG 1810" countermark found in a bundle of West Farleigh (Kent) Curate's licences.	PSM KAO P 143 2/1
		These countermarks together with the entry in the 1816 EGL suggest that John and Florence Young with some assistance from "Green" may have been the under–tenants for this period. Provisionally, therefore, it has been decided to show the succession at Cobtree as follows:–	
(1809)	*GREEN & YOUNG*		Countermark
(1810)	*JOHN YOUNG*		Countermark
1816	*Florence Young*		EGL 1816
1817–24	*Charles Wise*	The RB and Land Tax entries showed him there from 1817. The Dec. 1819 EGL also shows him at Cobtree mill. No.299. Shorter states that he was made bankrupt in 1824. (He was also in partnership with Charles Brenchley at Padsole mill in Maidstone and he was made bankrupt there as well as at Sandling in 1824).	Bib.7 (Apr).62
1825	*Allnutt & Co.*	This appears in the Land Tax Return. According to Shorter this must have been a partnership between Henry Allnutt (who was in partnership with Thomas Smith at Little Ivy and Great Ivy mills, Loose, Nos. 309, 308, to be extended to Lower Tovil mill, No.311 by 1832) and John Wilson (EGL June 1825).	KAO Q/RP1 42 Bib.7 (July) 60

[2] A Mrs. Young is shown as the Proprietor of some property in Boxley in the 1780 Land Tax Return (KAO Q/RP1 42) but beyond this nothing further has been discovered about the Youngs. A "J YOUNG Nestingens Fabrique" countermark has been recorded by Gravell, T.L. (Bib.16 19th January 1983) in a letter dated 1817 in Norfolk, Virginia but with no watermark date.

As to "GREEN" a George Green was assessed in the Hill Ward, Boxley in 1780 at £34 for some unspecified property; and a William Green in a varying capacity as an Occupant in both the Boxley RB and LT Returns between 1801–22, though he disappeared between 1810–1814. Some of the early assessments are large (£120–10) later falling to a much lower value (see Extracts Later in this Appendix). He also had property in Maidstone. It is conceivable then that ca.1809 he may have ventured briefly into financing papermaking.

Date			References
1826	*Allnutt & Wilson*	Paper mill and Land appears in the Land Tax. Shorter assumed that the mill was in their hands until it ceased work in 1838. This may have been so, in which case one has another instance of sub–letting, because the Land Tax Return for 1828 shows Lavender there.	KAO Q/RP1 42
1828	*John Lavender*	Paper mill and Land.	KAO Q/RP1 42
1830–38	*Smith & Allnutt*	The 1830 Land Tax Return merely shows Allnutt; but the 1832 EGL shows Smith & Allnutt and Shorter suggests that they remained the occupants until the mill ceased work in 1838. Smith & Allnutt both made Handmade *and* Machine made papers in colours *and* white. Colours are difficult to isolate in a mill that makes white paper. So it is possible that Allnutt's may have used a small mill like Cobtree to make one or the other, by hand that is.	Bib.7 (Apr). 62

COUNTERMARKS

As with Forstal Mill no countermarks, discovered so far, can be attributed with any certainty to this mill. Two Countermarks have been assigned tentatively:–

GREEN & YOUNG 1809

JOHN YOUNG 1810[29]

No "THOMAS" marks have been found yet: Charles Wise was active at Pratling Street in 1821 and at Forstal in 1824; so, if any countermarks of his are found between 1817–21, they might well belong to this mill. Allnutt[30] and Lavender were operating at other mills at the same time as this one.

---

[29] Previous page note 2 refers to another "J YOUNG" mark. This was found in an American document followed by "NESTINGENS FABRIQUE" possibly made for or in a French or Netherlandish mill.

[30] Henry Allnutt appeared in Maidstone in 1813 (Chap.VII TABLE XII). He may have come from Buckinghamshire, Weir House mill (Shorter, A. Bib.12 131). During the 19th C. the family had extensive papermaking interests in Surrey (inf.A. Crocker).

PRATLING STREET PAPER MILL, AYLESFORD (MR TQ 742593) (Excise No. 318)

In the Excise correspondence this mill was known only by its number and the anonymous title "NEW MILL". The first official reference to it is in the Excise General List for 1816 where Thomas Stroud is shown as Occupant. Because of Stroud's associations with both Hollingbourne and Leeds (see App.I & II)[31] it was first thought that this "New Mill" might have referred to a mill in that area such as the mention of a "New Mill at New England" in the Hollingbourne ratebook (1808). However, the discovery of Thomas Stroud's Will,[32] drawn up in 1813, made it clear that this "New Mill" was in fact the leasehold mill and premises at Pratling Street. But who built this mill and when has not been discovered.

The solution to these questions is not helped by the fact that no references to Thomas Stroud or to the mill can be found in the parish records of Aylesford, Boxley or Burham, only to his son, Thomas jnr., two years after his father's death, in the Aylesford ratebook.[33]

Although this mill seems to have slipped through the local records' net, there are one or two pointers to a previous occupant. It is now known that Stroud moved there in 1813 and there is a "STROUD 1814" countermark as well as a few countermarks of later occupants (see list following the synopsis of the mill's history), but in addition to these it is reasonable to attribute a number of Charles Brenchley's countermarks, those from at least 1804–12, to this mill. Shorter pointed out that Brenchley's name appears in the Boxley ratebook in the early 1800's;[34] he associated this with the discovery of "C BRENCHLEY 1807" and "1812" marks and suggested that the papers bearing these may have been made at Sandling mill (Excise No.513).[35] He went on to point out that Brenchley moved to Otham in 1812, his name appearing in the ratebook there for that year.[36] But there is at least one good reason which makes it impossible to validate Shorter's proposal. Sandling mill has a higher excise number than any appearing in the 1816 List (489) and, in fact, is only mentioned for the first time in Dec.1819 Excise Letter.

A more plausible alternative is to place Charles Brenchley in the Pratling Street mill which is situated astride the Boxley/Aylesford parish boundary. The Pratling Street Excise Number for one thing conforms to those of other established mills in the district. It is proposed here that recently discovered "C BRENCHLEY 1804" countermarks, together with any others up to 1812 should be assigned to Brenchley's operations at Pratling Street. Unless Brenchley had moved to Otham very early in 1812, it is unlikely that he would have inserted an 1812 date

---

[31] See App.I under Park mill; and App.II under Stroud.

[32] PRO PROB/11/1566 fo.159

[33] No entries referring either to this mill or to Stroud were found in the Aylesford or Burham Land Tax Returns (1799–1816/1820), only a reference to Cossington in the former, a property next to Pratling Street. Two references to a Thomas Stroud were found in the Aylesford RB (KAO P 12 12/6 1797–1820, but nothing in 12/7 1820-29); 1817 he was assessed for £50 and in 1819 £35. Since Thomas snr. d. 25:Feb:1815 these entries must refer to Thomas jnr. (Charles Wise took over in 1821).

Only one mill, the Corn mill, is mentioned in the Aylesford RB (see Map 10) a mill referred to in an indenture of 1655 KAO U 36 T906.

[34] Shorter, A. Bib.7 (April) 62.

[35] Sheets in the Springfield Collection, PSM. As will be seen earlier Brenchley countermarks (1804) have been found since.

[36] Shorter, A. Bib.7 (May) 54.

into his Otham production. Moreover, this suggestion accommodates Stroud leaving off his papermaking operations in Hollingbourne and moving them to Pratling Street in 1813.

Date			References
(1804–12)	*Charles Brenchley*	Suggested tentatively on the basis of 3 countermarks 1804, 1807 and 1812 and his presence in Boxley during the first decade of the 19th C.	PSM & KAO P 222b 5/2–8. Shorter Bib.7 (Apr) 62.
1813–15	*Thomas Stroud*	See Notes on Thomas Stroud (1767–1815).	App.II
1815–19	*Heirs of Thomas Stroud*	EGL for 1816 showed Thomas Stroud as the Occupant; but he died in 1815. It seems probable that, as in Leeds village, he was succeeded by Elizabeth, his widow; and, later, by his son, Thomas jnr., almost certainly a minor at the time of his death.	All SS. Maidstone PR
		The last reference to Thomas Stroud jnr. is an entry in the Aylesford RB for 1819.	KAO P/12 12/6
1821–24	*Charles Wise*	First mention is in an 1821 Excise Letter. Bankrupted in 1824. Had also been operating Cobtree mill during this period.	EGL Sept.1821 Boxley RB KAO P 40 12/7 and LT Q/RP1 42
1825–32	*John Lavender*	Countermark of 1825 found in a bill sent to Thomas Crispe by a William Kennard among Leeds Parish documents (For origin of LAVENDER see later in this Appendix)	KAO P 222B 12/34
1832	*Joseph Moreton*	Shorter quotes a reference to Moreton in the 1830's as a maker of White paper at Pratling Street.	Bib.7 (Apr.) 62
1846	*Samuel Salkeld Knight*	Millboard maker.	Shorter ibid.
1849	*Thomas Compers*	Bankrupt 1851.	Shorter ibid.
1851–60	*George Mason*	In 1851 the mill had only 1 beating Engine, so it must have been quite small.	Shorter ibid.
1860's/70's	*John Skelton Isaac* followed by *Warwick Isaac & Co.*	They made carriage panels; millboard; and, later, panels for railway carriages.	Shorter ibid.
1889		Reputed that mill was pulled down.	Shorter ibid.

COUNTERMARKS

C BRENCHLEY 1804[37]	Broomfield Document Hothfield Land Tax 1805	KAO P 222B 5/2–8 KAO Q/RP1 187
C BRENCHLEY 1807	Sheet of Laid Paper	Springfield Colln.
C BRENCHLEY 1812	Sheet of Laid Paper	Springfield Colln.
STROUD 1814	Harrietsham Document	KAO P 173 5/2 Pt.II.
J LAVENDER 1825	Leeds Document	KAO P 222 12/34.
LAVENDER 1825	Broomfield Document	KAO P 222B 12/11.

[37] An undated "C BRENCHLEY" mark has been found in a blank sheet of paper in American Archives (Gravell, T.L. Bib.68 113). This paper could have come from either Pratling Street or equally Otham mill after 1812.

NEW MILL, SANDLING WARD, BOXLEY   (Excise No.513 : later No.329)

The Excise Letter for Dec.1819 has the following entry "Sir, Entry having been made of a New Paper Mill in the undermentioned Collection, the Commissioners order that it be distinguished by the following Number" :–

ROCHESTER COLLECTION

513   William Peters and Thomas Chaplin   Sandling.

Dr. Shorter suggested that there might have been a paper mill here at an earlier date which had been occupied by Charles Brenchley, but on present evidence this proposal has been discounted (see Pratling Street). The location of this mill has also been discussed earlier in this Appendix and the proposal made that the Charles Family's Fulling mill was used for papermaking intermittently; and, further, that this later became the mill used for "sifting seeds" (see Map 10) with a modern Map Reference of TQ 752583 (quite possibly buried to-day under the spoil from the M20 Motorway ?).

Date			References
1819–25	*William Peters* *Thomas Chaplin*	EGL Dec.1819 Discontinued in 1825	Shorter Bib.7 (Apr.) 62
1839	*William Hunt* *Reuben Hunt*	This mill appears to have re–opened in 1839 under a new Excise Number; but by the end of the same year the Hunts had been declared bankrupt.	Shorter ibid.[1]
		There are no further references to this mill as a paper mill.	

COUNTERMARKS
None found to date.

[1]   The Excise No.329 was allocated to Sandling mill in 1839. This number had formerly belonged to Cray Valley mill, St. Paul's Cray (see 1816 EGL Charles Cripps). At some point, not investigated here, Cray Valley must have ceased production, possibly before 1828 when the Nash family were in occupation there and they were allocated Excise No.589 (Shears, W.S. "William Nash of St. Paul's Cray" 1950). As a matter of interest Medway (Monckton's) mill, Maidstone, was allocated the number 329 in 1860; this mill later became an adjunct of Springfield mill, Maidstone.

EXTRACTS FROM THE BOXLEY POOR RATEBOOKS & LAND TAX RETURNS

These extracts have been made to supplement information given in the main text on the existence and occupation of the Boxley paper mills and, where this is possible, to try to determine any change in their status. In addition they have been used to sort out the rapidly changing occupants, or sub-tenants, and proprietors of the Sandling mills and Pratling street at the beginning of the 19th C.

A Note on the first Boxley Poor Ratebook (KAO P 40 12/1).

The first entries in this book are for May 1691 and they are at *the back* of this book. The entries at the front of the book start with the Outdwellers for 1710, followed by the Assessments for 1711. It is probably as a result of this arrangement that the first entries were missed by both Dr. Shorter and Thomas Balston. The book ends with entries for 1723.

KAO P 40 12/1

DATE	£	TURKEY MILL Occupant	£	POLL MILL Occupant	£	FORSTAL MILL Occupant	£	COBTREE MILL Occupant
May 1691	16	George Gill			16	Peter Archer		
1692	16	do.			15	do.		
May 1693	16	do.			15	do.		
Nov.1693	16	do.			15	Goodman Hillary [2]		
1694-97	14	do.			15	John Hillary		
1699	14	do.			18	do.		
1700	14	do.			18	do.		
1701-07	16	do.			18	do.		
1710	16	do.			16	do.		
1715	16	do.			16	John Hylier		
Nov 1716	16	do. [3]			16	do.		[4]
May 1717	16	do.			16	do.	4	Gill for ye new paper mill [5]
May 1718	16	do.	31	Swinnock [6]	16	do.	4	do.
Oct 1718	16	do.	31	do.	16	do.	30	Mr.Deann for the paper mill [7]
May 1719	16	do.	16	do. [8]	16	do.	30	Mr.Dean
Oct 1719	16	do.	16	do.	16	do.	20	do.
May 1720	16	do.	16	do.	16	do.	14	do.
May 1721	16	do. [9]	16	do.	16	do.	15	do.
Oct 1723	16	Mr.William Gill [10]	14	do.	16	do.	15	do.

KAO P 40 12/2

DATE	£	TURKEY MILL Occupant	£	POLL MILL Occupant	£	FORSTAL MILL Occupant	£	COBTREE MILL Occupant
1724	16	Mr.William Gill	14	Swinnock	16	John Hillyer	15	David Dean
1725-27	16	do. [12]	14	do.	16	William Russell [11]	15	do.
May 1728	16	do.	30	John Swinnock now Mr.Gill [13]	16	do.	15	do.
Nov 1731	16	do.	30	Mr.Cordwell [14]	16	do.	15	do.
Jun 1732	40	(Cordwell) do. Gill mill [15]	30	do.	16	do.	15	do.
1733	40	do.	30	do.	16	Edward Russell [16]	15	Thomas Dean [16]
Jun 1735	40	Mr.Musgrove [17]	30	do.	16	do.	15	do.
Dec 1736	26	Mr.Harriss [18]	30	do.	16	do.	15	do.
May 1738	30	do. [19]	30	do.	16	do.	15	do.
Nov 1739	30	do. [20]	30	do.	16	do.	15	do.
Apr 1740	60	Mrs.Harris for mill & Land [21]	30	do.	16	do.	15	do.
Oct 1740	60	Mr.Wattman [22]	30	do.	16	do.	15	do.
Aug 1741	65	Mr.Whatman [23]	30	do.	16	do.	15	do.
1750	65	do.	30	do.	16	do.	15	Mrs. Dean [24]
Jun 1754	65	do.	30	do.	16	do.	15	Elizabeth Dean

KAO P 40 12/3 Ratable Values adjusted between June 1754 and Jan 1755.[25]

DATE	£	TURKEY MILL Occupant	£	POLL MILL Occupant	£	FORSTAL MILL Occupant	£	COBTREE MILL Occupant
Jan 1755	34	Mr.Whatman	37-10	Mr.Cordwell	22-10	Edward Russell	20	Elizabeth Dean
1756	34	do.	37-10	Mr.Fearon [26]	22-10	do.	20	do.
Dec 1759	34	Mrs.Whatman [27]	37-10	do.	22-10	do.	20	do.
May 1760	34	Mr.(sic) Whatman	37-10	do.	22-10	do.	20	do.
May 1761	34	Mrs.Whatman	37-10	Edward Fearon now Taylor [28]	22-10	do.	20	do.
Jun 1762	34	do.	37-10	Taylor	22-10	do.	20	do.
Jan 1763	34	do. [29]	37-10	do.	22-10	do.	20	do.

DATE	TURKEY MILL £	Occupant	POLL MILL £	Occupant	FORSTAL MILL £	Occupant	COBTREE MILL £	Occupant
May 1781	74 75 70 etc.	Whatman (Turkey Mill) James Whatman Esq. late Champney (Vinters) ?[31] ditto. house & land (other properties)	75	James Taylor[32]	45	Mr.(?) Russell[33]	40	Mr.Thomas late Dean[34]
Nov 1781	74	Whatman	75	do.	45	Mrs.Russell	40	do.
1787	74	do.	80	Whatman late Taylor[35]	45	do.	40	do.
Jul 1788	74 80	do. (Turkey mill) do. (Poll mill)	–	(assessment transferred to Whatman).	45	Mrs.Russell	40	Mrs.Thomas[36]
1790	74 80	do. do.	–		45	Mrs.Russell[37]	40	do.

KAO P 40 12/5 (May 1794–1800)

DATE	TURKEY MILL £	Occupant	POLL MILL £	Occupant	FORSTAL MILL £	Occupant	COBTREE MILL £	Occupant
Nov 1794	74 80	Whatman[38] do.[39]	–		45	Mrs.Russell	40	Mrs.Thomas
May 1795	138	Finch Hollingworth & Balston	–		45	do.	40	do.
1796	138	do.	–		45	do.[40]	40	do.[40]
1800	138	do.	–		45	do.[41]	40	do.

(TURKEY MILL & POLL MILL were in the WEAVERING WARD of Boxley : FORSTAL MILL & COBTREE MILL in the SANDLING WARD).

1 Peter Archer snr. left Forstal mill in May 1693. The destination suggested, Dartford, is discussed in Chap.IV p.140.

2 "Hillary" is synonymous with Hillyer (and var.). See App.IV under Hillyer.

3 In 1716 William Gill snr. effectively took over Turkey mill (see Chap.IV p.162 and Brooke's Insurance of Goods). Gill also insured Forstal.

4 No Ratebook entry between Jobe Potter/John Springate in 1716.

5 Gill entry for Cobtree mill appears for first time between Potter/Springate entries. Other Sandling assessments of a paper mill order such as John Powell (£13); Bartholomew Brawman (£13) and Ralph Covell (£11), who had been assessed since the 1690's carry on after Gill and Dean appear. They obviously had nothing to do with the Cobtree site. £22 in 1713 Mr. Sourls Goatley "for Cobtree" probably refers to the Manor.

6 The entry of £31 against Swinnock probably reflects the Old Fulling mill value. See Note 8 below.

7 First reference to David Dean snr. at Cobtree. For comment on the high initial assessments see this Appendix p.106.

8 The May 1719 entry against Swinnock reads £16 and then £27 John Swinnock now Mr. Moore. Swinnock had clearly shed something.

9 William Gill snr.'s mother, Susanna, died in 1720 and George, her father, evidently decided to leave Turkey Court in 1721 to live with his daughter, Ann Walter, at Ditton. The Boxley Ratebook entries reflect this change in 1722 (cf. Note 3 above).
   It is not known why (a) the ratable values of Turkey and Forstal mills fluctuated during this early period; and (b) why the values should have been of a comparable order. One would have thought that Turkey mill was much larger than Forstal and had Turkey Court attached to it as well.

10 First ratebook reference to William Gill snr. as such.

11 In 1725 William Russell took over Forstal. Who was he ? For discussion of this question see App.IV under Alexander Russell.

12 Between 1726–28 a "Mr. Harris" is assessed as an Outdweller, almost certainly Christopher (see Harris Genealogy App.II). He died in 1729 and was succeeded by "Widow Harris", probably Mary, who was assessed up to 1736 when she died. Not known whether any relation of Richard (see Note 18).

13 Swinnock was bankrupted 26:Sept:1727. The Ratable value was now more than doubled under William Gill when he took over Poll mill.

14 William Gill was finally bankrupted in Jan.1731. Cordwell taking over at Poll mill may have been a result of Gill's first bankruptcy in May 1729.

15 Cordwell took Turkey mill, 18 months after Gill's bankruptcy, with a ratable value of £40 now. What these increases were due to is not clear.

16 Thomas Dean took over Cobtree in May 1733; and Edward Russell Forstal in October 1733. Their ratable values remained unaltered.

17 "Mr. Musgrove", almost certainly Peter jnr., occupied Turkey mill for 1 year (see App.IV under Musgrove).

18 Richard Harris arrived at Turkey mill and insured his stock and storerooms (see App.1 Hollingbourne Old Mill).
   Note the drop in the Ratable value. The connection with Boxley of the Outdweller Harris' is probably fortuitous, no known connection with Richard.

19 The ratable value of Turkey mill had risen again to £30 in May 1738. Brooke had acquired the freehold 23:May:1738 and by 1st June had undertaken to convey to Harris, but conveyance not completed until 29:Sept:1738 when Harris started pulling down mill and modernizing it.

20 Date of Assessment 18:Nov:1739; Harris died 26th Nov; rebuilding was still in progress.

21 To judge from the £60 assessment against Harris' widow, Ann, the rebuilding appears to have been completed and the mill become operational.
   She was engaged to Whatman in Apr.1740; Marriage settlement 2:July; married 7:Aug.; and Whatman insured his mill 19:Aug:1740 (SFIP 85207).

22 James Whatman assessed for the first time (£60).

23 Ratable value of Turkey mill raised to £65 where it remained until Jan.1755.

24 Elizabeth Dean occupant continuing up to some unknown point before 1780 when a David Dean took over (see 1780 Land Tax Return).

25 The Boxley Ratable Values were adjusted in 1755; some values reduced e.g. Champneys, Bridgland and Whatman (£65 reduced to £34) others went up like Cordwell at Poll mill (£30–£37–10); Edward Austen, Edward Russell (£16–£22–10) and Elizabeth Dean (£15–£20), the last 3 in Sandling.

26 Fearon appears at Poll mill (Abraham, later Edward). See Chap.VII p.286 re possible installation of an Engine there.

27 The elder Whatman died 29:Jun:1759. Ann's interregnum ended technically when the property was transferred to James Whatman II on 28:Sept:1762.

28 Clement Taylor snr. arrives (see Chap.VII pp.286ff.; and his relationship to the Quelchs App.II).

29 The younger James Whatman took over the mill in the last quarter of 1762 (Balston, T. Bib.1 18).

30 The Boxley Ratebook for the second half of 1763 until the end of 1780 is missing, so one has to infer the changes that took place.

31 Whatman now has several assessments against his name; one for £75, late Champneys later appears as "Vinters". This and other acquisitions including the freehold of Poll mill are described in Balston, T. Bib.1 73, exchanges with the 4th Earl of Aylesford.

32 James was the s.o. Clement Taylor snr. who died in 1776 (see Chap.VII p.287).

33 The Land Tax Return for 1780 shows Mr. Gill/Hillier Russell £27 and this continues through the 1780's differing from the "Mrs. Russell" in the RB.

34 As above the LT differs from the RB. 1780 LT shows Mr. Gill/David Dean £24; 1781 LT missing; 1782 Mr. Gill/William Thomas £24.

35 In 1787 James Taylor was evicted by Whatman (Balston, T. Bib.1 86); RB also shows £165 (John) Taylor/(John) Russell (see Bib.1 87).

36 It is only in the 1790's that the entry becomes consistent as "Mrs. or Fanny Thomas" (see comment under Synopsis of History of Cobtree mill).

37 Hillier Russell was buried Aylesford 1790, but LT does not show Mrs. Russell at Forstal until 1792.

38 1793 LT shows James Whatman/late James Taylor £45 for first time replacing Lord Aylesford/James Taylor £45 (rather late in the day).

39 Whatman's mills, 3 of them, were sold to Finch and Thomas Hollingworth and William Balston on 10th Oct.1794. Whatman Estate assessed separately.

40 In the 1796 Boxley LT Returns "Mrs. Gill" replaces "Mr. Gill" as the Proprietor for Forstal and Cobtree mills; but "Mr. Gill" reappears in LT 1801–1810.

41 The first entries in the next Ratebook (P 40 12/6) 1801 show a "Mr. Russell" replacing "Mrs." see next TABLE p.116 note 1.

19TH C. EXTRACTS FROM BOXLEY RATEBOOKS AND LAND TAX RETURNS

19th C. Extracts from Boxley Ratebooks (KAO P 40 12/6,7) and Land Tax Returns (KAO Q/RP1/42) relating to FORSTAL and COBTREE Paper Mills and their Occupants. (As Poll mill and Turkey mill were by that time under the control of, initially, Hollingworth and Balston and, later, the Hollingworths only they have not been included as part of the rapidly changing scene around Maidstone in the early decades of the 19th Century. The demise of Poll mill has been covered in Chapter VII p.288. Turkey mill (see Spain, R.J. Bib.8 87) continued to operate under the Hollingworths until 1889 passing then to nieces, Mrs. Pitt and Lady George Gordon-Lennox, and thereafter under the control of Major William Pitt and the Cooke family. It finally ceased working in the mid-1970's.

| | FORSTAL PAPER MILL | | | | | | COBTREE PAPER MILL | | | | |
| | RATEBOOK | | | LAND TAX | | | RATEBOOK | | LAND TAX | | |
	YEAR	£	OCCUPANT	PROPRIETOR	OCCUPANT	£	£	OCCUPANT	PROPRIETOR	OCCUPANT	£
RB LT	1801-2	54	Mr.Russell [1]	Mr.Gill [2]	Mrs.Russell Mr.Russell	21	40	Mrs.Thomas [3]	Mr.Gill	William Thomas	20
RB	1803		Mr.Russell The Mill				40	Mrs.Thomas			
RB	1805-9	54	Hillyer Russell				40	Mrs.Thomas			
LT	1808			Mr.Gill	Hillyer Russell		[4]		Mr.Gill	William Thomas	20
RB	1810-11	54	Hillyer Russell				40	Fanny Thomas Mrs.Thomas			
LT	1810			Mr.Gill	Hillyer Russell	21			Mr.Gill	Fanny Thomas	20
RB	1812	54	Clement Russell				40	Mrs.Thomas			
RB	1813	54	Clement Russell				40	Stacey Wise late Thomas			
LT	1814-15			Mrs.Gill	Clement Russell	21			Mr.Stacey	Mr. Wise	20
RB	1814-16	50 +	Clement Taylor Russell				44 +	Wise			
RB	1817-18	50 +	C.T.Russell [5] or Occupier				44 +	Charles Wise			
LT	1817			Mrs.Gill	Clement Russell	21			Mr.Stacey	Charles Wise	20
RB	1819	50 +	C.T.Russell or Occupier (C.T.Russell				47 +	Charles Wise			
RB	1820	50 +	or Occupier deleted) TRINDALE				44 +	Charles Wise			
RB	1821	50 +	C.T.Russell & Pine				44 + 15	Charles Wise for House			
LT	1819-21			Mrs.Gill	Clement Russell	21			Mr.Stacey	Charles Wise	20
LT	1822			Mrs.Gill	John Pine	21			Mr.Stacey	Charles Wise	20
LT	1823-4			Mrs.Gill	C.T.Russell	21			Mr.Stacey	Charles Wise	20
LT	1825			Mrs.Gill	C.T.Russell	21			Mr.Stacey	Allnutt & Co.	20
LT	1826			Mrs.Gill	late C.T.Russell Paper mill & Land	21			William Stacey	Allnutt & Wilson Paper mill & Land	20
LT	1828			Geo. Fowle [6]	self	17			William & Henry & Courtenay Stacey	Lavender Papermill & Land	20
LT	1830			Geo.Fowle	self	17			ditto.	Allnutt	20

1. "Mr. Russell", who was this ? As mentioned in App.II p.61 the problems surrounding the genealogy of the Boxley Russells have not been resolved. Edward snr. occupied Forstal mill from 1733 to (probably) 1780. The Boxley LT Returns (but not the RB) for the 1780's show Hillyer Russell as successor, probably Edward snr.'s brother. Hillyer died in 1790 and the LT now shows Mrs. Russell as occupant. Meanwhile we know that John Russel (probably another brother) had 5 sons, 3 of whom became paper makers; Edward jnr. (bapt.1771, of age 1792); Clement Taylor (bapt.1774, of age 1796); and Hillyer jnr. (bapt.1781, of age 1802). Edward and C.T. Russell were in partnership at Upper Tovil 1797-99 (see this App. n.17); a Russell was also in partnership with Edmeads at Gurney's mill 1797-1802; and at Otham mill 1810-12. C.T. Russell does not appear in these Boxley records before 1812. It is suggested here that C.T. Russell was the Otham Russell (leaving there in 1812) and probably the Gurney's mill one also. Hillyer jnr. came of age in 1802 and makes his first recorded appearance at Forstal in 1805. The "Mr. Russell" of 1800 was either Edward jnr. or possibly (?) Hillyer jnr. as successor to Mrs. Russell.
2. Note that "Mr. Gill" and later "Mrs. Gill" were proprietors of both Forstal and Cobtree from 1780 to 1826 and 1808/9 respectively.
3. See comments on Mrs. Thomas as occupant in the synopsis of the History of Cobtree mill.
4. As mentioned in the synopsis of the history of Cobtree mill the "Youngs" (and a Green) were probably under-tenants of Mrs. Thomas and Stacey Wise between 1809 (at least) to 1816. The official records do not show this. Early in this period Wm. Thomas was otherwise engaged; and Charles Wise appears to terminate the Young occupation in 1817.
    Whether William Green was the Green concerned is uncertain. The Boxley RB shows that he was assessed as an Outdweller from at least 1800-9 for £120-10; the LT shows that for the same period he was occupant of property owned by Lady Calder (taxed @ £70); after 1814 and on into the 1820's Lord Aylesford was the Proprietor, still taxed at £70. But the RB assessments against Green drop dramatically after 1809 but he continues to make fitful appearances in the Sandling Ward assessed at relatively low sums. (Wm. Green also assessed for property in Maidstone).
5. As the synopsis of the History of Forstal mill shows, C.T. Russell was bankrupted in 1816. The 1816 EGL shows William Trindale as the occupant of this mill. This is not reflected either in the RB until 1820 or the LT at any time. C.T. Russell entries, however, continue until John Pine took over in 1821, though C.T. Russell's name does not make a final disappearance until 1825/6 (LT).
6. From 1800 (possibly earlier) Edward Fowle was assessed for £143.10 for Cobtree (manor). By 1814-16 this changes to George Fowle assessed for £208 with an additional £62 for Tythe. It is presumed that this is the George Fowle who later took over Forstal mill as a manufacturer of White paper.
7. Not shown in Table but Charles & Harris were assessed (RB) for a Fulling mill, certainly from 1800 into the 1820's (LT).

MISCELLANEOUS CONTEMPORARY CHARACTERS
WHO MAY HAVE BEEN ASSOCIATED WITH BOXLEY

In the foregoing pages the Boxley Ratebooks have been examined in relation to the paper mills in the Parish. The Parish bordered on the North–Eastern boundary of Maidstone so that by 1800 there would have been interchange with Maidstone inhabitants. Two have been singled out for separate attention and are dealt with below. There are, however, other figures who may or may not have played a significant role e.g. in the Maidstone RB for 1800/01 (KAO P 241 11/13) one finds EDWARD RUSSELL (see note 17 this App.) at the Bank in the Week Street Ward (and a neighbour of Catherine Vallance, William Balston's bride to be); in the West Boro' Ward there is WILLIAM GREEN assessed for £19–5, more for a house, a forge, land and 5 cottages; under the Stone Boro' Ward Hollingworth & Balston (which may have been part of Poll mill) £9 for a Drying Loft and £18 more for Lord Romney's Land (?); Hayes & Wise at Padsole mill in the Gabriel's Hill Ward; William Charles (see below) assessed for Chillington House Felt Factory: WILLIAM STROUD in Tovil Ward (see App.II); Finch Hollingworth in Holloway Road (and John Hollingworth under Boxley); in the Mill Lane Ward John Green snr. and jnr. thought to be the mould–makers. Finally, a William Crisp who disappears from the Boxley ratebooks in 1806 possibly the son of Edward Crispe of Leeds who died 1807 (see App.II p.68).

JOHN LAVENDER (Paper Maker)

John Lavender was the occupant of Pratling Street paper mill from 1825–1828; he appeared for a brief period at Cobtree mill also in 1828; as well as at Forstal mill in 1838–9 (see mill synopses). He has been included amongst the other Boxley ratepayers referred to in this Appendix solely on the grounds that, like Thomas Stroud, he may have had an association with Hollingbourne; it is only a possibility because the name Lavender is found elsewhere in the district, certainly in the early part of the 18th C.[38] The only John Lavender found in the course of the searches undertaken here who would accommodate this is a John who was baptised (base born) in Hollingbourne in 1777.[39] The name of the father is not known, but the presence of a John and Ann Lavender, contemporary and in the same parish, suggests that Sarah, the mother, may have been a sister of this "John". Shorter noted[40] that a John Lavender, whose name appears in the Smythe List of Apprentices in 1770, could well have been the husband of Ann and working at one of the Hollingbourne paper mills. This apprentice "JOHN" would have been approximately 67 in 1825, in other words too old to have taken over any of the Sandling and Aylesford paper mills. On the other hand, if he had been working as a paper maker in Hollingbourne, one could imagine the illegitimate John being brought up in a papermaking environment there. In support of this suggestion it may or may not be significant that the only Lavender countermarks discovered so far are found

---

[38] A John Lavender was baptised in Boxley in May 1694 (KAO P 40 1/2). Lavenders are also found in the early part of the 18th C. in the All Saints', Maidstone, PRs. No doubt there were other Lavender families in the district though not seen during searches.

[39] Hollingbourne PRs (KAO P 187 1/1).

Bapt. 1775	Edward (base born)	s.o. Sarah Lavender
Bapt. 1777	John (base born)	s.o. Sarah Lavender
Bapt. 1781	William	s.o. John & Ann Lavender

[40] Shorter, A. Bib.12 185 (p.no. corrected).

[handwritten note:] ✻ This reference is to an apprenticeship for William son of John the papermaker. John would be in his 90's in 1825!

in Leeds and Broomfield documents.[41] In addition to this Lavender took over Pratling Street paper mill only a few years after the Stroud family had relinquished it, Thomas Stroud formerly having had papermaking and farming interests in Hollingbourne and Leeds.[42]

WILLIAM CHARLES & FAMILY (Felt Manufacturer)

According to J.M. Russell[43] William Charles was baptised at Bradford, Wiltshire, and later educated for the Medical Profession. He married a Miss Arnold of Maidstone and settled in the district. He is of interest in the context of the present work as a Felt Manufacturer (see App.V.). His first appearance in Boxley is a reference in the Rate Book for 1784 as an Outdweller.[44] In that year he was described as Dr. Charles; later as William Charles. Russell states that he acquired Chillington House (now the Maidstone Museum) in St. Faith's Street, Maidstone, in 1800 where he set up a "business of felting and blanket–cleaning". The Universal British Director (1794 Vol.III), however, described William Charles (under Maidstone) as a "Felt Manufacturer" some six years before his acquisition of Chillington House. Presumably the property referred to in Boxley was a fulling mill there, "his Felt Factory in Sandling" where ca. 1826 he was described as "practically the only maker of Felts for Papermakers".[45] The Boxley Ratebook for 1800 confirms that he was assessed for a "Fuller's Mill" as well as for another property, probably that which was later described as "a cottage". In 1804 he was assessed for a third property, which in 1832 is itemized as a "cottage and fulling mill" at Sandling with other property in the Weavering Ward.

William Charles was assessed for the fulling mill up to 1835 (in the RB), but the business was not sold until 1840 (a Directory of 1839 gives the name of William Charles, Blanket Maker at Sandling; this would have been William jnr. his father having died in 1832).

Summarizing,

1784	Dr. William Charles first appearance in Boxley.
1794	Univ.Brit.Director described him as a Felt Manufacturer.
1800	Assessed for 2 properties in Boxley, one a fulling mill.
	Acquired Chillington House, Maidstone.
	In partnership with Robert Harris in a felting and blanket–cleaning business.
1832	William Charles snr. died.
1840	William Charles jnr. died and the Felt business was sold.
	William jnr. was a bachelor.
1855	Thomas Charles (another son and also a bachelor) died.
	He had been trained as a medical practitioner but retired into private life and collected antiquities, which he bequeathed to the town and which formed the nucleus of the Maidstone Museum's collection.

[41] See this Appendix under Pratling Street.
[42] See App.II Stroud.
[43] Russell, J.M. Bib.41 357.
[44] Boxley Land Tax Return for 1787 (KAO Q/RP1/42) shows Dr. Charles (O.D.) as proprietor of this property; Occupier late Arnold.
[45] Original Papermakers Recorder quoting from the Paper Trade Review Apr.1902.

APPENDIX IV

The Emergence of Maidstone as a Papermaking Centre

Preface to Appendix IV

This Appendix is mainly concerned in providing data to support the subject of Chapter II of this book, namely the emergence of Maidstone as a papermaking centre towards the end of the 17th C. It has not proved possible, however, to isolate the material thought to be relevant cleanly from that which may not be relevant; there are areas where the picture becomes blurred. Documentary evidence is very scarce almost to the point of being non–existent and in most cases the main source of evidence lies in the genealogy of the paper makers themselves, that is if one can demonstrate a connection between the names of individuals and papermaking and paper mills in a given district.

In order to try to simplify the very diverse material covered in this Appendix much of the less essential information has been relegated to "Working Papers". To give an example in the case of certain names, names known to be associated with papermaking but names very commonly found, it has been necessary to search the registers of more than 50 different parishes. It may be that in the majority of cases where the name has been found it has not been considered to have any bearing on the subject and the information in instances like this has been condensed to a statement such as "present; n/a". The reason for mentioning them at all is a consequence of the system of searching employed. At present it is out of the question to search every parish record in Kent so the searches carried out here have been limited to what have been designated as primary, secondary and tertiary search areas. There are obvious drawbacks in using this method, but in order to limit the search to reasonable proportions one has to take a chance.[1] Having adopted a system of this kind it has been thought essential that there should be a record of the registers or other records searched whether the results have been positive or not. A person who may wish to extend these searches may have noticed that a vital parish has been overlooked.[2]

The reasoning that lies behind this method of searching is that during the early period with which this Appendix is mainly concerned the paper mills in Kent were scattered very thinly on the ground and employing relatively few people. The chances are that anyone deciding to take up papermaking as a career would probably have lived somewhere in the vicinity of a paper mill. In other words if one is looking for the origin of such a person, one would be more likely to find his name in a papermaking district than elsewhere. Naturally there would have been exceptions to this; the prospective paper maker's home may not have been in that district. He may have stayed with relations near the paper mill as a child and decided that that was going to be his vocation. But, as mentioned, one has to take this chance.

Since the main subject of Chapter II is concerned with the emergence of Maidstone as a papermaking centre, one of the first objectives was to try to discover the origins of the paper

---

[1] This method, for example, would never have exposed or explained the presence of Thomas Chapman from Shropshire as a paper maker appearing at Dover in 1638.

[2] It was only some years after searching the records of the more prominent of the parishes surrounding Dartford that the little parish of Wilmington, tucked into the outskirts of Dartford, was "discovered", the PRs. revealing the marriage of William Quelch snr. and Richard Archer snr. there.

makers who worked the first mills there. How many of them came from that area ? Maidstone was, therefore, designated a Primary Search Area in this scheme and included parishes such as Boxley, Aylesford, Allington, Ditton, East Malling, Burham, Snodland etc. Based on the evidence available at the time when most of these searches were made there were only 5 paper mills active in Kent before the Maidstone district came into being. So, if a paper maker's origin could not be traced in the primary search area, then assuming he had a Kentish origin, the next place to search was the area surrounding and including the location of these 5 mills. These were designated secondary search areas.

Lastly, as mentioned in Chapter II (p.79) the question was asked whether there might not have been an early undiscovered paper mill, possibly in the High Weald, the forerunner perhaps of one only identified at a much later date, like the mills at Chafford, Benenden or Snodland. To take care of this eventuality areas such as these have been designated tertiary search areas.

Although App.IV is very largely concerned with the genealogies of the early Maidstone paper makers, it is not entirely. For convenience, some of these searches have been extended to cover later generations; and in other cases to the paper mills where they make their first appearance or which may have some associations with them. The following comments may be inserted here.

(1)   In certain cases like the Archers and the Gills the investigations have been carried beyond their 17th C. activities and these have been grouped together in this Appendix. There are exceptions to this, see (2) below.

(2)   Because some papermaking families were more closely associated with the elder Whatman and the developments at Hollingbourne, their genealogies will be found in App.II rather than IV. They include, for instance, the papermaking Harrises and the Quelchs.

(3)   Although Forstal paper mill played a contributory role in the early expansion of the Maidstone industry, it has not been included (as one might expect) in this Appendix, but treated instead as one of the Sandling mills in App.III. Nevertheless certain genealogical data relating to its first known paper makers will be found in this Appendix.

THE WILLARDS OF GOUDHURST

Thomas Willard is important in the present context as one of the first, together with Thomas Taylor and James West, identifiable paper makers in the Maidstone district.

The name Willard/Wellard was very common in Kent during the period under consideration. The implications of this will be considered later when the Willard genealogy is examined in more detail. At this point it is sufficient to know that after failing to find the baptism of a Thomas Willard in the Maidstone area whose dates fitted in with his tenure of Forstal mill, it was discovered that the Willards of Goudhurst provided the answer to the problem of his origin.

There was quite a considerable population of Willards in the High Weald[3] and the remote ancestry of our Thomas, the paper maker, has not been followed through. At Goudhurst the family probably goes back further than the middle of the 16th C.[4] There is, however, an apparent gap of a generation around the turn of the century so that one cannot follow through to Thomas (1) Willard, who was married there in 1632 and had Thomas (2), the future Forstal paper maker, in 1633. Thomas (1) had another son, Mathew (sic), baptised in 1639; the significance of "Mathew" will become clear later.

The search has shown, amongst other things, that a Thomas Willard, *paperman*, was in the district ca.1655.[5] This picture is complicated by the fact that at that date there were no fewer than four Thomas Willards living in the vicinity of the paper mill in Goudhurst. In Goudhurst itself there was Thomas (1) aged 43 or more by then and Thomas (2) aged 22; in addition there were two Thomas Willards in other villages close to the paper mill, one at Marden where the reference to the "paperman" is found; and one in Horsmonden on the opposite bank of the R. Teise to Goudhurst. It is clear from the parish registers of these two villages that neither of these last two Thomases could have been the same person as Thomas (2) of Goudhurst. In fact the evidence relating to the Horsmonden Thomas points to him still being there after Thomas (2) had moved to Forstal. The other family, at Marden, could have been the same as that which appeared in Aylesford in 1683 (see Willard genealogy later). Other evidence, to be presented later, supports the claim made earlier that it was Thomas (2) who eventually became the paper maker at Forstal; but this is not to say that there may not have been more than one Thomas Willard engaged in papermaking at Goudhurst.

Having made the claim that our Forstal Willard was Thomas (2) at Goudhurst, adding the proviso that there may have been more than one papermaking Willard in the area, one has now to consider where they and others to be named could have learnt their trade with so few identifiable paper mills in that part of England. The others include a Robert Wilson who was born in Goudhurst in 1624 (details later) and who was still described at his marriage in

---

[3] There were contemporary families of Willard at Goudhurst, Marden, Horsmonden and Benenden (where a Richard was married described as a Broad Weaver of Hawkhurst); but none in Cranbrook or Lamberhurst. These parishes effectively encircle Goudhurst.

[4] A William Willard had a family in Goudhurst between 1566–81 including a Richard (1569), John (1572) and a Thomas (1574).

[5] The Marden PR (still held by Incumbent) shows that an "unnamed" daughter of Thomas Willard, *paperman*, and Miss Sanders was bapt. 1655. The Registers show that a Thomas & Anne Willard were having a family there between 1661–1668 (at least) and thus a different Thomas to the Goudhurst one.

Maidstone in 1657 as "of Gowdhurst, papermaker"; and a Richard Roads, a paper maker of Bedhampton, Hampshire in 1666, but who was almost certainly born in Goudhurst (about 1 mile from the paper mill) in 1615.[6]

With no evidence for a founding date for this mill the question as to where these paper makers learnt their trade remains unsolved. In Chapter II two possible origins were suggested, a 16th C. one as a Press paper mill or built as a new mill ca. 1640–50, in either case founded by an immigrant. The second proposal would have enabled Thomas Willard (2) to have been trained on his home ground (suggested apprenticeship dates 1645–52), but this would not have accommodated either Wilson or Roads had either of them been trained in Goudhurst. The first proposal would have catered for them all. Since no evidence of either Wilson or Thomas (2) being present in any of the other Kent papermaking districts (1°, 2° or 3°) has been found yet, it is the first proposal that is favoured here. In view of the remoteness of the district, especially during the industrial decline of the area in the 17th C., a Cloth industry related founding is a more sensible idea than one that was quite unrelated to its environment.

On present evidence Robert Wilson is no more than a shadowy figure in this scene; his baptism and marriage have been discovered and after that he disappears. Perhaps he was a millhand or an under–papermaker or he may have left Kent altogether.[7] He is of interest only because he supplements the evidence of a paper mill at Goudhurst at that time.

In contrast Thomas Willard (2) is a more substantial papermaking figure. We know the date of his baptism (1633); the baptisms of four of his children, Robert (1660) who may have appeared later in Aylesford and Boxley (see below); and the last, Elizabeth (1666), noteworthy on two counts in that (i) this entry shows that Thomas (2)'s wife was named Dorothy, possibly the Dorothy buried in Aylesford (1696) and (ii) it is the last reference to Thomas (2) at Goudhurst. A child of Mathew's, Thomas (3), baptised in 1661 was later married to *Mical* Phillips of Cranbrook at Goudhurst in 1683, this last pair having their family in Boxley from 1684 onwards. Although there is the case of the Thomas "paperman" sowing a "wild oat" in Marden in 1655, a Thomas who could just as well have been the Marden Thomas Willard, we have no positive evidence of Thomas (2)'s presence in Goudhurst as an adult before 1660. It is conceivable, therefore, that between his baptism in 1633 and 1660 he might have gone elsewhere to learn his trade, but no evidence of this has come to light.[8] But in a proposal of this kind one has to provide a motive for him to go to some other place

---

[6] An Agreement (referred to again, later) dated 1666 (KAO U 1406) locates Richard Roads as a paper maker of Bedhampton, Hampshire. The Goudhurst PR (KAO P 157 28/1 Baptisms) records 2 Richard Rodes, one bapt. 21:Apr:1611 s.o. Edward Rode of Broad Ford (about 1 mile upstream on the Teise from the paper mill) evidently died; a second Richard was bapt. 30:Apr:1615. The Roads genealogy has not been pursued here, but there was a James Roads, for instance, of Mayfield and a man of property, some of this lying in Goudhurst (KAO U 383 T1). Other Documents (KAO U 769 T24) Probate of Richard Rode 1596 and (U 769 T131) Edward Rode of Goudhurst 1638 indicate that the Rode family owned property in the Broad Ford area, but neither of these documents add anything specific to our knowledge of the paper mill. But whether later they leased the paper mill from A.N. Other or owned it is not known. Certainly, Willard had to pay rent for the mill to Richard Roads in 1666.
It is perhaps worth noting that there is some evidence of a mill (function unknown) sited at one time at Broad Ford (approx: MR TQ 717397 inf. R.J. Spain). This site is exactly halfway between Goudhurst and Horsmonden and, if it had been a fulling mill, was probably in decline in Edward Rode's day.
[7] For further details see this App. under Robert Wilson.
[8] The distribution of Willards in 1° (Maidstone) and 2° search areas is given in the Willard/Wellard genealogy (this App. pp.134–135) but no signs of Thomas (2) having any papermaking link with any of these.

without there having been some sort of papermaking interest to prompt him. It was not as though paper mills were two a penny in that district or that his parents had moved away from there during his childhood; Thomas (1) was married and buried in Goudhurst. All these things point to Thomas (2) having learnt his trade at a paper mill there.

The first clear evidence that he was the Master Paper Maker there is not until 1666 and even this is by inference. The agreement of 1666, referred to in note 6, was between Thomas Willard of Goudhurst in the County of Kent, papermaker, and Richard Roads of Bedhampton in the County of Hampshire, papermaker. It appears that Thomas owed Richard the sum of £50, a debt relating to the paper mill and lands at Goudhurst. Thomas had to pay off the debt in kind[9] – by means of the sale of wheat and flax etc. from the paper mill lands, which demonstrates that the mill and lands were in his tenure and that he was *the* paper maker at Goudhurst.

Without the aid of speculation it is difficult to know how to place this piece of information in the context of the succession at Goudhurst paper mills. To begin with one might interpret this debt in two ways, that Richard Roads owned the paper mill and lands and that Willard was in arrears with his rental or alternatively, that Roads might have been the paper maker there before Willard and that the latter owed him money for the purchase of the leasehold. Added to this uncertainty one has the arrival of Alexander Manktellow in Goudhurst ca.1644 with Stephen Manktellow as potential successor to Willard at Goudhurst mill; and, further, the question as to why one finds Richard Roads making paper in Hampshire instead of what might have been his own mill.[10] The answers are not known. In the present context one can set Richard Roads on one side; if he was not the owner of this mill, then whoever leased it may have preferred to have a more up-to-date paper maker there and replaced him. It was suggested in Chapter II[11] that this hypothetical substitute might have been Alexander Manktellow invited to inject new life into an old mill. Roads, so far as we are concerned, had left the Kent papermaking scene.

Despite these uncertainties the evidence points to Thomas Willard (2) being trained at Goudhurst paper mill, possibly under Alexander Manktellow who installed Willard temporarily until Stephen was of age; and then between 1666 and 1671 he moved on (or was moved on) to Forstal.

GOUDHURST PAPER MILL   (MR probably TQ 726408)

Map 8 shows the site of Goudhurst paper mill. Before identifying the locality of this mill it may be helpful to give a brief account of the district illustrated. The River Teise (pron. Teez) flows from south to north and appears to divide at a point near to where the paper mill was sited, the western branch flowing into the Medway at Yalding and the eastern branch into the

---

[9]   The payments were to be made to Thomas Sharp of Hawkhurst and James Bayly of Cranbrook.

[10] Shorter, A. Bib.12 171 has labelled Bedhampton mill, Hampshire, as "doubtful". All the same it is clear from the wording of the Agreement and Richard's Will, proved in 1676, that a paper mill must have existed somewhere near there. Another Roades (sic), Thomas, was to follow later. Not far from Bedhampton was Warnford mill, active in 1617. Shorter states that the early paper makers there are not known; perhaps Roads was one of them ? The question as to whether there were Roads indigenous to that district has not been explored here.

[11] See Chap.II pp.72/73.

River Beult (pron. Belt) between Hunton and Yalding, where the latter also joins the Medway. Marden is situated in the Low Weald (alt.30 m.) with the Teise at this point about 19 m. above sea–level. Goudhurst (alt.123 m.) and Horsmonden (alt.55 m.) are situated in the High Weald with the Teise in the area of the paper mill at approx. 28 m. above sea–level. The parish boundaries of Marden, Horsmonden and Goudhurst meet at a point just north–east of the mill site, which lies in a position almost central in its relation to the three villages. To the west, south and east lies undulating hilly country intersected by numerous rivulets. Several other types of watermill were sited on the Teise and its tributaries, the paper mill being the lowest of these within the Goudhurst parish boundaries. (The subject of the quality of the water available to the mill has been discussed separately, see Chap.II note 105).

The sketch map also shows the position of Goudhurst in relation to other villages lying in the "Cranbrook District", for instance, those included in a secondary search area (Cranbrook, Horsmonden, Marden, Hawkhurst and Lamberhurst) and one in a tertiary (Benenden). It has already been noted that, with numerous 15–17th Cs. references to them and a large colony at Goudhurst, this was also Whatman country (App.II note 4).

With regard to the location of the paper mill, apportionments (2610, 2612 and 2614) of the 1842 Goudhurst Tithe map[12] record (respectively) Paper Mill Fields and the Upper Paper Mill. The map is not a precise one so that one or two additional landmarks were noted in order to help position the mill site as precisely as possible. The area in which these apportionments lay were, for example, to the north of Finchurst (Monk's Farm) and to the west of Winchet Hill and Curtisden Green. To judge from these details the paper mill fields appeared to lie along the right bank of the River Teise and, possibly, along a small tributary running off to the east with Upper Mill sited on the latter (approx. MR TQ 726408).[13] No earlier maps showing this mill have been discovered. However, Deeds of 1696 and 1699[14] refer to a "Paper–mill water–mill and appurtenances in the tenure of (Thomas) Ashedowne" all of it being described as being in the parish of Marden; but, in a more general sense, the documents with which these Deeds are associated all belong to Goudhurst. It has been assumed here that lying very close to the Marden parish boundary the property on which the mill was sited extended then into Marden and hence the description. Nothing has been

---

[12] Only tracings (in a rather damaged condition) of this area were seen; KAO 769 P 31; Apportionments P 157 27/1; and another tracing 769 P 21.

[13] The indications in these maps are not precise. No mill is marked in either of the tracings. On balance MR TQ 726408 seems the most probable site on what one would imagine to be an uncontaminated stream (except possibly from iron in the water). An alternative site has been suggested MR TQ 723407 at the division of the R. Teise. This might have meant that part of the mill lay in the parish of Horsmonden for which there is no evidence. Moreover, in view of the considerable industrial activity higher up the Teise, the latter might have been unacceptably contaminated with effluent.

[14] KAO U 769 T 55 Bundle 2 Nos.13, 14 concerns a lease for 1 year (13) by Charles Hawtry of Ruislip, drawn up in the 7th year of the reign of William III, of property including the paper mill to Jeffery Jeffery of London (there is also a Release Hawtry/Jeffrey of 1695, No.11, paper mill not specified); and a second lease for 1 year (14) by Jeffery Jeffery to Mr. Charles Booth dated 3:Apr:1699 of property including the paper mill. The first Deed merely designates the tenant as "Ashedowne"; the second "*Thomas* Ashedowne". Whether Hawtry's or Jeffery's interest was in the property as a whole or in the paper mill in particular has not been discovered. Ruislip, Hawtry's domicile, was close to a contemporary papermaking centre between Rickmansworth/Thames. How far back did Hawtry's interest extend? Were either Hawtry or Jeffery influenced at the time by (a) the COMPANY'S activities ? or (b) the stimulus given to White paper manufacture 1678–85?

Map 8 – Sketch Map Illustrating Site of Goudhurst Paper Mill

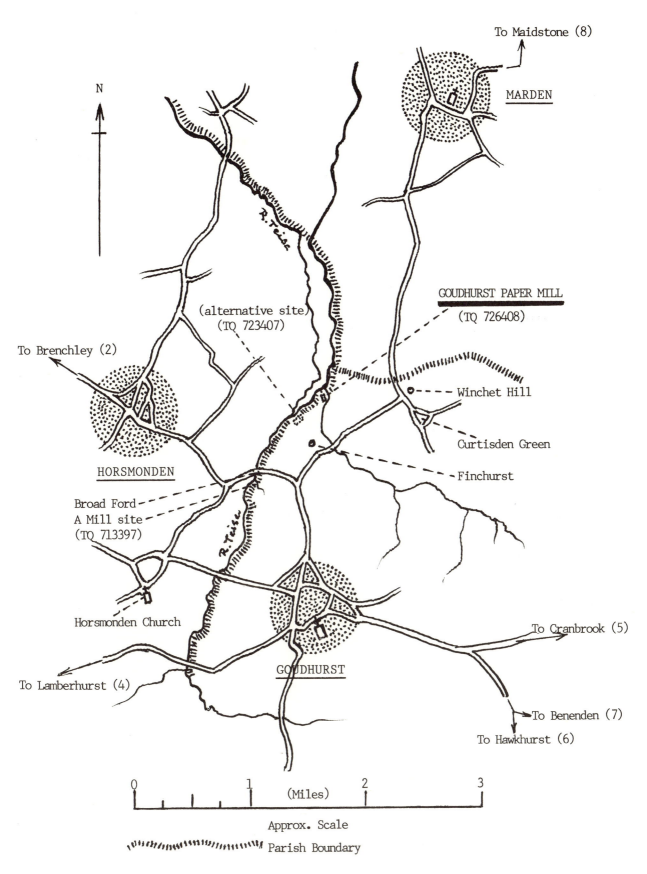

To Maidstone (8)

MARDEN

N

R. Teise

GOUDHURST PAPER MILL

(TQ 726408)

(alternative site)
(TQ 723407)

To Brenchley (2)

Winchet Hill

Curtisden Green

Finchurst

HORSMONDEN

Broad Ford
A Mill site
(TQ 713397)

R. Teise

Horsmonden Church

To Cranbrook (5)

GOUDHURST

To Lamberhurst (4)

To Benenden (7)

To Hawkhurst (6)

0   1   (Miles)   2   3

Approx. Scale

ʋʋʋʋʋʋʋʋʋʋʋʋ Parish Boundary

Figures in brackets following Village names not shown on
sketch map are in miles to the nearest village illustrated.

discovered yet to support the idea that there was another paper mill further down the River Teise.[15]

Earlier in this Appendix there have been references to "Thomas Willard, paperman" (1655), the Robert Wilson "of Gowdhurst, papermaker" (1657) and the Roads/Willard Agreement (1666) as items of evidence to show the existence of a paper mill in Goudhurst which are now confirmed by entries for the paper mill, actual or implicit, in the Ratebooks covering the years 1675–1697 and from 1711 onwards.[16]

The relevant entries are as follows:–

12/1	DATE	OCCUPANT	ASSESSMENT	
	1675–1679	Stephen Manktellow[17]	£10	(unspecified)
(corr:)	Jan. 1680	Peeter Archer late Manktellow	£10	(unspecified)
	Aug. 1683	Peeter Archer or Occupier[18]	£10	(unspecified)
(corr:)	Jan. 1684	George Beckett for the Papermills[19]	£ 7	
(corr:)	Jan. 1685	Richard Hover[20]	£10	(unspecified)
	1695–1696	do.	£15	(unspecified)
		do. for paper mill lands		(a small sum)
	1697	do. or ye Occupier		

12/2 (The first legible and intact entry is given below)

	1712	Mr. Charles Booth for the paper mill[21]	£4	

Apart from the fact that both Manktellow and Peter Archer are known to have been paper makers, who later moved to the Maidstone district, the point of greatest interest here is the reference (1684) to paper mills in the plural a point which fits in with the Tithe Map description of "the Upper Paper Mill". Further the fact that the assessment has remained of the same order from 1675 points also to the existence of an Upper and Lower mill at an earlier date. This notion gave rise to the suggestion made in Chapter II that perhaps originally there had been an old Press paper mill in the 16th C. and that the papermaking activity at this site had been given a new lease of life and a new mill when Alexander Manktellow arrived on the scene ca. 1644.

---

[15] No familiar paper makers' names (apart from Willard) have been found in the Marden KAO P 244 12/1 (starts 1703) or Horsmonden KAO P 192 12/1 (starts 1698) Ratebooks. The question of the tenant, Ashedowne, referred to in the 1696 Deed will be discussed later.

[16] Goudhurst KAO P 157 11/1 (1675–1697) : P 157 11/2 (starts) 1711. The latter piece in contrast to the former is in a very dilapidated state and, though restored, clearly has parts missing at, unfortunately, a very critical stage in the paper mill's history at a time when it obviously ceased operating.

[17] Stephen Manktellow moved to Upper Tovil mill, Maidstone, between Aug.1679/Jan.1680 and was first assessed for Upper Tovil late 1680.

[18] Peter Archer arrived in Boxley, nr Maidstone, at the end of 1683. (For details see this Appendix under Peter Archer snr.)

[19] Peter Archer had married Elizabeth Beckett in 1673, so George may have been his father– or brother–in–law.

[20] Nothing is known about the papermaking activity of Richard Hover although he was assessed for the paper mills for some 10 years. The 1697 entry does not show "Ashedowne" as one might expect, only "or ye Occupier". The Ratebook entries end here.

[21] To judge from this entry (2nd RB. and 15 years later) the paper mill had ceased operating.

The Succession at Goudhurst Paper Mill

After 1675 there is positive ratebook evidence of the occupants; but before this date no continuity emerges from the various references to paper makers in Goudhurst. There may have been 2 paper mills in this earlier period where Wilson had worked one and Willard the other. But the 1666 Roads/Willard agreement points to one Master Paper Maker in charge of the operations there. A purely conjectural "Succession" is proposed below compiled with a view to accommodating the suggestions and known facts mentioned in this Appendix. The bracketed dates are speculative; the ones without brackets documented.[22]

(−1630's)	(Occupants unknown)	(The assumption made here that the mill had an earlier existence: see Chap.II)
(ca. 1630's)	(Richard Roads)	(suggested apprenticeship dates ca. 1627–34. If the mill's earlier existence is accepted he could have served this at Goudhurst and moved to Hampshire later. If not, what else? He might also have been the Proprietor in 1666?)
(1636–1643)	(Robert Wilson)	(If ever apprenticed there, these are suggested dates; see also under 1657).
(1644 +)	(Alexander Manktellow)	(Almost certainly an immigrant. If a papermaker, he could have given the mill a new direction and perhaps founded the second mill).
(1645–1652)	(Thomas Willard 2)	(suggested apprenticeship dates. Could, in theory, have been apprenticed to Alexander Manktellow).
1655	Thomas Willard	"Paperman" reference at Marden. Thomas 2 would have been aged 22 then; but it could have referred to the Marden Thomas equally.
1657	Robert Wilson	Married at Maidstone and described as paper maker of Goudhurst. Might have been an under paper maker?
1660	Thomas Willard 2	Aged 27; first record of a child of his baptised in Goudhurst.
(1660's)	(Stephen Manktellow)	(may have been serving his apprenticeship at Goudhurst, if not served elsewhere e.g. in his native country).
1666	Thomas Willard 2	Aged 33; Master Paper Maker of Goudhurst paper mill and may have been so before this date, e.g. 1660 or earlier?
(1668–70)		Suggested dates for Thomas (2)'s move to Forstal, Boxley.
1672	Stephen Manktellow	Bapt. of a son, Thomas. First evidence of presence. (Could have taken mill over when Thomas (2) left).
1673	Peeter Archer	Married in Parish to Elizabeth Beckett. No other reference to him until 1680. (Did he move away for 7 years. For discussion see under Peter Archer, this Appendix p.140.)
1675	*RATEBOOK STARTS*	
1675–1769	Stephen Manktellow	Master Paper Maker at mill(s).
1680–1683	Peeter Archer	Master Paper Maker.
1684	George Beckett	Occupant of Paper mills at reduced rate. Possibly not fully operational. Evidently an in–law of Peter Archer's.
1685–1697	Richard Hover	The Master Paper maker (?). Rate increased in 1695. Mill probably in decline by 1697.
(1697 +)	Thomas Ashedowne	(Only evidence of his tenure is in Deeds of 1696/9 KAO U 769 T 55 Bundle 2 13/14. A Thomas Ashdown bur. Goudhurst 1709. For further discussion of Ashdown see below and later under Millhall mill, Ditton).
1712	Mr. Charles Booth	Assessed for "paper mill lands". The suggestion is that the mill had closed down by then.

[22] Mention has been made in Chap.II under the early history of Forstal and Millhall mills (and also in this Appendix under their names) of other paper makers who may have been associated with this mill though no positive evidence has been found for this. They have been omitted from the "Succession" set out below because of this; they include Thomas Taylor, James West, Alexander Russell. There were others also who may only have trained or worked here e.g. Peter Musgrove snr. and Thomas Gulvin (bur.Marden 1690).

NOTES ON OTHER PAPER MAKERS AT GOUDHURST PAPER MILL

ROBERT WILSON (1624–   )

Goudhurst PRs KAO TR P 157 28/1 (bapt) : 28/8 (marr) : 28/12 (bur).
1624 bapt. Robert s.o. Stephen Wilson

All SS. Maidstone
1657 marr. Robert Wilson of Gowdhurst, papermaker, and Joan Story of Loose.

The second entry suggests that he may have been familiar with the prevailing situation among the fulling mills in the Maidstone area, and along the River Loose in particular, which were probably by this time going through a difficult period. He could thus be seen as a source of information to the later Goudhurst paper makers who migrated to Maidstone some 20 years later.

It is also possible that the absence of his name later on in Goudhurst and other parish registers may have been the result of a change in religious persuasion. The Quakers were active in the High Weald in the 1650's and the Baptists in Loose before 1670. Alternatively, he may have left Kent; or moved to another as yet unidentified paper mill, a subject that was raised in Chapter II and again later in this Appendix under Peter Archer.

There were contemporary Wilsons in Horsmonden and Marden; not in Cranbrook. Other Wilsons in Allington, the only ones found in the Maidstone area. None in Dartford or Shoreham, but a family in Eynsford during the 1650's, but no Robert. Apart from the family of a Daniel Wilson at Goudhurst, Robert, so far as the searches made, disappears from the papermaking scene altogether.

STEPHEN MANKTELLOW (–1726)

"Manktellow" has been adopted here as the most convenient form of spelling his name; variants include Mantellow, Manckelow, Mantillim, Mankela, Manctellow, Manktelo (Goudhurst); Mankillo (East Farleigh) etc. It is an unusual name and Shorter, commenting on it in the context of the later Huguenot refugees, suggested that the origin might have been Dutch rather than French.[23] But as the name appears in Goudhurst much earlier than this, the thought occurred that, if Dutch, the Manktellows might have been connected with the Dutch Settlers who arrived in Kent in the 16th C. and found their way there in the course of time. Though on the face of it plausible, there is an alternative and more probable solution to this question.

First, the name is undoubtedly not English; the Society of Genealogists have no record of it before 1750 (over 100 years after its first appearance in Goudhurst).[24] Second, it does not appear to be a Dutch name either and the suggestion has been made that it had originally been French and took the form of Mantelleau.[25] This interpretation has been adopted here not because of any evidence demonstrating a French connection but solely on the grounds of the proximity of France to Kent and the fact that in spite of the Edict of Nantes (1598) the harassment of Protestants continued, especially under Richelieu and, later, under Louis XIV, throughout the 17th C. The supposition here is that families like Mantelleau decided to emigrate to England and find a sympathetic reception in the strongly Protestant district of the High Weald.[26] At present it is not possible to be more definite about the origin of the Manktellows other than to say that they were almost certainly not English.

The first reference to Manktellow (found anywhere in this search) is to Alexander, who had a family in Goudhurst between 1644–1669 and who was buried there in 1673 (Goudhurst PRs). His family appears to have consisted entirely of daughters, his first, Ales, baptised in 1644. There is no mention of a Stephen or of Alexander jnr., who may have

[23] Shorter, A. Bib.12 48 Note 77.
[24] Inf.Soc.Gen. 6:Jan:1984. The name does not appear in the Huguenot Society's list of denizations between 1509–1700; neither does it appear in the IGI; nor in their own Great Card Index of several million names before 1770.
[25] I am indebted to Mr. P.G. Hoftijzer, Sir Thomas Browne Institute, RijksUniversiteit Te Leiden, for this suggestion (9:Jan:1984); a Polish or German origin considered less likely. A 20th C. Manktelow has found not altogether convincing references to a somewhat similar name "Manekys/Manekilou" in Kent and Essex Books of Fees.
[26] Some support is lent to this notion by the fact that Stephen Manktellow was succeeded at the paper mill by Peter Archer who may also have been an immigrant (see this Appendix under Peter Archer snr.).

been Stephen's younger brother, marrying Ann Pont (1676) and who was buried in Goudhurst in 1709. There is no documentary evidence to show that either Alexander snr. or jnr. had any connection with the paper mill.

Later references indicate that Stephen must have been born between 1644–1652, although a date prior to 1644 is not impossible but it would have meant that he would have been over 82 when he died. His and Alexander jnr.'s baptisms may have been lost in the Commonwealth, but this is considered unlikely in view of the numerous entries for Alexander snr.'s daughters. An alternative is that Stephen and Alexander jnr. may have come from another branch of the family still living abroad. It was suggested in Chapter II that perhaps having no male heir Alexander snr. had summoned nephews or cousins from France to help him carry on whatever his business might have been in Goudhurst and seeing that Stephen later became the Master Paper Maker at the paper mill there this business could have been connected with papermaking. If they had in fact arrived in England after 1644, they would have been too young to have succeeded Alexander snr. in 1666, when Willard is known to have been the Master Paper Maker. The very fact that Stephen Manktellow *followed* Willard more or less proves this point.

Not until much later is the name found in any of the secondary or tertiary search areas; neither has it been found in any of the villages in the immediate neighbourhood.[27] It seems almost as if Alexander snr. had arrived in Goudhurst with some special purpose in mind and the only activity unusual to this area (perhaps not even established there yet, though good reasons to the contrary have already been put forward) would have been papermaking. Taking all things into consideration a perfectly plausible case can be made out for him being a paper maker himself either giving a declining business a new lease of life or else founding another mill in Goudhurst; and being followed by a relation of his *who, we know, was a paper maker*.

There are only 3 references to Stephen Manktellow in the Goudhurst PRs:-

Bapt.	1672	Thomas	s.o.	Stephen & Margarett Mantellow
Bapt.	1675	Stephen	s.o.	do.
Bur.	1676	Stephen	s.o.	do.

The Goudhurst Ratebook contains the only other entries relating to him showing him as the occupant of the paper mill from 1675–1679. Since this document only begins in 1675 and seeing that the first entry does not include "late A.N. Other", one might legitimately assume that Stephen was the occupant of the paper mill prior to this date (after all he was having a family in Goudhurst in 1672). Though we have no evidence of this it is reasonable to extend this assumption to a date ca.1668/9 and suggest that he was the successor to Willard there.

As with his predecessors one has to ask where had he learnt his papermaking? If he had not learnt it abroad, an idea that is perfectly feasible, he could have served an apprenticeship at Goudhurst which, at the latest, would have lasted from 1662–1670 but probably earlier than this. Part, if not all of this, might have been under Alexander snr. had he been the Master Paper Maker there since 1644.

No record of Stephen's first marriage has been found so Margaret's origin is not known. She died after their move and was buried at All Saints', Maidstone, in November 1682, Stephen remarrying in 1683 at Boxley, Anne Bonin.[28] He made this move between August 1679 and January 1680; he was not assessed for Upper Tovil mill until late 1680.[29] Motives for this are outlined below. Although the Manktellows were almost certainly Protestant refugees from Europe, to judge from the entries of baptisms, marriages and burials to be found in Church of England Registers, they were evidently staunchly orthodox in their faith and were not attracted initially to Tovil because of the strongly non-conformist element there despite their later association with the Pine family (see below).

27 For example, locally, not in Cranbrook, Horsmonden, Brenchley, Marden (though Manktellows there today), Lamberhurst or Benenden; nor in other secondary search areas which include (perhaps significantly) both Dover and Canterbury; similarly not in Dartford, Eynsford and all the tertiary search areas. The name only appears much later (18th C.) in Plaxtol, Wrotham, East Malling etc.

28 Boxley PRs. KAO P 40 1/2.

29 Spain, R.J. Bib.11 174. The last entry for Richard Parson, fuller, at Upper Tovil, May 1680. The Maidstone RB. entry for Jan.1681 "Mantiloe paper mill £13". But both Spain (loc. cit.) and Shorter, A. Bib.12 191, note that "Manktollo" was assessed for £12 in 1680, clearly while he was converting the mill.

During his tenure of Upper Tovil paper mill Stephen Manktellow took on a number of apprentices amongst whom, in 1698, was his son, Thomas.[30] This notice must have referred to the date of his freedom, since by that date he would have been aged 26 and, in fact, married Hannah Pine in the following year (1699) in East Farleigh.[31]

The question as to whether Stephen Manktellow ever made White paper has been discussed in Chapter II; as an immigrant one might well have expected him to do so. Goudhurst was not the ideal place for this and besides the main market for this class of paper was London. The embargo on French goods that came into force in 1678 would undoubtedly have been a powerful incentive to concentrate on this more profitable product line and the potential availability of mill sites in Maidstone coupled with improved shipping facilities must have been the principal motives for Manktellow's move, a move no doubt encouraged by news reaching him from his predecessor, Thomas Willard. So one can see him without too much difficulty setting up White paper manufacture initially at Upper Tovil, but that as soon as the embargo on French goods was lifted coupled with increasing competition from domestic sources like the COMPANY, he must have been an early casualty together with others in this venture. Whether one can relate this failure to the significant fall in the ratable value of Upper Tovil mill between 1680 (£12–£13) and 1721 (£8) is a matter for speculation; certainly when Peter Musgrove jnr. took the mill over in 1723 the ratable value was raised considerably (£20).[32] It was during this period, before 1716–19, that Dr. Harris remarked on the fact that *some* paper mills near Maidstone made a great deal of ordinary wrapping paper and one feels that Upper Tovil mill had become one of those he referred to. For instance, between 1703–1715 his son, Thomas, was working Great Ivy mill which from what is known of its later history was a Brown mill. By the time Stephen left off working Upper Tovil he could well have been in his 70's and no longer very active as a paper maker, another factor that might have contributed to the low ratable value of this mill.

PEETER ARCHER (For a full account of Peter Archer see later in this Appendix).

RICHARD HOVER (1662–)   (*THOMAS GULVIN*   –1690 Marden)

Absolutely nothing is known about Richard except that he was baptized in Goudhurst s.o. Richard. There is no record of his marriage there nor a burial entry before 1714 (where search ended). To judge from the Ratebook he was the Master Paper Maker at the mill between 1685–1696, by which time the assessment for the mill was lowered. As his name has not been discovered anywhere else and because his activities have no relevance to the development of the paper industry around Maidstone, the subject has not been pursued further nor that of Gulvin.[33]

THOMAS ASHEDOWNE (1658–1709) ?

The only certain reference is that in 1696/9 a "Paper-mill Water-mill and appurtenances were in the tenure of (Thomas) Ashedowne" a mill said to be in the parish of Marden, but surely Goudhurst paper mill. Since there is a vital gap in the Poor Rate Records for Goudhurst at this juncture one can only assume that the entry "Ye Occupier" that follows Hover's name in 1697 referred to Thomas Ashedowne. (See note 14).

On this basis the assumption is that (a) Ashedown was a paper maker because he was the tenant of a paper-mill; (b) he was the Thomas Ashdown buried in Goudhurst in 1709; and (c) he was the son of Abraham Ashedowne baptised in East Malling in 1658. Ashdown was a common name in Kent, but as will be seen later in this Appendix under the Succession at the East Malling paper mills and the notes on Millhall mill, Ditton, there is a case for believing that the same Ashedownes had associations with both of these mills. They had some interest in property in Millhall from 1677 and, coincidentally or not, this had ceased by 1698.

[30]   Shorter, A. Bib.12 191.

[31]   E.Farleigh PR. KAO TR 1042/7. In 1703 he took over Great Ivy paper mill from Richard Burnham (Maidstone & Loose RBs) and was succeeded there by Thomas Pine in 1715 (Spain, R.J. Bib.10 71 : see also Chapter VII TABLE XII.)

[32]   Ratable values given by Spain, R.J. Bib.11 174/5. The figure for £20 at this time is in line with those for Gurney's mill (£20), Lover Tovil (£25). These figures emphasise that £8 in 1721 was a very low figure indeed, perhaps indicating limited activity.

[33]   An Inventory (KAO PRC/11/54/216) of 1690 refers to Thomas Gulvin, paper maker of Marden, as a "very poor man"; clearly ex-Goudhurst mill.

EARLY FORSTAL PAPER MAKERS

The early paper makers at Forstal paper mill are discussed in this section; a general introduction to the early history of this mill is to be found in Chapter II; the location and succession at the mill in App.III; and information on the first paper maker known to be associated with this mill, Thomas Taylor, follows the notes below on Thomas Willard. The reason for placing Willard first here is due to the facts that (a) we know for certain that he was a paper maker; and (b) we know him definitely as the first paper maker to have migrated from Goudhurst to the Maidstone area; for paper makers like Taylor and West this is only a matter of surmise.

THOMAS WILLARD, Paper Maker (1633–1694)

A Deed of 1671 shows that Thomas Willard was the tenant of the paper mill in Boxley.[34] In App.III this mill was located and identified as Forstal. The wording of the Deed implies that Willard was at the mill before 1671 and the last entry for him at Goudhurst is 1666. Lastly, another Deed (KAO U 1823/24 T3 Part I) shows that Thomas Taylor was still described as in the occupation of this mill in 1668/9. It is thus reasonably certain that Willard had arrived in the parish ca.1668/9, although apart from the 1671 Deed and a sarcastic comment on him in the first parish register by Thomas Heymes at about this time,[35] there is no other record of any Willard being in the parish before 1684, 13 years after the Deed had been drawn up; a Thomas Willard had 4 children baptised in Boxley between 1684–93, this Thomas being the paper maker's nephew. How certain can we be of all this ? There is no need to retrace the many steps needed to piece this information together; they are summarised below.

Briefly, confronted initially with nothing more than the information given in the 1671 Deed, which fortunately contained the evidence enabling one to identify and locate the mill, the first question that had to be asked was whether he was a local man and unravel at the same time the three Thomas Willards buried in Boxley between 1693–96. Based on a number of assumptions at this stage it became clear that one would be looking for a birthdate for the paper maker of 1642 at the latest. At this stage there was absolutely *nothing to suggest* that he might have come from as unlikely a place as Goudhurst. A summary of the Willard genealogy which follows these notes shows that there were Willards in the 1° area, especially in Maidstone; and, seen in retrospect, one of possible interest in Allington. To cut a long story short the 1° search area was abandoned and 2° ones examined (Canterbury, for example, was a distinct possibility). In the event Thomas was traced to Goudhurst (bapt.1633). Initially there

---

[34] The whereabouts of this Deed is unknown, but the contents were published by Cock, F.W. "Notes & Queries" Vol.CLVIII (1930) 368.

[35] The first Boxley PR (KAO P 40 1/1 : bapt.marr. & bur.) has the following entry on the verso of the third last page among other memoranda. The date is unfortunately obscured by a blemish:–
"Goodman Wellard at the Paper Mill quia non frequentat Ecclesiam."
This is written in a block of fine handwriting in a golden coloured ink quite different in style from the Registers that follow (Incumbent Humphrey Lynde) and are characteristic of entries made from ca.1665 (Incumbent Thomas Heymes, 1646–77). Other names such as Goodman Saunders and Goodman Mitchell are also mentioned but do not assist in the dating. Judging from the frequency of these distinctive entries one might place the above comment nearer to 1665 than to the end of the register, in short nearer to Willard's arrival in the parish than later.

was nothing obvious to link this Thomas with Boxley beyond the fact that his estimated birthdate fell within the bracket set for him. The essential details for bridging this gap are:–

(i)   Both men were paper makers.

(ii)   Thomas (2) of Goudhurst (see genealogy) had a younger brother, Mathew (sic) baptised there in 1639. He in turn had a son, Thomas (3), baptised in Goudhurst in 1661, a date which in fact fits into the proposed birthdate range for the Thomas jnr. of Boxley (viz. 1663 or earlier).

(iii)   Thomas (3) married *Mical* Phillips of Cranbrook at Goudhurst in 1683. The first Thomas jnr. entry in the Boxley PRs. records the baptism of:–
        Matthew s.o. Thomas and *Michal* Willard in 1684. The name Mical/Michal is quite unusual. It is reasonable to assume that Thomas (3) and Thomas jnr. were one and the same person and the paper maker's nephew.

(iv)   Earlier in this Appendix under Goudhurst paper mill mention was made of the fact that there was a contemporary Thomas (and Anne) Willard in Marden. A Thomas and Ann also appear in the Aylesford PRs. with the baptism of a daughter there in 1683 (Ann was buried the following year). This is the first reference to any Willards in the Aylesford registers, so that it is self-evident that they must have come from outside. The chances are even that this Thomas and Ann were identical with the Marden Willards and that they may have been related to the Goudhurst ones.

        Another reason to support this notion is that Thomas (2), the paper maker, was married to Dorothy (see the entry of their daughter Elizabeth's baptism, 1666, in the Goudhurst PR.). Goodman Willard was buried in Boxley in 1694 and the family came to an end there in 1696. It would not be surprising, therefore, to find that his widow had moved to Aylesford, if the Willards there were related (see para above), and that she was the Dorothy buried there in 1696. No other Dorothys have been identified in the district.

(v)   More doubtful, a Robert Willard made a brief appearance in Aylesford in 1685 (the baptism and burial of a son Robert). A Robert Wellard, householder, was buried in Boxley in the following year (1686 corr:date). Whether either of them was the Robert Willard (bapt. Goudhurst 1660) believed to have been a son of the paper maker is a matter of conjecture, plausible on the grounds that there were no previous Willards in Boxley or Aylesford before the paper maker and that this perhaps represented another Willard migrant to the area from Goudhurst.

All these points strongly (one might say indisputably) support the proposal that the Goudhurst Thomas (2) and the Boxley paper maker were the same person. Having made this discovery, an extension of the examination of the Goudhurst records (see earlier in this Appendix) has provided all the evidence needed to demonstrate the migration that took place of paper makers from there to the Maidstone area in the decades following Willard's move. As said, he may not have been the first paper maker to make this move, but he is the first for whom there is positive evidence.

WILLARD/WELLARD GENEALOGY

Some 30 Kent PRs. were searched for Thomas Willard (1°, 2° and 3°) but once his origin had been located the majority of these results were no longer relevant (those discarded will be found in working papers). The remaining results are presented below (a) to confirm the conclusions reached in the text; (b) to see whether any of those round Maidstone might have encouraged him to leave Goudhurst; and (c) to see whether any of the others provided any possible links with the contemporary papermaking Canterbury Wellards. Although the latter were active in their papermaking at a rather later date than Thomas (2) from Goudhurst, it does seem to be rather more than a coincidence that there should have been two papermaking Willard/Wellard families operating relatively close in date to each other.

MAIDSTONE DISTRICT

	PARISH	ENTRY	EVENT		DETAIL
	MAIDSTONE				Willards date back to Elizabethan times. (Russell, J.M. Bib.41 280). There were 6 families producing offspring between 1632-77, but no obvious link with the Goudhurst family. William, shoemaker, act.1632-46; Thomas & Alice 1637-40; John & Elizabeth 1660's; George & Martha ditto; others, a William and Richard, later.
1	ALLINGTON	1650, 1651	marr. x 2		In 1651 George to Elizabeth Farbie of Horsmonden.
2	BOXLEY	1684	bapt. Matthew	s.o.	Thomas & Michal Willard (see Goudhurst)
		1686 (corr:date)	bur. Robert		Wellard (householder) (see Goudhurst)
		1688	bur. Ann		Wellard (widow)
		1689	bapt. Sarah	d.o.	Thomas & Ann (Michal appears to have died).
		1691-93	bapt./bur. x 3	c.o.	Thomas & Ann
		1694 (11:Dec)	bur. Thomas		(probably Thomas & Ann's last child)
		(11:Dec)	bur.		GOODMAN WILLARD (clearly the paper maker)
		(24:Dec)	bur. Mary		(probably Thomas & Ann's child)
		1696	bur. Thomas		(Probably Ann's husband)
3	AYLESFORD	1683	bapt. Elizabeth	d.o.	Thomas & Ann Willard (see Marden)
		1684	bur. Ann		Willard (must have been Thomas' wife)
		1685	bapt./bur. Robert	s.o.	Robert & Keturah Willard
		1689	marr.		Thomas to Mary Stow (both of Aylesford : 2nd. marr.)
		1690-1701	bapt. x 6	c.o.	Thomas & Mary
		1692	bur. Elizabeth		Willard
4		1696	bur. Dorothy		Willard (paper maker's widow ? see Goudhurst)

WEST KENT

	PARISH	ENTRY	EVENT		DETAIL
	DARTFORD	1582	bapt. -	d.o.	Richard Willard
		1666	bur.		Richard Willard (see origins of Wellards at Canterbury)
	SHOREHAM	1623	marr.		Richard to Anne Mosier (1 child bapt./bur.1624)

EAST KENT

	PARISH	ENTRY	EVENT		DETAIL
5	DOVER AREA				Willards present in parish of St.Mary's Dover
		1665	bapt. Margaret	d.o.	Richard Willard, St.James, Dover (and others)
		1665-73			No Willards in parishes of either Buckland or River, possibly both with paper mills by then.
	CANTERBURY				The PRs of 10 Churches in Canterbury searched. A few Wellards seen must most seem n/a.
					The main interest is in trying to trace the origin of 2 Richard Wellards, paper makers at Barton mill, father and son (1693-ca.1714). The confirmed details of these two are:-
					Richard snr. purchased freedom 1693
		1698	marr.		Mary Newsome at St.Alphege's in Canterbury
		1704	bur.		Richard snr.
					Richard jnr. apprenticed to Richard snr. 1700 made free by apprenticeship in 1704
		1707	marr.		Martha Knock (widow) at St.Alphege's
6					Are there any clues to their origin ?

	PARISH	ENTRY	EVENT				DETAIL
	GOUDHURST						The earlier generations of Willard have been omitted here. (appears to have been a generation gap)
		1632	marr.				Thomas (1) Willard to Joane Carpenter
7		1633	bapt.	Thomas (2)	s.o.		Thomas (1) Willard (mother not named here)
		1635	bapt.	Nicholas	s.o.		Thomas (1) & Joane
		1639	bapt.	Mathew(sic)	s.o.		ditto (see bapts. below)
		1646	bapt.	John	s.o.		ditto
		1649	bur.	Richard			father of William and Alice bapt.1620/22 (?)
		1660 (corr:date)	bur.	Thomas (1)			died of the poxe (8 Mar).
8		1660	bapt.	Robert	s.o.		almost certainly Thomas (2) (23:Sept)
		1661	bapt.	Thomas (3)	s.o.		Mathew (sic) Willard.
		1663	bapt.	Elizabeth	d.o.		Thomas (2) Willard
		1663	bapt.	Elizabeth	d.o.		Mathew Willard
		1665	bapt.	Mary	d.o.		Thomas (2) Willard
		1666	bapt.	Margaret	d.o.		Mathew & Ann Willard
9		1666	bapt.	Elizabeth	d.o.		Thomas (2) & Dorothy Willard· (see Aylesford)
		1669	bapt.	Richard	s.o.		Mathew & Ann.
10		1671	bur.	Richard	s.o.		ditto
		1675	bapt.	John	s.o.		ditto
11		1683	marr.				Thomas (3) Willard to Mical Phillips (of Cranbrook)
	MARDEN	1640	bur.	Richard			Willard
12		1655	bapt.	Unnamed	d.o.		Thomas Willard, Paperman & Miss Saunders
13		1661-68	bapt. x 4 (1 bur).		c.o.		Thomas & Anne Willard (see Aylesford)
14	HORSMONDEN	1653-62	bapt. x 5		c.o.		Thomas & Susan Willard
	CRANBROOK						Willards but n/a.
	LAMBERHURST						No Willards for the period, 1630 +
	BENENDEN	1601	bapt.	Francis	s.o.		Richard Willard
		1602	bur.	Henry	s.o.		ditto
		1608	bur.	Thomas			Willard (could have been Thomas [1]'s father ?)
							Family of Laurence Willard also
		1654	marr.				Richard Willard, a Broad Weaver from Hawkhurst

1   For possible links with the papermaking activities at Goudhurst, George Willard and a Horsmonden link. See also, same dates, the Richardsons of Allington (this Appendix), Alexander Richardson and Marden.

2   Two points to note (i) No Willards in Boxley for the period prior to Thomas (2)'s arrival; and (ii) the arrival of Michal Willard from Goudhurst.

3   No Willards in Aylesford for the period prior to the arrival of Thomas & Anne from Marden in 1683.

4   No other Dorothys seen other than Thomas (2)'s wife; see 1666 baptism entry at Goudhurst.

5   No known paper mill connection for this Richard Willard; not seen at Buckland or River between 1665-73.

6   Are there any clues to Richard snr.'s origin ? In theory Richard snr., who appears in 1693 as the paper maker at Barton mill and husband of Mary Newsome from Wingham, could have been married a first time to Ursula. A Richard & Ursula Wellard had 2 daughters bapt. in the parish of St. Mary Northgate, Canterbury (believed to be the paper mill parish) in 1671, 1672 (corr:dates : result of search undertaken by Miss A.M. Oakley, Archivist, Cathedral, City and Diocesan R.O.). There was also a Wellud family who married there at the same time from the neighbouring parish of Sturry. Richard jnr. must have been born ca.1685 i.e. by a previous marriage. Even supposing Richard snr. was the husband of Ursula in 1671, he could still have been the son of the Dover Richard (St. James'). (There is evidence of considerable interchange between the Canterbury and Dover areas in terms of papermaking apprentices. See Chap.II and VII under Barton mill; and Chap.VII River mill).

   Alternatively, there were other Richard Willards as will be seen from the other Regions listed in the Table including one from the Dartford area and some in South Kent, the latter could have been related to the Goudhurst Willards as well (see both Goudhurst and Benenden).

7   This Thomas (2) was the future paper maker at Forstal.

8   The father, Thomas (1), would have been at least 50 by this time. He could in theory have been the father of this Robert, but there would have been a gap of 14 years between this and his previous child. It is thus thought that the father in this case was Thomas (2), but whether it was his first child or not is not known.

9   This is the last Thomas (2) entry at Goudhurst, 1666.

10 Even if Mathew and Ann had had a second "Richard", it is not likely to have been the Canterbury one.

11 See first Boxley entry.

12 As discussed in the text, the father of this base-born child might have been Thomas (2) who, we know, was a paper maker; but it is also possible that the Marden Thomas might have been a paper maker as well seeing they appear to have moved to Aylesford later (see note below).

13 In Marden Thomas & Anne lost a daughter in 1665 (aged 3) and although there is an 18-year gap their first child in Aylesford might have been a second try. Ann died in the following year.

14 The Horsmonden Thomas cannot have been Thomas (2); their wives had different names and Susan(nah) was having a family at the same time as Thomas (2)'s wife in Goudhurst. There were other Willards in Horsmonden after 1675.

FURTHER NOTES ON THE EARLY PAPER MAKERS AT FORSTAL MILL

The succession, as far as it is known, at Forstal mill has been outlined in the synopsis of the history of this paper mill in App.III (q.v.). Due to the absence of a ratebook before 1691 there is uncertainty as to who precisely occupied the mill before this date and for how long. The position of Thomas Willard has already been considered; he is known to have been a paper maker and his origin has also been determined, but it is not known when he finished papermaking there. It is known that Peter Archer was the occupant in 1691 and it has been proposed that as a Master Paper Maker from Goudhurst he took over the mill late 1683 when he arrived in Boxley. Peter Musgrove (q.v.) may initially either have worked for Willard between 1678–83 or assisted in the conversion of Turkey mill. Other paper makers like Thomas Newman have been assigned to Turkey mill as employees. In the pages that follow consideration will be given to what little is known about Thomas Taylor, the first known occupant, Peter Archer and Thomas Hillyer. For convenience, Taylor and Hillyer will be dealt with before the Archers, the latter being seen in a wider papermaking context than that of the Maidstone district.

THOMAS TAYLOR

In the Notes on Forstal mill (App.III) 2 Deeds were cited, one of 1665 and the other of 1668/9, relating to the Estate of George Duke of Cossington, Aylesford, both of these showing that the paper mill at that time was occupied by Thomas Taylor. Though Taylor is not described as a paper maker in either document neither is Willard in a similar Deed of 1671; but since Willard was one it is fair to assume that Taylor, as occupant of a paper mill, was one also. The 2 Deeds which name him indicate that Taylor was associated with other Duke property, but in what capacity is not known. As both documents refer to Taylor in the same terms this gives one some latitude in thinking that he was not necessarily a new tenant but one who could have been there some time. So, in attempting to trace his origin, one has to consider a possible birthdate of not later than, say, 1644 and, in view of his responsibilities, one that might have been much earlier. The Will (1640) of Sir Edward Duke, father of George, shows that he left "mills" to his son, a fact which gives one some scope for the length of Taylor's tenure or that of his predecessor. Finally, as the paper mill was sited in the parish of Boxley, certainly all the later tenants were assessed for their rates in it, one might assume that Taylor would have been domiciled there also, albeit the mill was much closer to Aylesford (3/4 mile) than Boxley (3 miles). For the significance of this in terms of Taylors in the area see below.

In the Aylesford PR (starts 1654), one finds a strong contingent of Taylors there. Baptisms from this date show 3 families producing offspring but apart from 1 Thomas (s.o. John Taylor, Yeoman, and Mary bapt.1658 and died as an infant) and 2 families, Nicholas snr. and jnr. there is no sign of the paper maker; all that one can say is that Taylors were present in the area who might have produced the paper maker and someone who had learnt his trade locally on the assumption that an early paper mill existed there. The Boxley PRs were searched from 1630 but show no evidence of Taylor families there though it is clear that Taylors raided the parish from time to time in search of brides, a Stephen marrying Frances (1630) and a Peter, interestingly, a Susanna Richardson (1642), see also this Appendix under "The Richardsons". It has to be said, however, that under the Puritan incumbent, Thomas Heymes (1648–78), the Register is virtually empty during his ministry, 1 or 2 entries per annum, sometimes none at all (e.g. 1651, 1667–78) vital years for discovering whether Thomas was there, was married and having a family. From other sources[36] one learns that a Thomas Taylor was born in 1657 s.o. a Thomas who became the first baronet in 1664 but died in 1665, showing that they could not be linked to the Thomas of the 1668/9 Deed.[37]

Further afield the most promising finds come from Goudhurst, a Thomas Taylour (b.1612) and a Thomas Taylor (b.1640); the first being 53 in 1665, the second 25. If in fact Thomas came from Goudhurst then one might favour

---

[36] Burke's Extinct Baronetage (1844). See also Russell, J.M. Bib.41 199ff. and 336.

[37] Though no connection with the paper mills can be shown, it may be noted that Thomas Duke, the first of this family to live at Cossington, had married Elizabeth Culpeper of Wakehurst, Sussex; and there were Culpepers at Preston Hall, the last of whom, Alicia, married Thomas Taylor (1650's).

the first i.e. born earlier than Willard, his successor. A training in Goudhurst, taken in conjunction with the fact that he was succeeded by Willard, would certainly fit the general picture; but beyond this there is nothing to go on. Born in Aylesford would strongly indicate the earlier existence of Forstal; born in Goudhurst the same could be said unless one sees the elder of the two founding a mill during the closing years of Sir Edward's regime. In view of the widespread occurrence of the name the search has not been extended in any depth.[38] No other papermaking Taylors appear until the 1740's, Clement Taylor at Basted mill (see App.II).

JOHN HYLLIER alias HILLER (1686), HILLARY (1693), HILLIER (1697), HYLIER (1715)[39]

Goodman Hillary followed Peter Archer at Forstal in 1693 and remained there until 1724. He had married Constant Russell in Boxley in 1686 but with no entries for children before 1697 (bapt.Elizabeth). He has been included here as a borderline case among the early Maidstone paper makers, possibly serving his apprenticeship there, say, 1677–84(?).

His origin has not been traced. Eleven parishes in the Maidstone area, 10 in the West Kent Region and 5 in the South have drawn a blank (a Hillbye and Hilder found in Penshurst). The Great Card Index (Soc.Genealogists) show some Hillyers in Kent, but the main concentrations lay at this time in London, Middlesex, Wiltshire, Hampshire and Buckinghamshire with a few scattered elsewhere. Overend (Bib.66 196) refers to a 1666 Patent issued to Charles Hyllyerd Esq. for making "blew paper used by sugar bakers and others" and (loc.cit.205) to a Thomas Hyllyard as one of several persons summoned by Lord Sunderland in 1687 to answer charges of infringing the COMPANY's monopoly for the manufacture of White paper; they included paper makers from Surrey, Middlesex and Buckinghamshire though Hyllyard is not specifically referred to as one. Nevertheless, there seem to have been papermaking associations with this name and possibly related to our Hyllier, the inference from this being that he might also have had anti–COMPANY leanings. Finally, Shorter[40] refers to Hillyers who were proprietors of Down mill, Cobham (Surrey), early in the 18th C. In 1733 Mary Hillyer, widow, insured her new–built mill; the previous one had just been burnt down. This information suggests that Mary's husband may have been the previous proprietor. The mill is mentioned in 1687, but the PRs for Cobham and nearby Fleet have no Hillyer entries.[41] As proprietors the Hillyers need not have lived in the neighbourhood. This Surrey association of the name with papermaking is regarded here as a promising source for our John.

The Hillyer/Russell marriage and the perpetuation of the name "Hillier Russell" is referred to under "The Russells".

[38] Seven families at Dartford between 1614–41, a Thomas father but no Thomas sons; Eynsford none before 1642; Goudhurst, Thomas s.o. Moses Taylour bapt.1612 and Thomas s.o. Thomas & Elenor Taylor bapt.1640 (no Thomas Taylo(u)r burials up to 1665).

[39] Boxley PRs KAO P 40 1/2 marr. John Hiller to Constant Russell 1686; 1697 Elizabeth d.o. John & Constance Hillier. Other var. P 40 12/1 (RB).

[40] Shorter, A. Bib.12 237 occupant in 1733 Richard Hinton (propr. Mary Hillyer) occupant 1687 William Berrey/Borroy (propr. not known). Could the latter have been the French refugee paper maker "Berreau" summoned as a witness against the COMPANY's Bill in 1690 ? (Overend Bib.66.209). This would strengthen the comment made above concerning anti–COMPANY feelings.

[41] Search kindly made by Staff, Surrey R.O. Kingston–upon–Thames (16:Feb:84); Cobham starts 1655 and Byfleet 1698.

THE ARCHERS : THEIR ORIGIN AND RELATIONSHIP DISCUSSED

There are references to one papermaking Archer or another throughout this work from Chapter II onwards.[42] In some ways one might regard the activities of these Archers as nothing more than a foil to the main theme of the work, the elder Whatman, his background, investment and influence on contemporary papermaking. However, they all played their part in the progress of these events and in their eventual fruition and, though as individuals none of them appears to have made a major contribution to the advancement of the domestic White paper industry, cumulatively they must be reckoned as an important element in its development. Looked at the other way, if the assumptions made are correct, one might ask whether the sequence of events covered in the text would have followed the course they did, or would have been considerably delayed, had the Archers been absent from this scene ? For instance, if the introduction of an early form of Engine into the papermaking operations at Dartford No.2 mill in the 1690's is seen as a starting point for the growth and spread of the Quelch empire on the one hand and the appearance of Engines at Chartham mill on the other as stemming from Peter Archer snr.'s suggested management of the Dartford mill, then one has an important source of influence to consider even though the operations themselves may not have been especially noteworthy at that stage in establishing the manufacture of White paper on a secure and profitable basis. In short the Archers had a status that warrants a more detailed study than has been devoted to many of their contemporaries. The problems surrounding this study, their origin and relationship, have not however been resolved definitively.

From 1673 onwards various Archers connected with papermaking, almost exclusively in Kent, appear and disappear from the papermaking scene making it very difficult to establish any continuity in their careers and determine whether they were related to one another or not. Vital information at certain critical points in their histories is missing. In order to present a reasonably coherent and credible picture of their activities, activities that may have been an important source of influence on Harris and Whatman when preparing for their Hollingbourne investment in 1733, certain assumptions were proposed for this purpose[43] in order to account for the developments that were taking place at that time. The text was not the place, however, to discuss alternatives or the reasons why the assumptions proposed were adopted.

There were seven Archers who were connected with some certainty in one way or another with paper and papermaking during the period covered by the latter part of the 17th C. and the first half of the 18th. An eighth Archer, Richard jnr., one feels must be added to this list too. So far as we know he was the *eldest* son of Richard snr., the paper maker at Dartford No.2 mill from 1701–40; he was still in Dartford and having a family there in 1740 when his father was made bankrupt. One imagines that prior to this he had worked in the paper mill; but after 1740 he appears to have made the neighbouring parish of Sutton–at–Hone his home.

---

[42] There are casual mentions of Peter Archer snr. in Chap.II.
      But the main discussion of the Archers is set out in Chap.IV, principally in the context of possible influence on Whatman and Harris. (Reference is also made there to Robert Archer's continued operations at Sittingbourne).
      In Chap.VII there are references to Richard snr. at Dartford; his sons at Hawley mill; the type of Engine installed at Sittingbourne either by Peter snr. or jnr.; and, finally, Peter jnr. and the succession at Chartham mill, and the suggested link between Dartford and Chartham based on a Richard/Peter jnr. relationship.
      Other references in Chap.VIII; App.II under the Whatmans formerly at Chartham; and App.III Boxley RB entries.
[43] See Chap.IV pp.140–143 + notes.

It was suggested that later he may have found work at Hawley paper mill where his younger brother, William, became the occupant in 1762.[44]

Finally, was there another related Archer, Elizabeth ? An Elizabeth Archer married a John Walter at Aylesford in 1703; this could have been the John Walter and his wife Elizabeth who became the occupants of River No.1 mill, near Dover, some time prior to 1717 and remaining there maybe as late as 1733 (see Chap.VII and this Appendix under The Walters).

Up to this point it has been possible to identify two families of Archers amongst these six paper makers; but it would be of much greater significance and make better sense to find evidence, so far undiscovered, that combined these two into one. Out of the seven Archers Peter snr. was the eldest and the first to be identified with papermaking.

Like so many of the names encountered in this work the name Archer occurs widely and frequently in the records of this period; the spelling of the name was sometimes variable, changing from Archer to Archal(l) or Archel(l) for the same family within a given parish. To show that these variations were not isolated or due to some quirk of the incumbent they are to be found in parishes as far apart as Aylesford and Shoreham.[45]

A short summary of what is known about each of these Archers follows with a discussion which attempts to resolve the problems they pose. An attempt to unify the relationship may not seem to be an important matter; but it has a considerable effect on the kind of conclusions that may be drawn regarding their respective papermaking activities. Also, when searching 1°, 2° and 3° papermaking districts, not having basic information on their birthplaces and birthdates, it becomes a matter of conjecture as to whether any of them were related to the various families of Archer one finds already domiciled in these areas or whether our paper makers came, perhaps, from a totally different place or even country.

To judge from the results of the searches there appear to be three credible solutions to this situation:–

(i)   that all these Archers were related.
(ii)   that they were of immigrant origin.
(iii) that they had an English origin.

One thing they all had in common was the fact that all the Archers to be considered were connected with that somewhat unusual and rural occupation, papermaking. This is at least one point in favour of seeing them as a unified family.

---

[44] See Chap.VII p.301.
[45] There is even a "Larshal", possibly a variant of "Arshal" who may have been connected with papermaking at St Mary Cray in 1756. (Chap.VII).

1. *THE FAMILY RELATIONSHIP* ?

PETER ARCHER snr.

Though well over 50 different Kent parish records have been searched (as well as the IGI for Kent, Sussex and Surrey) Peter snr.'s birthdate and birthplace have not been discovered. The first documentary evidence that we have of him is the entry for his marriage to Elizabeth Beckett in 1673 at Goudhurst.[46] The entry designates him as "Peeter" Archer, as indeed do the later references to him in the Goudhurst Poor Ratebook between 1680–83.[47] The spelling of the name Peter (in two different records) may or may not have special significance. It is perhaps worth noting that there is a baptismal entry in the same register for a "Peeter" West in 1651, quite distinct from another "Peter" baptised there in the same year.[48] It may also be noted that no other Archers have been found in that area at that time;[49] our Peter clearly came to Goudhurst from some other place.

One gets the impression that after his marriage Peter and Elizabeth may have moved to another area. Between 1673–80 there are no references to them having a family in Goudhurst[50] or being assessed (in this case between 1675, when the ratebook starts, and 1680). It is possible, to explain all this, that Peter and Elizabeth had been living with a member of her family e.g. George Beckett.[51] On the other hand the absence of any reference to offspring is puzzling because on and after their arrival in Boxley in 1684 they produced a family of three. There may have been a good reason for this apparent "barrenness" during the first seven years of their married life, that is to say if one is to judge this state by the Goudhurst records alone.

The idea that they might have moved somewhere else (because there was perhaps no place for them at the Goudhurst mills)[52] was one of the reasons for creating the 3° search areas; but no sign (in Kent) of them living elsewhere has come to light yet. One important reason for making this additional search is that it so happens that this period, 1673–80, would accommodate Richard snr.'s birthdate very conveniently; and that of an Elizabeth (named after her mother) mentioned earlier. To establish either that Richard or Elizabeth were Peter's children would not only help us locate Peter's whereabouts at that time, but be a major step in resolving the problems surrounding these Archers.[53]

[46] KAO P 157 28/8.

[47] See this Appendix p.127 and KAO P 157 11/1.

[48] This reference to a "West" is referred to again later in this Appendix under James West.

[49] This includes the parishes of Goudhurst, Cranbrook, Horsmonden, Lamberhurst, Marden, Benenden (not until (1697) and Brenchley (see also note 50 below)

[50] There is an isolated reference to an Archer, the burial of a Mary Archer, in Brenchley in 1679, but nothing to indicate her age or origin (see note 53 below). She could quite easily have been a daughter of Peter's.

[51] See this Appendix note 19.

[52] See this Appendix for the suggested succession at these mills p.128; and discussed in Chap.II.

[53] There is a limit to the number of parishes that can be searched within reason for this kind of work. It has been suggested that in an area like the High Weald, particularly in a period of decline with many fulling mills becoming redundant, there might have been small paper mills almost anywhere whose existence has never come to light. If, for example, the Mary Archer buried at Brenchley had been Peter's daughter, one only has to look at the map to see that this large parish would have afforded many potential sites for paper mills (Matfield and Paddock Wood are modern parishes); but there are no ratebooks or other documents from which evidence can be obtained.

The Goudhurst ratebook indicates that Peter Archer had already left Goudhurst by August 1683.[54] The next reference to Peter and Elizabeth is that they had a daughter, Margaret, baptised in Boxley on 1st January 1684 (corr:date).[55] It is also clear from the Boxley ratebook (starts 1691) that by 1691 he was the occupant of Forstal mill.[56] The only other paper mill in Boxley at that time was Turkey mill so it is most unlikely that Peter, a Master Paper Maker at Goudhurst, would have made the move to work under somebody else e.g. George Gill and then make another move shortly after. It is more probable that, following in Stephen Manktellow's footsteps, he moved to Maidstone because a vacancy in a more promising area had arisen, e.g. that Thomas Willard had by 1683 ceased to be the occupant of Forstal mill and that Peter Archer had taken his place.[57]

There is continuity in the evidence, baptisms and ratebook, to show that Peter snr. was resident in Boxley from the end of 1683–May 1693, after which he disappears from the records and no further trace of him has been discovered to date. One of the baptisms recorded in the Boxley period, that of his son, Peter, in 1683, without doubt refers to the Peter Archer, the paper maker from Sittingbourne who was married in Canterbury in 1708 and who moved later to Chartham (see below).[58] Another son, John, was also baptised in Boxley, in 1689, and will be discussed later.

RICHARD ARCHER snr.

The next papermaking Archer in age was Richard snr. who bought Dartford No.2 paper mill in 1701 and sold it again to William Quelch snr. at the end of 1736.[59] He remained an occupant of this mill until made bankrupt in 1740. His birthdate, birthplace and his parents (suggested as having been Peter snr. and Elizabeth) have not been found. He married Sarah Whitewood in the neighbouring parish of Wilmington (though *both described as of Dartford*) in 1703 (corr:date),[60] following the example of his opposite number, William Quelch snr. at Dartford No.1 (see Quelch genealogy in App.II) also married in Wilmington but a year earlier. Richard snr. was buried in nearby Darenth in 1746.

Richard snr. and Sarah had 4 sons and 4 daughters baptised in Dartford between 1704–21 and all of them, except a second Elizabeth (bapt.1718), were buried like their parents in Darenth. Three of the sons survived as adults, John, Richard and William (2) and will be referred to below. The fact that they were all buried in Darenth points to them as being a closely knit family.

---

[54] The entry in the Ratebook for Aug.1683 "Peter Archer or Occupier" and succeeded by George Beckett in Jan.1684.

[55] See later under Archer genealogy.

[56] See Boxley Ratebook entries App.III under Forstal mill.

[57] News may also have reached Peter snr. via Thomas Willard ex–Goudhurst and lately at Forstal mill. For Peter Musgrove as another possible candidate see this Appendix under Peter Musgrove snr.

[58] Regarding other possible claimants the subject will be discussed below under "Peter Archer jnr."; there was, for example, another contemporary Peter Archer born in Aylesford in 1679.

[59] The occupant immediately preceding Richard snr. was John Quelch, but it is not known whether Richard bought the mill off him or not. (See App.II under the Quelch genealogy: and Chap.VII under St Mary Cray pp.305/306).

[60] KAO P 397 1/1 (Licence Bishop of Rochester). It may be noted here that in addition to the uncertainty of Richard snr.'s origin, Sarah Whitewood was not baptised in Dartford either, nor is there any mention of her family.

Up to this point particulars have been given of the two principal Archers; Peter snr. from his marriage in 1673 to his disappearance in 1693; and Richard from the purchase of Dartford No.2 mill in 1701, by which time he must have been at least 21 years old (with a baptism date during or *before* 1680), to his burial in 1746. Richard snr. was always described in the records either as Mr. Archer of Dartford or as "Gentleman". It may be noted here that before Richard snr.'s appearance in Dartford in 1701, the only other Archers found in this district before this date were a John or Johan Archer who had 8 children baptised in Dartford between 1604–1620; none earlier at Wilmington, Darenth, Sutton–at–Hone or Eynsford. The conclusion that presents itself is that Richard originated from somewhere outside the Dartford district.

Were the elder Peter and Richard Archers related ?

One might argue that there were sufficient Archer families in Kent to have produced two separate papermaking families and, if we knew it, it might well prove to be the case. Indeed it is possible that our Archers were linked in some way with some of the families who have been found in parishes close to those where Peter, for one, worked e.g. the Archer/ Archals of Aylesford.

However, based on the scattered evidence presented above it makes better sense to assume that Peter and Richard were related; how closely is another matter, but in terms of birthdate and age brackets there are no obstacles to the proposal that Richard was Peter's son.

There is, for example, a case for examining the apparent barrenness of Peter and Elizabeth between 1673–80, a period in which Richard could have been born, making him between 21–26 by the time he came to buy Dartford No.2 mill in 1701 and at the same time be of marriageable age to fit in with the actual date of his marriage in 1703. One also has to take into account the facts that (a) Richard appears to have come to Dartford from somewhere else i.e. no Archer families in the district; (b) in buying and operating a paper mill he must have been trained as a paper maker by a paper maker somewhere; and (c) no Richard Archers have been found in any of the 1°, 2° or 3° search areas who survived from infancy or whose dates would fit in with our Richard snr. On the basis of this reasoning alone there is a good case for believing that Richard was probably Peter snr.'s eldest son, there was certainly a gap of a generation between them.[61]

The above agrees well with the two proposals made in Chapter IV, namely, that Peter snr. moved from Forstal mill, Boxley, in 1693 to manage Dartford No.2 mill for the COMPANY, accounting not only for certain unexplained subscript "A's" to COMPANY countermarks but

---

[61] Mention should be made here of 2 families of Archers who were active in Aylesford between ca.1661–1707. From the ratebook they appear to have lived in the "Millale" ward of the parish, that is very close to Millhall paper mill in Ditton and it is just conceivable that they may have worked in Millhall mill. They are mentioned here on three counts: (i) the older of the 2 families was headed by a William and the other by a Richard. The latter had a son, Richard, b.1672–bur.1673; there was no gap amongst his other children to allow for another Richard to fit in with the Dartford one's dates. There was, however, quite a noteworthy gap in William's family between 1664–71 but no entry of a Richard baptism; had there been it would have made him 31–36 at the time the Dartford mill was acquired. Neither of these families appear to have been prosperous enough to have bought a paper mill. (ii) In 1679 both families changed their name from Archer to Archell (or var.). (iii) Richard & Ann Archell had a son, Peter, baptised in 1679, discussed below under Peter Archer jnr.

also providing a rationale for Richard's subsequent interest in this mill (and the occurrence of an "RA" countermark); and a proposal that with the demise of the COMPANY'S operations at Dartford, Peter snr. moved to Sittingbourne, which in turn would account for Peter jnr.'s presence there in 1708, all accommodated within the age brackets of the people concerned and with the few facts that we know about their movements.

If one assumes that (a) Peter and Richard came from entirely different families and areas, and (b) that Peter may never have gone to Dartford (and there is an alternative) etc. one is left, as indicated in Chapter IV, with unexplained evidence difficult to relate to other circumstances associated with them e.g. why should Richard have appeared at Dartford for no apparent reason ? and how does one explain Peter jnr.'s presence at Sittingbourne ?[62]

The next question, does Elizabeth, as a daughter of Peter snr. and his wife Elizabeth, fit into this picture as well ? No Elizabeth Archer baptisms have been found that remotely approach the date bracket that one would be looking for, if she was married in 1703 (probably ca.1677–83). On this basis she would just qualify for a baptism during Peter snr.'s "missing" period. There is no record of an Elizabeth being baptised in the Aylesford district where the marriage took place; not in Aylesford, Boxley, Ditton, Allington or East Malling. Who then was this Elizabeth Archer who married John Walter there in 1703 ? Peter snr. had left Boxley in 1693. If this Elizabeth was his daughter then she must have returned to the district for her marriage there, maybe because Walter had worked for or been apprenticed to her father there earlier ? This subject, to gain any credibility, really has to be seen in association with an equally indeterminate one concerning a Thomas Radford who like Walter appeared in River from another papermaking district, albeit a generation later (see this Appendix under "the Walters" and "Thomas Radford").

All in all, the most promising solution to the above is to accept that Richard snr. and Elizabeth were Peter's eldest children and that there might also have been a Mary (bur. at Brenchley).

PETER ARCHER jnr. (b.Boxley 1685–bur.Chartham 1737 corr:date)

If the above conclusion is accepted then Peter jnr. and Richard snr. were brothers. The notion of this relationship is strengthened by the links that, on the face of it, existed between the Dartford and Chartham mills, links that, once forged, lasted into the 19th C.[63] It is also obvious that the dateable events in Peter jnr.'s life distinguish him completely from his father

---

[62] An indenture of 1655 (KAO U 36 T906) between Sir William Smedley Bart. and Thomas Lucy and Robert Tomkins which links Milton mill, presumably corn (see below), and "Alisford" mill which was a corn mill. Later, Harris, J. (Bib.4 207) ca.1715 refers to two very good springs about half a mile from the town, Sittingbourne (Milton being part of it) that drive 4 corn mills, but Harris makes no mention of a paper mill there. That we know there was one is shown by Peter Archer jnr's marriage licence (1708) and the Insurance Policy (SFIP 61335 10:July:1733). The point to note here is that there may have been a link between the corn mills at Sittingbourne and Aylesford that provided information to Peter snr. during his residence in Boxley regarding a paper mill in Sittingbourne or the potential for one. Though somewhat far-fetched this information could furnish one with a reason whereby Peter snr. and his family moved direct from Boxley to Sittingbourne in 1693 thus by-passing the proposed move to Dartford. It leaves one, however, with the need to explain Richard's presence at Dartford instead of taking his father's place at Sittingbourne.

[63] Chap.VII see under Chartham mill.

(a distinction that was not clear to Shorter Bib.12.57). The events of Peter jnr.'s career have already been set out in Chap.IV, his marriage in Canterbury,[64] his move to Chartham in 1732, his conversion of the mill there by 1733 and his early death in 1737.

Mention has been made earlier of another Peter Archell (family formerly Archer) baptised in Aylesford in 1679. Could this have been the Chartham Peter ? This is a very unlikely supposition. The Archer/Archells, even if they had been engaged in papermaking, were certainly not Master Paper Makers like the Boxley Peter snr. It seems quite reasonable then to dismiss Peter Archell as a possible candidate. Neither of the younger Peters were buried in the district, so there is no obvious objection to Peter snr.'s son being the one who was buried in Chartham.

JOHN ARCHER   (b.Boxley 1689–bur.Hatfield 1764 ?)

That this John Archer was a paper maker is much more a matter of surmise than one of any certainty. So far as we know John was the youngest of Peter snr.'s family. Apart from his baptism nothing more is known about him. However, a John Archer appeared as a paper maker at Hatfield paper mill in Hertfordshire in 1726 and in view of the Boxley John's baptism date it is feasible to suggest that he may have been the Hatfield paper maker. No other likely candidate has been found yet, certainly none with a known papermaking background.

---

[64] This is a convenient point for reviewing the significance of various references made in the text to Peter jnr.'s association with the Dean(e) family of Canterbury whilst still in Sittingbourne and after his move to Chartham. The basic fact underlying these situations is Peter's marriage to Margaret Deane at Holy Cross, 18:Nov:1708, the eldest d.o. Robert Deane of Holy Cross, Westgate, Canterbury, millwright (d.1729), and Margaret (formerly Spencer of Selling or Boughton–under–Blean 7 miles west of Canterbury and 4 from Chartham, d.1753) and whose youngest son, John, was the miller who later insured Barton Mill (1736). In the marriage licence (1708) Peter is described as "of Sittingbourne, paper maker" and in the first Insurance policy for Chartham (1733) as a "paper maker from Sittingbourne". The only Sittingbourne connection in this context is an entry in St. Peter's, Canterbury, PR for the marriage of his daughter, Ann Archer of Milton, Sittingbourne, to James Green in 1726. The only other Archer/Sittingbourne reference is to his brother(?) Robert (q.v.). No records have survived for Sittingbourne itself and the Milton PRs. (searched 1647–53, 1695–1732) and the RBs. (1690–1710) make no mention of Archer (KAO TR 1042/15 and P 253 12/2, 12/3). The Dean connection raises three issues :–

(i) *Interregional Movement* (referred to in Chap.VII)
   In this case more correctly termed connection. Other instances have been cited where the possibility of a Maidstone district/Canterbury connection has been mooted e.g. Willard/Wellard and (mentioned later) Blunden paper makers in both camps. The question here is whether the Deanes of Boxley, roughly contemporary with the Peter Archers' presence there, and the Deanes of Canterbury, who although not paper makers were either millwrights, millers or married to a paper maker, were related. The Archer/Canterbury connection was clearly quite strong; Robert Deane left his daughter £10 for mourning; and his widow, Margaret, left certain property to her daughter after Peter's death; to Ann her grd.–dtr.; and to Elizabeth her gt.grd.–dtr. (PRC 17/94 f.472 1747). No documentary evidence, however, has surfaced yet which shows a positive relationship between the two Deane families; though it may be noted that David Deane of Boxley and Mary Spice of Allington were married at Christchurch, Canterbury in July 1697. At the moment the Archers are the only known common link between Boxley and Canterbury.

(ii) *A reciprocal Influence between Chartham and Barton Mills* discussed Chap.VII p.331.
   Namely, the possibility of Barton mill installing an Engine at an early date because of the family connection with Chartham.

(iii) *Alternative or additional reasons for Peter Archer's move to Chartham.*
   Questioning whether in fact the Engines he introduced at Chartham really were more advanced or more efficiently operated than the one at Sittingbourne, and that the move might have been made for other, family, reasons.

The facts are these. Thomas Frewen converted the fulling mill at Mill Green, Hatfield, to a Brown paper mill in 1663.[65] Frewen was succeeded there sometime prior to 1691 by Isaac Moore, whose daughter, Sarah, married a John Archer in 1726, the latter succeeding Moore at this mill in the same year (Moore d.Nov.1726). The PRs. and the Wills of both John and Sarah throw no light at all on the origins of this John Archer.[66]

There were numerous families of Archer living at Hatfield between the 16th–18th Cs., including two early Peter Archers.[67] The Mormon microfiche for Hertfordshire includes several John Archers baptised during the period ca.1689–1710;[68] but there is no evidence that any of them were connected with papermaking. The only John Archer baptised in Hatfield at this time was the son of Ralph and Sarah in 1704, on the young side for marrying Sarah Moore, who would have been 34 in 1726 (she was baptised Dec.1692).[69] A more feasible match would have been a marriage between the John Archer of Boxley, who would have been 37 then (and 75 at the time of the Hatfield papermaker's death in 1764), and Sarah Moore. The Boxley candidate would undoubtedly have had the added advantage of papermaking experience. It is also conceivable that his father might have come from Hatfield in the first place; this suggestion will be discussed later. The Kent and the Hertfordshire Archers may have been related in some other way ? But the absence of Sittingbourne records prevents us from seeing whether the Boxley John Archer ever moved there or had even been buried there. So the assignment to the Hatfield mill remains unresolved.[70]

RICHARD ARCHER SNR'S FAMILY

JOHN ARCHER   (b.Dartford 1704–bur.Darenth 1741)

John Archer was the eldest son, so far as we know, of Richard Snr.[71] Apart from the dates of his baptism and burial only two other facts have been discovered about him. In 1717 John s.o. Richard Archer of Dartford, Kent, *Gentleman*, was apprenticed to Charles Walkden, Citizen and Stationer (presumably of London).[72] It appears he did not take up the Livery, but evidently continued to live in London. He was buried with all the other members of the

---

[65] Hatfield WEA Publ. "Families and Trades" (1960's) p.69.

[66] Inf. Mr. Henry W. Gray of Hatfield, an authority on the families there, to whom I am indebted (letter 27:Sept:1983).

[67] Ibid. Letter 29:Jan:84. The Hatfield PRs. only begin in 1653. A Peter Archer was buried there 24:Oct:1656. He may *perhaps* have been the son of a Yeoman Peter Archer of Hatfield whose Will, dated 1593, was proved in 1605.

[68] Inf. Hertfordshire R.O. (Mrs. Catchpole, letter 9:Aug:1983).

[69] Inf. Henry W. Gray, letter 27:Sept:1983.

[70] As a matter of interest it may be noted that 24 years after John Archer's death, this mill twice rebuilt passed into the hands of Thomas Vallance, father of Catherine who married William Balston in 1806 at Maidstone.

[71] John is the first-born of Richard and Sarah's family, but whether Richard had had a previous marriage or not, is not known. The reason for raising this point is that the Sutton-at-Hone PRs. record the marriage of Thomas Archer to Mary Hinton in 1721, both being described as "of Dartford", when there is no record of any Archers at Dartford other than Richard. This Thomas could quite easily have come from outside the area and settled for a short time in Dartford; he does not necessarily have to be related to Richard. Certainly, there is no further mention of him in the area. There was also an unexplained "Henry" Archer bapt. 1719 in Darenth and buried there in 1720, no parents given.

[72] Soc.Genealogists' Index of Apprentices of Great Britain Book 6 fo.35.

family at Darenth in 1741 and the entry describes him as "John of St. Saviour's, London"– This piece of information might indicate that he was engaged in selling paper for the family in London; the fact that he was buried in Darenth demonstrates that he had at least kept up the family connection and that his basic training had been aimed at selling paper for them. (See also under William Archer, his younger brother, below).

RICHARD ARCHER jnr. (b.Dartford 1706–bur.Darenth 1773)

Little is known about Richard apart from his date of birth, that he married sometime before 1732, in which year he had a daughter baptised in Wilmington;[73] that he may have continued to live in Dartford up to ca.1740;[74] and that his burial entry at Darenth in the year 1773 describes him as aged 68 (which ties in with his birthdate) and that he came from Sutton-at-Hone. (See William Archer below).

WILLIAM ARCHER (2)   (b.Dartford 1721–bur.Darenth 1781)

Richard snr. had two goes at a "William", the first was b.1716 and bur.1719 at Darenth s.o. Mr. Archer of Dartford. The second William can only have been 19 when his father was made bankrupt. His next appearance is at Sutton-at-Hone where there are 3 entries under baptisms, 1758 (William), 1761 (John) and 1762 (Elizabeth), children of William and Lucinda. That they were connected with Richard snr. is made clear from their burial entries at Darenth, William in 1781 (aged 59) and described as "of Southwark" and Lucinda in 1784, widow of William Archer (aged 62) described as of St. Giles in the Fields, Middlesex. From this data one imagines that William and Lucinda had been living perhaps in London, possibly having succeeded John there selling paper after 1741.[75] As suggested in the account of Hawley mill (see Chap.VII) in the course of his work in London he could have discovered a profitable new line in paper, not White but maybe something like Cartridge, and persuaded Mr. Jarrett to add a paper mill to his leather dressing mill in 1758. It was suggested in this account that Richard jnr., who lived only about a mile away, may have been working at Hawley and taken over the papermaking after 1758, until Richard died in 1773 when the Archer connection with this mill ceased, though William back in London again may have continued to serve it.

---

[73] KAO P 397 1/1 the baptism of Elizabeth d.o. Richard Archer; the mother's name not given, but we know it must have been Richard jnr. because Richard snr. had already had a daughter, Elizabeth, baptized in Dartford in 1718 and no corresponding burial at Darenth.

[74] The later Richard Archer baptisms at Dartford present a confusing picture partly because there are no Archer marriage entries in the Register for this period. Following William Archer's baptism in 1721, there is a long gap until one comes to Richard s.o. Richard & Sarah, which obviously refers to the grandson of Richard snr; he was bapt. 1733. But almost immediately following in 1734 is Nicholas s.o. Richard & Diana. Finally in 1740 one has Sarah d.o. Richard & Sarah. It does appear however that Richard jnr. and Anne were domiciled in Dartford up to 1740 (after 1740 other Archers appear on the scene e.g. there are baptisms from 1742 to Stephen & Jane).

[75] Apart from the fact that William was later described as "from Southwark" there are two other reasons for thinking that he worked in London prior to his interest in Hawley mill: (i) he can have had little experience, if any, of papermaking at Dartford No.2 before his father's bankruptcy. In other words it is more likely that he would have worked with his elder brother, John, in London. And (ii) Lucinda would have been aged 36 when she had William at Sutton-at-Hone and 40 when her daughter, Elizabeth, was born. One would have expected William and Lucinda to have had other children earlier than this and there is no record of this happening locally.

ROBERT ARCHER OF SITTINGBOURNE (d.1737)

The only information discovered about Robert is that he was the occupant of Sittingbourne paper mill at the time of his death in 1737, information derived from an inventory that included the contents of the mill.[76] It is not known what relation he was to the other Archers and his presence there throws open the whole question as to why Peter jnr. (and/or Peter snr.) moved to Sittingbourne. He may have been, as suggested, Peter jnr.'s elder brother whom one would have to include with Richard snr., Elizabeth and Mary (?) as children of Peter snr. born during his "missing" period[77]; or he may have been a cousin or other close relation. There is no evidence that he was Peter jnr.'s son. If he was a brother of Peter jnr.'s, then he would have been the senior paper maker at Sittingbourne. As there are no records for Sittingbourne nor of Robert's baptism the point remains unresolved. The papermaking contents of the paper mill have already been discussed in Chapter IV note 25. It may be noted that the inventory also recorded that upstairs in the paper mill (as distinct from the dwelling house) there were musical instruments, a violin, dulcimer, hautboy and flutes.

THE ARCHER GENEALOGY (see Table overleaf)

A very much shortened version of this follows. Initially the searches in some registers covered a much wider range of dates than some later ones simply because so little was known about Peter snr. beyond the fact that he appeared in Boxley in 1684. Once his marriage was discovered it was possible to narrow the ranges. The narrower ranges were not only aimed at discovering Peeter snr.'s baptism but where he might have gone between 1673–80. Other searches were directed at trying to piece together the movements of his descendants; and there are some also that refer to quite different families or to later generations outside the limits of the search brackets.

---

[76] KAO PRC 11/81/166 Nov.1737.
[77] See Chap.II p.79. There would have been time for four children to have been born.

ARCHER GENEALOGY SEARCHES

MAIDSTONE DISTRICT

PARISH	SEARCH BRACKET	ENTRY	EVENT	DETAIL
Maidstone	from 1632			Archers present
		1635	bapt. Susan	d.o. William & Mildred
		1661-76	bapt. x 6	c.o. Benjamin & Mary incl. Richard 1673-9
		1678	bur.	Benjamin
		1664	bur. Isabell	Widow (family unknown)
		1670	bur. Joane	w.o. John (possibly E.Malling)
		1681	bur. Mary	d.o. ye Widow Archer (?)
Boxley	1650-1710	1684 (1:Jan)	bapt. Margaret	d.o. Peter snr. & Elizabeth (corr:date)
		1685 (1:Jul)	bapt. Peter Jnr.	s.o. Peter & Elizabeth
		1689	bapt. John	s.o. Peter & Elizabeth
		1710	bur.	No Archer burials up to 1710
Aylesford	1654-1741	1660	marr.	William to Mary Edmonds
		1661 +	bapt. x 9	c.o. William & Mary
				too late for Peter snr.
				first bapt. Allington no vicar Aylesford
				no Richard or Elizabeth
				name changed to Archell/Archall in 1679
		1702	bur.	William Archall
		1664	marr.	Richard to Frances Kingswood
				Richard described as from Burham (? q.v.)
		1664-70	bapt. x 2	changed name in 1679
		1672 (?)	remarried	Anne Pearson
		1673-9	bapt. x 4	
		1673	bapt./bur.Richard	(lived 3 weeks)
		1679	bapt. Peter	Archell (no bur. found to 1732)
		1707	bur.	Richard Archel
East Malling	from 1570	1628	bur. John	possibly h.o Joane (Maidstone q.v.) ?
		1640	bur. Margaret	possibly a dtr. ? or widow ?
Burham	from 1626			No Archers
Allington	from 1630		bapt. William	No Archers (cf.Aylesford)
Ditton	from 1663			No Archers
Barming	from 1541			No Archers
East Farleigh	1650-87			No Archers
Loose	1660-83			No Archers
Otham	-			Registers too damaged to inspect
Bearsted	(transcript)			No Archers
Snodland	from 1559			No Archers
West Malling	-			Starts too late
Wateringbury	-			Starts too late

WEST KENT REGION

PARISH	SEARCH BRACKET	ENTRY	EVENT	DETAIL
Dartford	from 1561	1604-20	bapt. x 6	c.o. John/Johan Archer
				1 son Thomas surviving
		1704-21	bapt. x 8	c.o. Richard snr. & Sarah (see Wilmington)
		1704	bapt. John	
		1706	bapt. Richard jr.	(see Darenth for family burials)
		1716	bapt. William	
		1721	bapt. William	
		1733-40	bapt. x 2 or 3 ?	c.o. Richard jnr. & Anne (see Wilmington)
Wilmington	from 1686	1703	marr.	Richard snr. to Sarah Whitewood
		1732	bapt. Elizabeth	d.o. Richard jnr.
Sutton-at-Hone	from 1607	1721	marr.	Thomas to Mary Hinton (both of Dartford) ?
		1758	bapt. William	s.o. William & Lucinda
		1761	bapt. John	s.o. William & Lucinda
		1762	bapt. Elizabeth	d.o. William & Lucinda
Darenth	from 1678	1713	bur. Sarah	d.o. Mr.Archer of Dartford
		1716	bur. Mary	d.o. Richard Archer "att the paper mills Dartford"
		1716	bur. Elizabeth	
		1719	bur. William	s.o. Mr.Archer of Dartford
		1741	bur. Mr.John	Archer of St.Saviour's, London
		1746	bur. Mr.Richard	Archer of Dartford (snr.)
		1757	bur. Sarah	w.o. Mr.Richard Archer of Dartford
		1766	bur. Sarah	
		1768	bur. Mary	
		1773	bur. Richard	Archer from Sutton-at-Hone aged 68 (jr.)
		1775	bur. John	Archer of Dartford aged 45
		1781	bur. William	Archer from Southwark aged 59
		1784	bur. Lucinda	w.o. William, aged 62 and from St.Giles
				in the Fields, Middlesex

WEST KENT REGION (ctd.)

PARISH	SEARCH BRACKET	ENTRY	EVENT	DETAIL
Eynsford	from 1530			No Archers (a mid-18th C.marr.)
Shoreham	from 1558	1625	marr.	William Archall to Elizabeth (see under)
		1626-39	bapt. x 7	first-born "Marie"; second Alexander
Otford	from 1630	1654-75	bapt. x 4	c.o. Alexander & Jane (see Shoreham)
Wrotham	from 1558	1698-1702	misc.	present but n/a.
Ightham	from 1558	1664	marr.	Guillelmus to Anna Baker n/a.
Plaxtol	from 1648			18th /19th Cs. only
West Peckham	from 1561			n.s.
Tonbridge	-	1649 +	bapts.	Thomas Archer's family
				no Peter (1640-53)

NORTH KENT

Sittingboure	no records	1708	marr.	Peter jnr. (known from marriage licence)
Milton	(transcript)	1647-53		no Peter snr.(no Archers)
		1695-1732		no Peter jnr.(no Archers)

EAST KENT

Dover (St.Mary's)	from 1652	1660-67	bapts.	c.o. Allenn & Susan	
(St.James')				No Archers	
Buckland	1640-53			No Archers	1
River	1640-53			No Archers	1
Canterbury				9 parishes out of 17 examined	
St.Aphege's	from 1558	1638-42	bapt. x 2	c.o. William & Alice (see Cathedral)	
		1666	banns	Margaret	
St.George's	from 1538			No Archers	
St.Paul's	from 1652	1624	marr.	Richard to Elizabeth Tomson	
		1634	bapt. Thomas	s.o. William (see St.Alphege's)	
St.Dunstan's	from 1559			1 or 2 early 17th C.	
Cathedral	from 1564	1629-31	bapts.	c.o. William (see SS.Alphege & Paul)	
St.Peter's	from 1560	1726	marr.	James Green to Ann Archer (of Milton)	
St.Mary Magdalene	from 1560			only Archers mentioned from Deal	
St.Mary Bredin	1650-1690			Archers present : none relevant	2
St.Mary Northgate	1650-1690			No Archers	2
Sturry	1650-1690			No Archers	2.
Chartham	from late 16th C.	1737	bur.	Peter jnr. (corr:date)	
Boughton-under-					
Blean	from late 16th C.			n.s.	
Ospringe	17th C.			n.s.	

SOUTH KENT

Goudhurst	from 16th C.	1673	marr.	Peeter snr. to Elizabeth Beckett	
Cranbrook	(transcripts)			No Archers	
Horsmonden	17th C.			No Archers	
Lamberhurst	17th C.			No Archers	
Marden	from 1648			No Archers	
Benenden	from 1558	1697-1745	misc.	No Archers until 1697	
Brenchley	17th C.	1679	bur. Mary	no indication of age	
Penshurst	from 1558	1628	bur. Robert	s.o. Robert (only entry)	
Chiddingstone	1631-64			No Archers	
	1673-90			No Archers	
Ashurst	(start too late)				
Speldhurst	(start too late)				

Footnotes

1 Search carried out by Mr. Michael Heenan, Genealogist.

2 Searches carried out by Miss A.M. Oakley, Archivist, Cathedral, City & Diocesan R.O. who pointed out (letter 16:Feb:1984) that Barton paper mill probably lay in the parish of St. Mary Northgate. Archer marriages before 1670 included Barnabe to Elenor Pen and Edward to Martha Dunkin in the parish of St. Mary Bredin.

Additional Information

1 No Wills or Inventories have been found for Peter Archer snr. at Canterbury (both Prerogative and Consistory Courts); ditto Rochester; ditto Peculiar of Shoreham.

2 Benjamin Archer of Maidstone (q.v.) is mentioned in "Maidstone Records" Monckton, L. et al. (1926).

3 As mentioned above no Records for Sittingbourne have survived so it has not been included in the Table above. But it is known from other records that Peter Archer jnr. was a paper maker there between 1708-1732; and from an Inventory that Robert Archer, also a paper maker, was at the paper mill when he died in 1737. The contents of this Inventory have been discussed in Chap. IV and earlier in this section under Robert.

2. *THE CASE FOR AN IMMIGRANT ORIGIN*

An inevitable question will be, why should Peter Archer snr.'s origin ever have been considered as foreign when Archer families were two a penny in this country? Moreover, one would expect to find numerous Peters linked with a name as common as this. Indeed the idea never cropped up until an exhaustive search of parishes in the Kent 1°, 2° and 3° search areas had been undertaken, more than 50 parishes, and not a single Peter Archer baptism found in the relevant period (or before it). There was, and still is, the possibility that his name might be found just outside one of these areas; or that his baptism had fallen foul of the Commonwealth which started in 1649, a date within his estimated birthdate bracket.[77]

The notion that Peeter may have had an immigrant origin was first conceived when his marriage to Elizabeth Beckett, 1673, was found at Goudhurst. There are three points to be noted about this discovery (i) there were no Archer families to be found in any of the surrounding parishes and only isolated and very scattered ones further afield; the main concentrations were in Maidstone, Aylesford and Canterbury[78] and in none of these were there any Peter baptisms; (ii) within 7 years of his marriage Peter snr. was the Master Paper Maker (of 2 paper mills) at Goudhurst, following directly in the shoes of Stephen Manktellow who was undoubtedly an immigrant; and (iii) in the Goudhurst records his name is spelt consistently as "Peeter" when, as pointed out earlier, there were other parishioners whose Christian name was spelt normally as "Peter".

Peter had appeared, like the Manktellows, from out of the blue and whereas he could have come from somewhere else in England (to be discussed later) he could equally have arrived as an immigrant, as we believe both Alexander and Stephen Manktellow had done. With such an English sounding name was there anything to justify the idea that Peter Archer might have been a foreigner?

Reaney's Dictionary of British Surnames[79] shows that the names Archer, Larcher and L'Archer probably have a Norman origin and date in this country from 1199. The more recent records of the Huguenot Society of London also contain the names of *Archer*/Arché and Larcher/L'Archer. The registers of the French Church in London (Threadneedle Street), which date from 1600, contain entries of Larcher and Larché, including a Pierre Larcher baptised in 1696, the son of Francois Larcher from the Ardennes.[80] Further to this, the Records of the Huguenot Society also show that a *John Archer* was naturalized in 1656/7 but that nothing more is known about him; and a Jacobus Larcher was among the denizations for 1681/2 (denizations could date years, even decades, after immigration).[81] These citations are merely to indicate that our Peter Archer could easily have been an immigrant and like the Manktellows found refuge in the High Weald without necessarily appearing in a list of

[77] The searches included not only the search for Peter's baptism but for the presence of an Archer family who might have had a Peter but whose baptism had not been recorded.

[78] The results of these searches are set out in the Archer Genealogy (this App.). It may be noted that the Archal/Archer form was used at Shoreham, Otford and Aylesford; and that there were no Archers in the St. Mary Northgate parish, Canterbury, in which Barton mill was situated.

[79] Reaney, P.H. (Routledge, Kegan & Paul, 1976)

[80] Information Huguenot Soc. London.

[81] Huguenot Soc.Publ. Vol.XVIII.

denizations. Indeed, one can envisage him following directly in the footsteps of the Manktellows.

That his family might have come over as paper makers with the Netherlandish refugees who settled in Maidstone in the 16th C. is another possibility, albeit a very slender one; even if they had there is no evidence of any connection between the Archers who were domiciled in Maidstone when Peter might have been baptised, and our Peeter.[82] If these settlers had subsequently moved to Aylesford because Forstal or Ditton might have provided suitable sites for paper mills there is still no evidence of any connection with our Peeter. The (surviving) Parish Registers do not begin before 1654 and even so there is nothing to suggest the presence of a family before William's family, the earliest to be found, which starts with his marriage in 1660;[83] there are no unexplained burials, for instance, that might suggest that other families had been there earlier.

If Peeter Archer was indeed an immigrant, and possibly a man of substance as well, then this would accord with the proposals already made in the text (see Chapter IV), namely, that he had managed the COMPANY'S operations for them at Dartford No.2 (1693–1697/8), and used countermarks. Collectively, a credible case can be made out for this theory.

3. *THE CASE FOR AN ENGLISH ORIGIN*

The provisos referred to above for searches in Kent, such as missing records or the conditions prevailing during the Commonwealth, can be used in a negative way to support the case for an English origin. But even within a wider search spectrum, e.g. the IGI for Kent, Sussex and Surrey and, yet more extensive, Shorter's searches have all failed to reveal a Peter.

Once outside Kent one has the possibility that Peter may have been baptised almost anywhere in the country but, even if examples are discovered, there are certain problems that will accompany them. *First, could we demonstrate any connection between one of these "Peters" and papermaking ? and second, if we could, is there anything to show why a particular papermaking "Peter" should want to move to a district where there were no other contemporary Archers ?*

With searches of this nature with all the deficiencies one meets one has to draw the line somewhere. So in answer to the questions above the literature to date has not come up with *any* Archer connections with the paper industry in this country prior to Peeter's arrival in Goudhurst in 1673. That there were other Peter Archers has already been demonstrated when the subject of a John Archer appearing at Hatfield paper mill was considered earlier (see John

---

[82] There are no 17th C. indications of Peter being connected with the Maidstone Archers, the name Archer (or var.) has not been found in any of the lists of foreign tradesmen in Maidstone e.g. lists of 1585, 1622 or that made in the reign of Charles II; nor for that matter in Faversham, Canterbury (the main centre) and Sandwich (inf. Huguenot Soc. London). Against this it may be said that paper makers usually lived in rural areas and might not have figured in such lists. An even more far-fetched proposal, namely, that Peter's baptism was lost at Maidstone in the Commonwealth and that a chance encounter there with Robert Wilson from Goudhurst might have stimulated his interest in Goudhurst, does not hold water on several counts.

[83] The head of the other family, Richard, first appears in 1664 with his marriage; in the entry he was described as "from Burham", but the Burham PRs. which date back to 1626 show no Archers.

Archer b.Boxley 1689). The investigation showed that a Peter Archer had been buried in Hatfield in 1656. It is unfortunate that the parish registers there do not begin until 1653, so we have no means of knowing whether this Peter had a son bearing his name or not. If we assume that one was baptised between, say, 1640–53, then one could imagine him, perhaps, undergoing training in papermaking at either Hatfield mill (converted 1663) or at an older mill at Sopwell (founded 1649),[84] which would have allowed him time to appear at Goudhurst as a trained paper maker. But there is no evidence to support either of these suppositions. A point that favours this suggestion is that, a Peter born in Hatfield could provide a reason for his son, John, returning there at a later date; but it leaves the awkward question as to why this Peter should have left Hatfield and moved to a remote area in the Weald of Kent ? This sort of move was not unknown; e.g. Thomas Chapman as a young man from Shropshire appearing as a paper maker near Dover in 1638; and one might ask what prompted Richard Roads to leave Goudhurst and settle in Hampshire ?

If one pursues the idea of Peter having an English origin, one can see that following in the footsteps of Manktellow he could have learnt something about White paper manufacture there (which he could not have learnt at Hatfield) and, possibly, the practice of using countermarks. But beyond that the notion that he might have managed the COMPANY'S operations at Dartford is now less easy to accept and without that one wonders from what source did Peter jnr. learn about the Engine and how did Richard come to acquire Dartford No.2 mill ?

Whereas the idea of an English origin for Peter has its attractions it has more drawbacks to it than the proposal that he followed the Manktellows over to this country from abroad. In neither case is there any positive evidence to clinch the matter one way or the other. At the beginning of this section on the Archers it was said that their activities can be seen as a foil to the main theme of this book, so it may not matter a great deal which of these two solutions was the real one; but the idea that Peter snr. had an immigrant origin can be turned to advantage to explain a number of events and facts which are difficult to account for credibly in any other way.

---

[84] Finerty, E.T. Bib.69 (Apr.) 309 states that Sopwell mill (about 5 miles to the West of Hatfield on the River Ver) had been in existence from at least 1649 and belonged to a Robert Sadleir, who leased it to John Philbie in 1660 for 21 years. In 1675 it was leased to John Hammerton; and by 1691 it was described as "a decayed house and paper mill to be rebuilt as a corn mill".

MILLHALL MILL, DITTON (MR TQ 715589 : see also Map 10)

There have been several earlier references to this mill more especially in Chapters II,[85] IV and VII, the first of these being the most important in the present context. It was suggested there that because of the similarity of its site and construction its history might parallel that of its near neighbour, Forstal, in going back to a date prior to 1660, possibly prior to 1640 and might even have been founded in the 16th C. In the event there is no solid documentary evidence of its existence before the early 1680's, a map of 1684[86] that illustrates an established site with a (i) Papermill Farm; (ii) Papermill Pond; and (iii) Papermill Field, the inference being that the papermill had been operating there before this date. What this does not tell one is whether it was a used or disused site at that time. There are names of one or two families connected with papermaking living in the area that would be difficult to associate for all, or even part, of the time with the only other paper mill in the district, Forstal, already occupied. 1663 is the earliest entry in the case of the Ditton parish registers and 1677 for the rates;[87] and in neither of these documents does one find initially the names of any of the paper makers who were active at a later date.

Fuller in his book on "The Watermills of the East Malling Stream"[88] mentions 2 watermills in Ditton (a) Church mill, a grist or corn mill (MR TQ 709582); and (b) a Domesday mill, function unknown, which he believes might have existed on or near the site later occupied by the paper mill. Because this last site lay within the flood plain of the River Medway it may have been abandoned at some stage early in its history, but re-used later; it was less than 7.5m. above m.s.l. compared to Forstal's marginally superior position at ca.7.5m. Fuller has suggested[89] that the paper mill could have been built as a paper mill i.e. not converted from some other use. Because of the low level of the site extensive banking was needed to provide the mill pond with a sufficient head of water for papermaking. It may be that at an early date conditions for manufacturing paper were less demanding, particularly if Brown or Press papers were being made or a rag fermentation process in use, than at a later date when the mill may have turned to the lower qualities of White; even so both mills survived to ca.1840 and Forstal was certainly making White paper before it closed.[90]

[85] See Chap.II under Forstal and Millhall paper mills.
[86] The map is of Court Lodge Farm, A. Walter KAO U 49 P4 dated 1684.
[87] No Poor Ratebook for Ditton has survived for this period. Kent Archives, however, have a document described as the Ditton Church Book (KAO P 118/5 which gives accounts between 1677-1758 kept initially by the Churchwarden, William Richardson; the Book was rebound in the 19th C.). Only a record of the actual amount of the Rate paid was kept; these were based on a rate of sixpence in the pound of the assessment, so that by using a multiplier of 40 one can arrive at the assessments and thus make comparisons with those of other parishes. This method of recording the sums paid was maintained certainly up to the time of William Harris jnr. who occupied the mill in 1724, so one has a double check as it were on the value of the early assessments.
[88] Fuller, M.J. Bib.13
[89] M.J. Fuller verbal communication.
[90] Forstal mill was certainly making White paper of a kind as late as 1832; but nothing is known about the later manufacture at Millhall mill, although to judge from the paper makers who occupied it one would imagine that they could have managed to produce a quality at least equivalent to Forstal's. Both had underlying gravel beds with ample supplies of clean water; but as a cheaper and perhaps more practical alternative in the 17th C. it has been assumed from the lie of the land (but not checked) that spring water was available to both mills. An insufficiency of water power might have been a limiting factor in their later development.

It is clear from the Church Book that Ditton must have had a very small population in the 17th C. In 1677, for instance, only 8 residents were assessed for rates together with ca. 14 Outdwellers (forraioners).[91] In the absence of any reference to the paper mill the small number of entries assumes a special significance in this investigation. By working backwards from a known paper maker such as William Harris jnr. (1st entry 1724) to another known paper maker, Edward Middleton, (1st entry 1695) one comes to assessments prior to this made, for the same value and in the same position as Middleton, against James West, not previously identified as a paper maker in Kent and now believed to have been the Ancient Paper Maker of that name not hitherto located. West's tenure only went back to 1688, the ratable value falling from £12 (1694) to £8. Further back than this one enters speculative territory.

With so few residents it is possible to eliminate a number of them on the grounds that their assessments extend more or less unchanged well beyond the appearance at the paper mill of West and Middleton; in the case of others because the much higher ratable values obviously refer to more substantial property e.g. the corn mill. By this means one can reduce possible candidates for the occupation of the mill to the Richardsons occupying the same position as West, John assessed for £4 from 1682–87 and William £4–£6 between 1677 (book starts)–1681. A further point to note here is that William died in 1681 and John in 1687 clearly pointing to continuity in the succession to the lease of this property.[92]

In view of the low value of the assessments made against the Richardsons questions arise as to whether (a) the paper mill was operating under them as a paper mill; (b) anyone else may have contributed to the financing of it; and (c) what James West and others might have been doing at this time.

Taking these points in the same order, was £4 a rate representative of a paper mill at this time? In the normal way one would think not; but, first, it has to be remembered that Ditton was a very small parish where one might expect the rate to have been low; and, second, though mills like Dartford No.2 paid an average rate of £40, there were *mills* in a much bigger community like Goudhurst only rated at £10. Even the newly converted mills in Maidstone like Lower Tovil were rated at £12 (1685); Upper Tovil £12 (1780); Great Ivy at £16 (1686) and Gurney's mill probably ca. £8, assessed in 2 parishes, (1689).

If this is considered to have been too low a value, apart from drawing the conclusion that the paper mill may not have been operating as such, there is an alternative, namely, was anyone else contributing to this cost ? For example, it was noticed that among the Outdwellers the name Ashedowne occurs, first Abraham (1677) followed by Thomas (1678–1697) assessed at ca. £4–13–0. Thomas was born in East Malling and it has been suggested that he might have been a paper maker at Middle mill (see later under the Succession at the Paper Mills of East Malling), because ca.1696/7 a Thomas Ashdowne appears as a paper maker at Goudhurst

---

[91] The principal Outdwellers included Sir Thomas Twisden Bart., Lady Helen Culpeper and Sir Oliver Boteler. One wonders whether any of these owned the paper mill ? Though there is no evidence that Thomas Duke of Cossington, the first of the Duke family to live in Aylesford, ever owned a mill, he married Elizabeth Culpeper from Wakehurst in Sussex. Could the Culpepers of Preston Hall also in Aylesford have owned Millhall mill ? On the other hand among the residents there was a John Tomlin who might conceivably have had an interest in the mill (for discussion see note 93 opposite).

[92] More details relating to the Richardsons and West follow this account.

and simultaneously one finds the Ditton Outdweller assessment ceases. Can one read into this the fact that the Ashedownes had a papermaking interest in Millhall mill ? Certainly, one finds that the assessment made against Middleton, the resident paper maker, of £12 in 1796 rises by £4 to £16 after this. Although no documents have survived to confirm this suggestion, a case can be put forward to support the idea that there might have been some form of partnership between the Richardsons and the Ashedownes; they may have been related, for instance.[93]

Finally, the third question, were the Richardsons paper makers ? or did they employ others to make their paper for them ? The answers are not known. It is possible to suggest a number of permutations of the situation to allow for both of these; but whether the Richardsons actually made paper or not but farmed the land around is not the important issue here. The main interest is in trying to account for James West's activities in the neighbourhood between 1668 to the time he became occupant of the paper mill in 1688.

The proposals and arguments for and against him working as a papermaker for the Richardsons are discussed in some detail in the notes that follow this section. It has been suggested there that West may have come from the West of London trained or partially trained as a paper maker to Aylesford via Goudhurst. It seems unlikely in the circumstances that prevailed that his services would have been needed for long, if at all, at Forstal with the newly arrived Willard there. But something must have attracted him to the area and papermaking at Millhall mill is as good a reason as any. Moreover, at a later date, there are features that suggest a fairly close working relationship between John Richardson and West, the latter stepping immediately into the former's shoes and possibly acquiring Richardson property in Aylesford when John died.

Out of the early Ditton residents there would seem to be only a John Smith who might qualify as an assistant paper maker at the mill really on the grounds that a James Smith, paper maker, was buried at Snodland (the next parish downstream on the left bank of the River Medway) in 1705. In the absence of any firm evidence for a paper mill there before 1755 it has been suggested that James might have been a paper maker, living perhaps with John, working at Millhall mill. The ratable value of his property is in the same category as William

---

[93] This subject has not been investigated here, but it may be noted (i) there had been a Taylor/Richardson marriage in Boxley (1642); (ii) there were a lot of Middletons in East Malling and Middletons figure in both Millhall and Middle mill (E.Malling) paper mills; (iii) as a later example there was a Tomlyn/Gowlett partnership at Middle mill (see under Succession there). In other words there could have been relationships and partnerships between the 2 parishes in other mill or paper mill enterprises. Tomlyns feature prominently in both parishes. It is not beyond the bounds of possibility, therefore, that the Ashedownes and Richardsons had a common papermaking interest on a family basis ?

Richardson (£4) and at least two others.[94] The assessments against him drop in value to £1–6–8 in 1682, remaining at approximately that level and disappearing ca.1693/4.[95]

Summing up, the succession proposed here is as follows:–

Prior to 1677	Not known	possibly William Richardson, Smith & West?
1677–1681	William Richardson	assisted by Smith & West
1682–1687	John Richardson	ditto
1688–1694	James West	
1695–1724	Edward Middleton	(Thomas Ashedown gone by 1696/7)
1724–1741	William Harris jnr.	

Although there were at least two Tomlins resident in the parish and others as Outdwellers, there is nothing to suggest that they were involved with the paper mill, more probably to judge from the size of the assessment made against them (£26–13–4) with some other business, perhaps the corn mill.[96] The notion that the Culpeper family at Preston Hall might have been the proprietors of the mill has not been investigated.

---

[94] The first list of resident ratepayers in the Ditton Church Book (1677) is as follows:

James Garfar	£26–13–4	(d.1678);	Thomas Hoad (1682);
Thomas Miller	£13–6–8		John Miller snr. & jnr. (1682) combined £6. Still there 1693.
Jasper Boorman	£14		£6 in 1693.
John Tomlin	£26–13–4		£40 in 1693. Continues well into Middleton's time.
John Smith	£4	(d.1703)	£2 in 1693.
Nicholas Hammond	£6		Replaced by James Gilbert (1682) ?
William Richardson	£4 (£6 in 1678)		Replaced by John Richardson at £4 in 1682 : replaced by West at £8 in 1688.
John Austin	£3–13–4	(d.1679)	

[95] John s.o. John Smith bapt. 1665.
John Smith bur. 1703.
There was also a John Smith of Boxley bur. 1663.

[96] There were many Tomlin entries in the Ditton PRs. A William married in 1680; John Tomlin seems to have married again in 1689 (The Goodwife of John Tomlin having died in 1679); and another, Sarah, wife of John Tomlin of this parish was buried in East Malling in 1676 etc. East Malling was seething with Tomlins but, curiously, Roger Tomlyn's baptism not found there; Roger was the proprietor of Middle mill (see The Succession at the Paper Mills of East Malling).

THE EARLY SUCCESSION AT MILLHALL PAPER MILL, DITTON

Because the early history of this mill is shrouded in the mists of uncertainty so too are the names and origins of the people who may have worked it in these times. One cannot be certain, for instance, whether the Richardsons were among the early paper makers there or not; or whether they leased the mill and James West operated it for them. The later occupants are covered elsewhere in this work and in the literature.[97] The limited information relating to the earlier ones is given below.

THE RICHARDSONS

The Ditton PRs do not start until 1663 so that one cannot trace their parentage in this parish, but it seems likely that they were born in the next–door parish of Allington, William (1649) and John (1651) ss.o. John and Frances.[98] The next reference to them, which fits in well with the above, is to be found under Ditton;–

1677	bur.	Joane w.o. William Richardson	
1681	bur.	William Richardson	(this ties in with the ratebook entries)
1683–6	bapt.x3	Mary, John and William c.o. of John and Mary Richardson	
1687	bur.	John Richardson	(this ties in with the first James West RB. entry in 1688)
1694	bur.	Mary Richardson	

The Registers of 25 other Kent parishes were searched (1°, 2° and 3°) but produced no information on or connections with the Allington parents. There were early 17th C. Richardsons at Dartford and a 1625 marriage, John to Margaret Combdenn, at Goudhurst. One can only speculate as to whether this John was connected in any way with the Allington ones.

To judge from the records these Richardsons came of local stock and may never have left the district. If one assumes that they were papermakers, they would have had to learn their papermaking locally i.e. at Forstal or a conjectural early Millhall or possibly an East Malling paper mill. On the other hand they may not have been paper makers but employed James West (see below) to operate the paper mill.[99] In this context it may be noted that William also appears as an Outdweller in the Aylesford Churchwarden's accounts in 1680 (assessed for £1–10), in other words he either owned or leased other property; he was also Churchwarden for Ditton.

JAMES WEST (ANCIENT PAPER MAKER)

To understand the problems surrounding the documentary evidence connected with James West it is an advantage to have an idea of the local geography. Map 10 shows that both "Millhall" and "Preston Hall" (to the East) lie within the parish of Aylesford but on the opposite bank of the river to the village. The parish forms a deep wedge on the south bank of the River Medway separating Allington to the East from Ditton to the West. Millhall (or Millale as it was sometimes described) was a small settlement in the parish of Aylesford only a stone's throw from Millhall paper

---

[97] The early history of the mill has been discussed in Chap.II. The succession after James West and the mill's further development under the Harris family in Chap.IV and in App.II under the Harris family genealogy.

[98] That there were other Richardsons in the area who may have had contacts with papermaking might be seen in (a) an Alexander Richardson of Hollingbourne who married Deborah Howman of Marden *in Allington* in 1651; and (b) a John Richardson mentioned in connection with land adjoining Turkey mill in 1658 (Spain R.J., Bib.84). (Apropos of Alexander/Deborah cf. Willard/Farbie marriage at Allington 1651).
Another remote possibility of an indirect connection with papermaking is the marriage of a Peter Taylor and Susanna Richardson in Boxley in 1642 (Richardsons were present in the parish at least as far back as the 1630's).

[99] The Smith(s) were tentatively suggested as early occupants of Millhall mill in Table III (see Chap.II) but on no other basis than a Smith resident in Ditton and another buried in Snodland as a "paperman" (see this App.).

mill in the parish of Ditton. In other words James West, who appears in the Aylesford parish records over a period of many years, may have, indeed probably, lived much closer to Millhall mill than to Forstal.

James West's birthplace and birthdate have not been discovered. The first fact known about him is his marriage:-

| 1668 | marr. | James West to Margarett Austen (in Aylesford) |

The Registers for Aylesford start (1654); Ditton (1663); it is not possible therefore to check whether James was born in either of these parishes. The Aylesford RB starts earlier, but with no reference to James before 1673. As no West families appear among the early entries in any of these records, it is clear that he must have come there from outside. As to the Austens/Austins since Margaret was married in Aylesford, though found in both parishes, an Aylesford Austen is favoured here.

1669–85	10 of their children	bapt. Aylesford
1689	a son John	bapt. Ditton
1694	both parents	bur. Aylesford

The baptismal evidence presented above indicates that once he had arrived he never left the parish again except to move into Ditton where the Church Book shows that for the last 6 years of his life he occupied the paper mill, accounting for only 6 out of the 26 years he was in the area. Before considering the various permutations that might account for his activities during the first 20 years, RB evidence for Aylesford is set out below, but this does not show the continuity one might expect from the regular nature of the baptisms.[100]

1673	£1–10	No previous mention of West.[101]
1675	do.	
1679–82		No West entries.
1684	£1–6–8	
(1688–94)	£1–10	An Aylesford Outdweller assessment

The last Outdweller entry covers the period when he was working in the paper mill and coupled with John's baptism shows that he was domiciled in Ditton. What then did this Aylesford assessment represent?

Aylesford Austen/Austin entries occur before 1668 and continue until after West's death; but this is not the case in Ditton (Church Book) where John Austin entries are shown from the start of the book (1667) up to his death in 1679. The only other data which may or may not be significant is the fact that William Richardson (see above) was assessed as an Outdweller in the Aylesford RB in 1680 for £1–10.

Nothing conclusive can be deduced from this data; much depends on the interpretation placed on the status of James West as a papermaker (see below). For example, if he arrived as a trained paper maker, then one might see him working for the Richardsons at Millhall mill and that they found accommodation for him some of the time. One might postulate that he lived with his in-laws initially, but by 1674 there had been 5 births and 2 burials and James may have been forced to live in separate accommodation which belonged to the Richardsons (certainly later they were assessed as Outdwellers for property in Aylesford). A Boxley option (no records) though not impossible is a less likely one. Neither proposal precludes James from working in either Forstal or Millhall mills, though the latter on balance is favoured. The period 1679–88 is puzzling. John Austin of Ditton died that year and one might assume James moving from Millale Ward into his property subsidized by the Richardsons. That there seems to have been some link between James and the Richardsons is indicated (a) by the fact that the former stepped directly into the latter's shoes when John died; and (b) John's widow who survived until 1694 can be seen to occupy what was now

---

[100] The Aylesford Churchwardens' Accounts (KAO P 12 5/1) were discovered in a dilapidated state in the Guildhall, Rochester and rebound. Although the claim is made that they date from the early part of the 17th C. the first complete assessment is for the year 1657; only fragments precede this. Even after 1657 the assessments do not follow a regular pattern. For instance, the next ones are 1661, 1668, 1669, 1670 1672 etc.

[101] As suggested in the text it is conceivable that West, if he had been working at Forstal mill, might have started married life in Boxley, but we have no means of discovering this. With Willard installed at this mill this proposal would only have an element of credibility, if one assumes that West had arrived untrained and was learning his trade there.

West property in Aylesford from 1688–94. The intention of this hypothesis is not so much one of concern with James' housing problem but rather to lend support to the proposal that he arrived in Aylesford as a trained paper maker.

The important question then is, was James trained or untrained when he arrived in Aylesford? Both cases are feasible. He could have arrived from elsewhere in search of work and trained under Taylor, and later, Willard at Forstal. The mill was just over half a mile from the centre of the village. Equally, if Millhall mill existed then (and the continuity of the Church Book entries indicate that it may have been) he could have learnt his papermaking under a Richardson/Ashedowne association, possibly including a Smith "paperman".[102] There is nothing positive to rule either solution out. In support of the former one might argue that as more paper makers started arriving in Boxley ca. 1680, James felt that better opportunities were open to him at Ditton and moved accordingly.

The alternative is to see him at an age of at least 21 arriving as a trained paper maker, in which case Taylor might not have wanted his services, whereas the Richardsons, who may not have been experienced, perhaps merely leasing the property, engaged James to make paper for them. This proposal leads to the questions, where had he trained ? and why should he have gone to Aylesford?

There can be little doubt that for the last 6 years of his life James was a paper maker. As one of the Ancient Paper Makers was a "James West" (untraced as yet), it is considered legitimate to assume that he was the missing APM. To make him eligible (see Russell, this Appendix) from what has been said he must either have had ca. 20 years of papermaking in his own right; or, failing this, he came from a family with a long tradition in this trade. If previously trained, his 26 years in the district would have qualified him for this distinction. Over and above this, is there any other evidence to support this claim?

Reverting to the questions asked above, as there appear to have been no contemporary West families in that area[103], one has to search for an origin and for a papermaking district where he might have learnt his trade. Without going into all the possible permutations that this situation creates two proposals that would satisfy all of these questions emerge.

First, there were 4 West families in Goudhurst (none in Marden or Horsmonden).[104] No James was found there and one must postulate that either his baptism was lost in the Commonwealth or that he may have been a relation sent there to learn his trade or, if already trained, find work there in the paper mill. Whichever proposal is considered, finding a surfeit of paper makers there, he could have followed Thomas Taylor's example (see Taylor) and moved to Aylesford, where there was at least one paper mill, a state of affairs that may have prompted Thomas Willard to move a couple of years later.

Second, and perhaps more important, there were two other West Ancient Paper Makers, John of Horton Mill, Buckinghamshire and Richard of Poyle Mill, Middlesex and these Wests had a long family background in papermaking (see TABLE overleaf). It is unfortunate that no "James Wests" baptisms have been found in these parishes. One has to resort to the notion that he might have belonged to another West family, not in one of these parishes, but related, and found work in one or other of these mills. If not a direct descendant of one of the earlier Wests, he may have gone in search of a papermaking job elsewhere. The Table shows that the Wests had widespread interests. In theory James could have gone direct to Aylesford, but this course of action is most unlikely with no other Wests in that area. A more plausible solution is to suggest West relations in Goudhurst, which can be seen as a staging post in his journey to Aylesford.

If one accepts that the Aylesford "James West" was the Ancient Paper Maker, then the solution favoured here is that he came to this district trained; that he was related in some way to the other two APMs and to explain his movements

---

[102] Shorter A., Bib.12 184.

[103] Twenty-four parishes in Kent were searched for West families (in a wider selection e.g. Willard and Archer, Wests not consciously seen); 8 of these were in the Maidstone area, one of these with an early marriage entry in East Malling, 1613, John West to Mary Cusham; 10 in West Kent; 16th C. Wests in Eynsford peter out in the 1590's; another family 1637–50 in Ightham but no James; 6 plus in South Kent with no Wests in Marden or Horsmonden but with isolated ones at Benenden, Penshurst and, of course, a large colony at Goudhurst but with no James.

[104] The 4 families at Goudhurst were Peter in the 1620's; John and Bridget in the 1630's; and a Peeter and William in the 1640's/50's. In 1645 Peeter had a son, John, and in 1651 William had a son, Peeter (other Peters, except Archer, were spelt normally as "Peter").

THE PAPERMAKING WESTS OF THE WEST LONDON REGION

Information extracted from Shorter, A. Bib.12 with page
numbers given in the Table below and further information
added by the author.

Bib.12 p.No.	DATE	PAPER MILL	COUNTY	PAPER MAKERS		
123	1631	Sutton Courtenay	Berkshire	s.o. Richard West, paperman bapt. Later this was Rice Watkins' mill making White paper.		
146	1649	Horton mill	Buckinghamshire	Richard West. paper maker bur. 1657 (Will proved)		
	1663			marr. Thomas West, paper maker.		
	1680			Timothy West (Will proved)		
	1687			JOHN WEST, ANCIENT PAPER MAKER	*	
214	1636	Poyle mill	Middlesex	Shorter has suggested that this mill was probably held by Richard West (Author: presumably the Richard that died in 1657, see Horton mill above).		
	1684			RICHARD WEST, ANCIENT PAPER MAKER bur. 1690 (but not clear whether at Poyle or Wraysbury)	*	1
240	1658	Eashing mill	Surrey	Corn mills conveyed to William West, paper maker of Wraysbury, Buckinghamshire. (Only reference to him).		
170	1698/9	Bramshott mill	Hampshire	bur. George West from ye Paper mill (in other words active at Bramshott before this date).		

GEOGRAPHY   Sutton Courtenay (Berkshire) is about 7 miles SSW of Oxford and about 35 miles WNW of Horton.

Horton, Wraysbury and Poyle form a triangle between Staines and Slough on the Buckinghamshire/ Middlesex border; they are ca.1–2 miles apart.

Eashing (Surrey) is about 7–8 miles from Bramshott (Hampshire) and about 18–20 miles SSW of Horton.

FOOTNOTE
1.   The County Archivist (Mr. H.A. Hanley), Buckinghamshire R.O. kindly pointed out (letter 6:Oct:1983) that Horton, rather than Wraysbury (Shorter's main source), may have been the principle domicile of the West family.

The Horton PRs. start in 1571 and show that Edmund s.o. *Richard West* was baptised in Jan.1641 (corr:date); and in 1643 Henry, another son, was baptised and buried there, these two demonstrating that Richard West was in Horton for part of the search period (1638–1650). The search (kindly carried out by the Rev. S.J. Newell, Vicar, 10:Mar:1984) failed to reveal any sign of a James West.

The Wraysbury PRs. only begin in 1734, but Bishops' Transcripts have survived for an earlier period, in particular 1634, but the important years 1641–47 are missing for James' estimated baptism bracket 1638–49). No James West found here.

he travelled via Goudhurst. There is some support for the theory that once in Aylesford he worked at Millhall mill; was Master Paper Maker there from at least 1688–94, the date of his death;[105] was succeeded there by Edward Middleton (see below); and finally, had been familiar enough with the Richardsons to have acquired or leased property that had belonged to them.

EDWARD MIDDLETON

Strictly speaking Edward Middleton is not one of the early paper makers of the Maidstone district; he is a borderline case. The evidence for his presence at Millhall mill is to be found in Mr. M.J. Fuller's work (Bib.13); the following additional information is given here because it throws a little more light on relationships in this area:–

(i)   Although there were numerous Middletons living in East Malling Edward's baptism has not been found there. As he does not appear in either the Ditton or the Aylesford PRs. or Church accounts before 1695, one must conclude that he came from outside. It seems likely that his apprenticeship was served during the 1680's and could have been completed within the Maidstone district.

(ii)   Information from the Ditton PRs.:–

1695	bapt.	William s.o.	Edward Middleton	(This was almost certainly the future paper maker at East Malling)
1705	bur.	Edward s.o.	Edward Middleton	
1706	bur.	Mary d.o.	do.	
1738	bur.	Mary Middleton		
1748	bur.	William Middleton		
1749	bur.	Edward Middleton		(for further discussion see "The Succession at the Paper Mills in East Malling" in the pages that follow)

THOMAS ASHEDOWNE (see also this Appendix p.131)

(i)   1658   bapt.   Thomas s.o.   Abraham Ashdowne (East Malling, Kent)

(ii)   Abraham Ashdowne appears as an Outdweller in the Ditton Church Book for 1677; and Thomas Ashdowne from 1688–1697. Their suggested role in Ditton has been discussed under "Millhall Mill"; and at Goudhurst earlier in this Appendix.

---

[105] For a possible relationship between the Thomas Wests at Conyer mill, East Kent, and James West see Chap VII p.326.

THE SUCCESSION AT THE PAPER MILLS IN EAST MALLING, KENT

A recent account of the Watermills on the East Malling Stream was published by M.J. Fuller in 1973 (Bib.13). Earlier and less complete accounts of the paper mills are given by Shorter (Bib.12 185 and Bib.7[Mar] 62/3) in 1957/60 respectively. None of these accounts gives an adequate description of the succession of paper makers in the three paper mills there nor a definitive allocation of Excise Numbers to them. As a consequence a detailed examination of both the Churchwardens' Receipts (KAO P 242 12/1–4) and Assessments and Disbursements (KAO p 242 5/1–2) has been undertaken to sort this matter out. 12/1 starts in 1728 and 5/1 in 1752. By combining the relevant entries in these pieces with extracts from the Land Tax Returns from 1780 (KAO Q/RP1/245) a picture of the Succession at the three paper mills emerges, Middle (Ex.No.315), Upper (Ex.No.314) and Lower (Ex.No.313); all three paper mills were found to come under the "Upper Division" of the parish (the "Lower" bordering on the parish of Ditton with its Millhall paper mill discussed earlier). The results of this search are summarised below (details will be found in working papers). In places it has been necessary to make certain assumptions to explain, for example, the presence of a paper maker in the parish not obviously related at that time to any one of the paper mills. This applies particularly to the first appearance of George Blunden and, later, John Larking. In the context of the subject of this book it is not thought that any of the East Malling mills have any significance before the post–1740 developments. The possibility that there might have been a reverse movement from this district, in the form of Thomas Ashedowne, to Goudhurst at the end of the 17th C. is of interest.

The Succession at Middle Mill

Middle mill was the oldest of the 3 paper mills and the succession there is mostly straightforward. What has not been discovered yet is how far back the existence of this paper mill extends. The first documentary reference to it is in Roger Tomlyn's Will of 1704,[106] a reference that clearly implies that the paper mill existed before that date. It has been tentatively suggested here that Thomas Ashedowne (bapt. East Malling 1658–bur.possibly in Goudhurst 1709 q.v.) might have been the occupant of Tomlyn's mill during the last decade of the 17th C. but not later than 1696.[107] Whether his father, Abraham, had been there before him is an even more conjectural suggestion.

If Thomas Ashedowne had in fact worked this mill up to 1696, a gap of at least 20 years follows before the earliest date when this mill could have been occupied by William Middleton (1695–1748 bapt. and bur. Ditton). An advertisement shows that he was there in 1723[108] possibly taking the mill over from another Middleton or relative).[109] The ratebook, which starts in 1728, confirms his presence at the mill with half–yearly assessments of £15 to £17-10, comparable values to Forstal and Cobtree mills); the mill remained Tomlyn property. In 1741 Middleton was succeeded first by his wife and then by Mr. (Richard) Gowlett.[110] From then on the succession is seemingly without complications, the assessment appearing to double (as for all other paper mills) in 1757. Later, the property looks as if it had been shared between James Tomlyn and Mrs. Mary Gowlett, Gowlett's widow and possibly

[106]   Fuller, M.J. Bib.13 47.
[107]   Other references see this Appendix.
[108]   Shorter, A. Bib.12 185.
[109]   There were a number of Middletons living in East Malling in the late 17th/early 18th Cs. Edward Middleton, William's father, occupant of the nearby Millhall mill, Ditton, 1695–1724, after retirement was assessed for some small property including a field (£2 + £1) in East Malling, late 1720's–1730's; buried in Ditton a year after his son, and probably related to the East Malling Middletons, who may have had a papermaking relation preceding William at Middle mill.
[110]   Richard Gowlett's origin not known (not E. Malling, Ditton, Aylesford or Allington; Boxley not searched). He was married to Mary and had a daughter 1741, Mary, and a son, Richard, 1745. He took apprentices 1755; and insured a paper mill in 1770 (SFIP 280432 23:Jan:1770).

a relation of the Tomlyn's (?), (1779-1797)[111] after which it was acquired by George Blunden (1798), remaining in the hands of his descendants well into the 19th C. (See summary below).

The Succession at Upper Mill

The succession at this mill, the only White mill, is much more complicated than in the other two. The first documentary evidence as a paper mill is an insurance policy taken out by James Brooks, papermaker, who insured his utensils and stock in *his* paper mill on 12:Nov:1763 (SFIP 204448; no mention of any tenant). Fuller has reported that a mill is marked on this site in a map of 1706[112] where it is shown simply as "a mill" in contrast to Middle mill marked as a "paper mill". Mr. Brooks (possibly John Brooks) had appeared in the ratebooks by 1738 (if not earlier) assessed for £6 (half yearly).[113] By 1743 this ratable value had risen dramatically to £59-10 (£61 in 1751) and it remained at these high values, slightly increased in the years following, up to 1786. It is noteworthy then that there was no increase in 1763 when James insured his paper mill. 1763 is the year in which both James Brooks[114] and George Blunden make their appearance, the latter being assessed at £15 (a figure that might be regarded as a paper mill value) replacing a George Nokes in the same position and at the same ratable value.[115]

James Brooks continued to be assessed at a high figure, with an additional entry in 1790 "£20 more for Mr. Thornton's", and these remain until ca.1803/4 when shortly after his name disappears from the records (gone by 1806); but James was evidently regarded as a paper maker as late as 1803 when he attended a meeting of paper makers at the George & Vulture (London).

The point of mentioning these facts is to explain the presence of George Blunden, assessed at £15 for the 16 years following 1763 and ask the question was he working for James Brooks at Upper Mill during this period ? and, if so, did George Nokes, his predecessor there as far back as 1745 (2 years after John Brooks' dramatic increase in ratable value), also work a paper mill for the Brooks ? In addition to this there is the question, was Blunden replaced by John Larking who made his first appearance in the ratebooks in 1779/80 at a time when George Blunden moved to new premises, Tomlyn property, which proved to possess, among other things, the new paper mill, Lower mill ? Larking, when he arrived, was assessed at £20 initially, but in 1783 this was increased substantially to £73, falling again in 1789 to £43, an assessment that was thereafter described as "his own Estate". As noted above, after this date (1789) James Brooks had to absorb an additional "£20 for Mr. Thornton's". Both Blunden and Larking were nominally, if not actually, paper makers although there is no record of their apprenticeships.[116]

A solution of these problems stems from the fact that an insurance policy taken out by James Brooks in 1801 describes him as a paper maker *and* a miller insuring a paper mill *and* a corn mill in East Malling (SFIP 713345 2:Jan:1801), again with no mention of a tenant.[117] It is claimed here that the consistent and high assessments made against the Brooks from 1743 referred to their corn mill and associated property and that the paper mill was assessed separately. (Alternatives do not accommodate Larking). Further it is almost certain that none of these properties

---

[111] Gowlett may have learnt his papermaking under William Middleton and married into the Tomlyn family, an event that would explain the joint proprietorship of Middle mill for 18 years.

[112] Map of the Manor of East Malling, I Gostling 1706.

[113] The Brooks, like the Tomlyns, figure in the E. Malling PR.s from the early 17th C. The 2 Brooks families that dominate the PRs. during the 18th C. are the first family of Henry Brooks followed by Robert Brooks. John may have been a son of Henry (?) though there were other Brooks families in the district. James was the s.o. Robert & Elizabeth bapt.1735.

[114] In 1763 James Brooks would have been aged 28, and 68 in 1803. For Blunden see below.

[115] George Blunden takes the place of George Nokes in the RB. (1763) and was assessed at the same value, £15. Nokes' assessments go back to 1745 (halving in value, as all others, at £7-10). This suggests that John Brooks may have started Upper Mill as early as 1745? Brooks' position in the RB does not alter between 1738 (£6)-1743 (£59-10); he is placed above Allchin, a place Nokes occupies in 1745.

[116] George Blunden's origin is uncertain; there were Blundens in the neighbourhood, a George bapt.1729 in Offham (3 miles West of E. Malling) making him 79 in 1808 when succeeded by William; another Blunden m. in Rochester (1729), family in Aylesford (1738-43) and more in E. Malling from 1746, too late for this George; another alternative, paper maker Blundens at Barton mill, Canterbury e.g. Ralph Blunden (d.1750) followed by other Blundens (1754- late 1760's), but no record of a George as an apprentice there.

[117] Guildhall MS. 11937/37 County Department Policy Registers.

belonged to the Brooks. Although lumbered with high ratable values Brooks does not appear in the Land Tax Returns except as an occupant of property owned by Mr. John Thornton and taxed at £8.

We have a situation then where (i) James Brooks was described as a paper maker in 1763 and a paper maker and miller in 1801; (ii) the Brooks could have delegated early papermaking operations (1745-1763) to George Nokes about whom we know nothing; and (1763-1779) to George Blunden who, we know from later evidence, was a paper maker and who took an apprentice in 1798; and (iii) in later papermaking operations (1780-89) the Brooks may have delegated these to John Larking, who for want of any other explanation of his presence must have been engaged in papermaking as soon as he arrived in East Malling, a proposal supported by (a) his replacing Blunden in 1780, but with a slightly higher ratebook assessment against him (£20) and (b) the fact that during his first years he lodged in the Reverend Daniel Hill's house (1780-4) a quite separate charge to the one allegedly for the paper mill.

Though Larking is not described specifically at this stage as a "paper maker", there are a number of facts about his career, some difficult to accommodate, that indicate that this was so. As evidence of his papermaking activity we have (i) his involvement with Whatman in 1791 concerning proposed alterations to the Excise Act;[118] (ii) his experience in the early 1790's with the bleaching of rags;[119] (iii) the production of White paper bearing his countermark from at least 1794-1816 (there is an undated countermarked sheet in the Springfield collection that could have been made before 1794); and, finally, (iv) Larking appeared with Brooks and Blunden at a meeting of Master Paper Makers on 13:June:1803.

In contrast to the above one finds that in 1789 Larking appears to have shed the assessment believed to relate to the paper mill and that he was assessed for the next 10 years at a reduced sum of "for his own Estate". At the same time precisely the charges against James Brooks were raised by the sum of "£20 more for Mr. Thornton's". By this date Larking would have been aged 34 and was obviously building a career and estate for himself far beyond the running of a small paper mill;[120] the roles of Larking and Brooks were clearly reversed at this juncture; it was now a case of Larking delegating the papermaking to Brooks. Support is given to this notion by the fact that when Brooks made his last appearance in the ratebooks (1803) he was succeeded in the following years by John Morrice assessed at £26, Morrice either then or later became Larking's partner.

Larking's responsibilities both domestic[121] and in business were increasing rapidly during the 1790's and there is evidence that Morrice was already acting as Agent for Larking in various matters from 1799. With Larking's apparent demise in 1816 (a) the partnership was dissolved and (b) the 1816 Excise General List shows Morrice as occupant of the two East Malling mills, Upper and Lower (see below), but not for long, being replaced at Upper Mill by William Blunden in 1821. The succession at Upper mill thereafter is straightforward, the mill passing into the Busbridges' hands by 1850 and continuing under their name until 1932.

The only problem that remains is the question as to how early Upper mill became White mill. It seems doubtful that it became one that made first class papers before Larking's time. If Blunden was his predecessor and had come from Canterbury then it is possible that he was only capable of producing Brown paper (q.v. Chap.VII). Even if he had come from the Rochester or Offham Blundens, it is doubtful that he would have had any experience of making

---

[118] Shorter, A. Bib.12 185.

[119] Balston, T. Bib.1 104.

[120] John Larking (1755-1838) was during his lifetime a paper maker (in name at least), a timber merchant, banker and in the early 19th C. Deputy lieutenant of Kent and High Sheriff of the County in 1808, clearly a man with wide interests and therefore likely to delegate his papermaking activities to another person. Certainly from 1799 John Morrice was assessed for the first time in the RB. (£26) rising to £167 in 1816-18, but not under Land Tax. 1803 may have been a date when Larking entered into a partnership with Morrice, because a partnership between these two was dissolved in 1816 (London Gazette); but whether James Brooks acted as paper maker for Larking before 1803 or not e.g. 1789-1803 is a point that has not been resolved, but is a likely solution.

[121] The ratable value of Larking's property increased from £43/44 in the 1790's to £290 in 1799/1800. This must reflect the acquisition of land for the park surrounding Clare House which Larking had built in 1793 and where he had 12 children in 16 years by Dorothy d.o. Sir Charles Style Bart. of Wateringbury Place. From 1803 the main assessment stood at £285-10 but with numerous additional ones, 13 for example in 1810 including "Brooks and Blunden £155-10". In 1816 Larking's name ceases to appear and the entry is Alexander Randall in Trust £618 with an additional item of £30; was this the paper mill? The Land Tax merely shows "The Trustees John Larking/themselves £214" for 1817. It is not known what happened to Larking between 1816 and the date of his death, 1838.

Class I papers. On the other hand for James Brooks to have insured his mill in 1763 it may have been for the reason that they made some sorts of White paper there, the rags for and the stocks of White paper being worth much more than the equivalent Brown paper and therefore worth insuring; in addition he may have taken this opportunity, if the mill had an earlier history, to modernise it. The site of the mill was admirably suited to the manufacture of White paper and, on balance, 1763 would seem to be a reasonable date for its inception with Larking raising its status to Class I in the 1780's/90's, the 1790's the most likely.

The Succession at Lower Mill

This is really only of importance in helping to unravel the succession at the other two mills. The first suggestion of Lower mill's existence as a paper mill is George Blunden's move there in 1780. The ratebook shows that his assessment, formerly £15, rose to £79–£83 during the period 1780–99 after which John Larking, in name at least, took the place over. For one year Blunden was his tenant and then he moved to Middle mill, vacated by Mary Gowlett. The Blundens remained there for the next 30 years.

The premises that George Blunden moved to in 1780 seem to have been much more than a Brown paper mill. The ratable value and the Land Tax show no significant change between 1789–1800 and though never insured by Blunden we know that he took an apprentice in 1798 and that he was described as a paper maker. This property, which at that time belonged to Tomlyn, must have been some sort of mill complex probably similar to the Brooks' which contained a paper mill and would explain the high ratable value. Whether Blunden managed it all by himself we do not know, but the evidence of the apprenticeship shows that at least he looked after the papermaking side of it, which was certainly the case later at Middle mill. John Larking took over this complex in 1800, but to whom he delegated the papermaking is not known; Morrice was the occupant in 1816.

THE SUCCESSION SUMMARISED

DATE	LOWER MILL (No. 313)		MIDDLE MILL (No. 315)		UPPER MILL (No. 314)	
17th C.			Thomas Ashedowne ?	but not after 1697		
1723			William Middleton	after 1716		
1738			do.		(John Brooks)	(appears in RB)
1741			Mrs. Middleton			
1741/2			Mr. (Richard) Gowlett			
1745			do.		(George Nokes)	(appears in RB)
1763			do.		James Brooks	insures paper mill
					George Blunden	replaces George Nokes
1779/80	George Blunden	propr. Tomlyn	Mrs. Mary Gowlett	propr. Jas. Tomlyn and Mary Gowlett	John Larking	replaces Geo. Blunden
1798/99	George Blunden	propr. Larking	George Blunden	propr. Blunden	John Larking	in name
1800	John Larking	in name	do.			
1807	do.		William Blunden	propr. Blunden	do.	(Morrice ?)
1816	John Morrice	(Larking Estate)	do.	do.	John Morrice	(Larking Estate)
1821	Robert Tassell	(1821–1838)	do.	do.	William Blunden	
1824	do.		George Blunden[1]		Robert Tassell	(1824–1838)
1836	do.		Robert Tassell		do.	
1848	MILL CLOSED		T. & G. Busbridge		T. & G. Busbridge	
1932			MILL CLOSED		MILL CLOSED	

NOTES TO THE TABLE

1. From this point onwards the Table is an approximation e.g. Shorter, A. Bib.7 reports that in 1829 George Blunden was bankrupt but went into partnership with Francis Collins, eventually Tassell taking over the mill in 1836. In 1838 Henry Smith appears and seems to have partnered Tasell in all 3 mills.

Comment

Purely out of interest a William Manktellow was buried in E. Malling 1778; and in the 1770's Matthias Manktellow appears in the RB. In the Land Tax John Golding appears as propr./occ.; and in 1796 Mr. Clement Taylor as an occupant, £4.

THE GILL FAMILY

JOHN GYLL	of Widdial, Hertfordshire	d.1547	
(7 daughters & 5 sons)			
MICHAEL GYLL	settled in Dartford area in Kent	ca.1570	
(5th.son)			
JOHN GYLL	buried Sutton-at-Hone near Dartford	6:Apr:1624	(1)
	widow also buried at Sutton-at-Hone	d.1626	(2)
JOHN GYLL	baptism not found		
	m. Ursula d.o. ―― Langridge Esq. Sutton-at-Hone	1611	(2)
	John bur.Dartford	1646	(3)
THOMAS GYLL	baptism not found		(4)
	noted as holding various offices in Dartford parish		
	m. Alice (d.1672)		(5)
	Thomas bur. Dartford	1667	
GEORGE GILL	baptism not found		
	m. Susanna d.o. Thomas Cox Esq. of Dartford	1676	(6)
	issue several; referred to in text. almost certainly apprenticed Dartford and obviously before 1676.		
	first documented appearance in Boxley, nr.Maidstone moved to Ditton	1680 1721	
	bur. All Saints', Maidstone	1726	
WILLIAM GILL snr.	baptised Boxley	1686	
	m. Elizabeth d.o. John Laurence Esq. of Kent (niece of James Brooke Esq. of Lewisham, High Sheriff of Kent in 1731; London Stationer & Bookseller; and mortgagee of Turkey mill for a time).	1713	(7)
	issue several, but domicile not found and consequently baptisms; see under Wm.Gill snr.		
	Elizabeth d.1750		
	William snr. bur. All Saints', Maidstone	1754	
WILLIAM GILL jnr.	baptism not found (estimated)	ca.1725	
	apprenticed to James Brooke (see above) 5th.July took up Livery of the Stationers	1737 1747	
	m.(1) Elizabeth d.o. Robert Prowse Hassel (Bucks & Surrey).	1751	(8)
	partner of Thomas Wright (brother-in-law) as Stationers of (1) London Bridge (2) Abchurch Lane	1753-59 1761-98	
	m.(2) Mary d.o. John Broome Esq. of Ludlow.	1773	(8)
	Lord Mayor	1788	
	bur.	1798	

---

1   Not in Sutton-at-Hone PRs.
2   Sutton-at-Hone PRs. (KAO TR 2322/69)
3   Dartford PRs. (KAO TR 1303/2)
4   Not in Dartford or Sutton-at-Hone PRs.
5   Only 1 positive reference to Thomas' family in Dartford PRs. (no George)
         bur. 1664 Thomas s.o. Thomas Gill
    There is another contemporary burial of An Gill 1667 no indication of relationship.
6   Not in Dartford PRs or Sutton-at-Hone (Darenth too late, starts 1678).
7   See Chap.IV pp.163-166, 168-170.
8   From these 2 marriages only 1 son, William (b.1774), survived. By 1794 this son held the rank of Captain (2nd Regt.Life-Guards and Equerry to H.R.H. the Duke of Sussex). His family were born after 1794 and therefore clearly not connected directly with Mr. Gill the Proprietor of Forstal and Cobtree paper mills.

THE GILL FAMILY

The pedigree of the Gill family is illustrated opposite. For John Gyll's antecedents see Burke's "Landed Gentry" (1853 Ed. pp.521-3).

GEORGE GILL (ca.1650's–1726) (ANCIENT PAPER MAKER)

[handwritten annotation: Not so: - Court Baron of 1680 refers to Dorothy Cripps assessed for 2 mills + 7 acres in occupation of Gill]

So far as is known George Gill was the first member of his family to become a paper maker.[122] The entry for the burial of his daughter in 1681 is the earliest evidence discovered indicating him in this occupation.[123] Another entry, 1683, for the burial of his son, George, also describes him as "paperman". As the Table opposite shows he came from Dartford, where, it is believed, he served his apprenticeship.[124]

His first appearance in the Maidstone district is to be found in the Boxley PRs., an entry for the baptism of his daughter, Elizabeth, in 1680.[125] There are no Gill entries prior to this date. Entries continue up to 1693 and these include William (1686 s.o. George and Susanna).

There was no other mill in the Maidstone district, and Boxley in particular, where George might have been a Master Paper Maker other than Turkey mill. The inference is that he must have converted Turkey mill from a fulling mill sometime prior to 1681, by which time he was described as a "paperman" in the area. (It is known from a lease that this mill was converted sometime between 1671–1693[126] and this could therefore have taken place at any time between 1676–1680). George Gill must surely have supervised the conversion himself; the recent construction of Dartford No.2 paper mill (obviously before 1679, when it was first assessed) may have been the inspiration for his move to Boxley coupled with the current incentive given to the manufacture of White paper as a result of the embargo on French goods (1678–85). One might legitimately assume then that the manufacture of White paper was in hand at Turkey mill by 1680 at the latest.

The Gill genealogy has not been pursued here in any depth, but William (b.1686) seems to have been George's only surviving son; and yet in the first 2 decades of the 18th C. three other Gill families may be identified in the PRs. of All Saints', Maidstone; and another in the

---

[122] Keyes, S.K. "Dartford, Historical Notes" (1933, 1938) refers to Thomas Gill as a tailor in the High Street, who made his mark when signing a document. Whether this was George's father or not is not known; no other Thomas Gills are mentioned.

[123] PRs. All Saints', Maidstone (held by Incumbent).

[124] The Petition of the Ancient Paper Makers (1690) implies that they, including George Gill and other paper makers opposing the COMPANY, had all served their apprenticeship in papermaking. There is no documentary evidence of George having served one in Dartford, but the fact that he was born and married there and later became a noted manufacturer of White paper all support the notion that he learnt his trade there. Dunkin, J. (Bib.46 306 footnote) specifically states that George Gill, s.o. Thomas Gill of Dartford (d.1667), was apprenticed to the Spilmans or Mr. Blackwall (sic), but provides no evidence to support this statement. William Blackwell, traditionally held to have been apprenticed to the Spilmans, was the occupant of Dartford No.1 paper mill by 1670 and a Ralph Blackwell, paper maker, there in 1666/7 (Shorter, A. Bib.12 180). The name of William Blackwell appears in the Dartford PRs. as early as 1651. All this lends credence to Dunkin's statement. In any case where else could he have learnt his papermaking ?

[125] Boxley PRs. KAO P 40 1/2.

[126] Spain, R.J. Bib.8 85 (see also in this Appendix under Peter Musgrove).

1720's.[127]   No other Gills have been noticed in the records of the district prior to George's arrival. It is possible that these other Gills came from the Dartford area following in George's wake and were perhaps cousins.[128]   They are mentioned solely because the ownership of Forstal and, later, Cobtree paper mills passed into the hands of a "Mr. and Mrs. Gill" at some unknown point in the 18th C., Forstal (and possibly Cobtree) having belonged formerly to William snr. and later William jnr., George Gill's grandson. From what we know of William jnr.'s family (see Pedigree; *Landed Gentry* 1853) this "Mr. Gill" cannot have been directly related, but was perhaps a descendant of a collateral of George's e.g. one of the Maidstone families referred to.

It is not known where George Gill was domiciled when he first arrived in the Maidstone area. Five baptisms and 1 burial are recorded in the Boxley PRs. between 1680–93;[129] whilst 3 burials are found in the All SS'. Maidstone PRs. all connected with George's family for this period. His own burial is recorded in Maidstone also.[130] Turkey mill lies less than a mile due East from the centre of Maidstone and 2.75 miles South of Boxley Church. Living at Turkey Court might account for his use of two churches for baptisms and burials, though Thomas Balston reckoned that Turkey Court, the house adjoining the mill, may only have been built ca.1688.[131] The occupation of the latter cannot be distinguished from that of the paper mill when the first assessments appear in the ratebook (1691). These assessments (see App.III for Extracts from RB) are almost identical in value to those of Forstal mill for the initial period, a surprising fact when one considers the different size of the two mills; it is unlikely that once converted George Gill made any further major alterations to the supply of water power, though in 1695 he rebuilt the drying lofts and carried out miscellaneous improvements to the value of £200.[132]

Although William snr. seems to have taken over the mill in 1716 (see below) his father and mother probably remained at Turkey Court until 1720 at least, the year Susanna died;[133] the assessments in the Boxley RB remain in George's name up to and including 1721, after which he moved to Ditton. His Will, made in 1725, begins "I, George Gill of Ditten in the County of Kent paper maker.......".[134] George must have gone to live with his only surviving daughter, Ann, and her husband, Thomas Walter, who were resident there.[135] Whether Ditton had any special significance for George is not known; his son had been the owner of

[127] All SS' Maidstone PRs. show a John & Mary Gill from 1707; a Thomas & Frances from 1708; a Robert & Elizabeth from 1717; and John & Anne from 1721.

[128] Dartford PRs. (KAO TR 1303/2) no Gill baptisms 1640–78 : Sutton-at-Hone (TR 2322/69) no Gill in PRs. other than 2 shown in pedigree (1611, 1626) : Darenth PRs. (TR 2322/20) only begin 1678, 11 yrs after Thomas' death. There may have been Gills in Darenth earlier because there is a "Gill's Farm" marked on modern O.S. maps (MR TQ 576698) and this quite probably goes back to the 17th C. since the Darenth PRs. show that a John & Sarah Walter were having a family at "Gill's Farm" from 1715, possibly a connection with Thomas and Ann Walter at Ditton (?) George's son-in-law.

[129] Boxley PRs. KAO P 40 1/2.

[130] All SS', Maidstone, PRs.

[131] Balston, T. Bib.1 11 and Plate IIa.

[132] Ibid. 4.

[133] The only evidence of the date of Susanna's death found is in Dunkin's description of William Gill snr.'s tomb in the churchyard of All SS', Maidstone (Dunkin, J. Bib.46).

[134] KAO PRC 32/60 fo.176 (1725).

[135] Ditton Churchbook KAO P 118/5 shows Thomas Walter among those assessed in 1717 (there are no entries for Gill). The PRs. show that Susan d.o. Thomas Walter was bapt. 1717 and Thomas s.o. Thomas & Ann Walter bapt. 1725 (see also Note in this appendix on the "WALTERS").

nearby Forstal mill since 1716 and there may have been Ancient Paper Maker associations there as well (?), viz. a Walter connection with Court Lodge Farm on which Millhall paper mill was sited.

It has been said that Wills of that period rarely contained any mention of real estate and this certainly seems to have been the case in George's Will. He was only the tenant of Turkey mill, so it is difficult to know whether one should attempt to read anything untoward into the terms since one does not know what his possessions were. Had he quarrelled with his son, for instance ? He left £40 to 2 grand–daughters, Susan and Ann, when they came of age; £30, his silver tankard and books to his grandson George; and to William, his son, 5 shillings only to be paid 12 months and a day after his decease. Everything else was left to his daughter, possibly because her need was greater than William's ?

George Gill was buried at All Saints', Maidstone, in 1726. His contribution to English papermaking and to the manufacture of White paper in particular has been discussed in the text (see Chaps.II & IV). In retrospect he must be considered as one of the most important paper makers of his time; Dr. Harris in his account of the mill evidently thought likewise. He can justifiably be seen then as a significant influence on the next generation of White paper makers in Kent.[136]

WILLIAM GILL snr. (1686–1754) (PAPER MAKER)

William was baptised in 1686 in Boxley,[137] (MAP 2) in whose Weavering Ward lay Turkey mill and in the Sandling Ward Forstal and Cobtree paper mills. He appears to have been the only surviving son of George and Susanna; no other sons are mentioned in his father's Will.[138]

Whereas George Gill seems to have spent his life single–mindedly establishing and perfecting his White paper manufacture at Turkey mill, his son had more ambitious ideas which when put into practice resembled a reversion to the entrepreneurial style of an earlier age. Possibly through his wife, Elizabeth, daughter and co–heir of John Laurence Esq. he may have married himself into a more wealthy circle than that of his father. From 1716 onwards, 3 years after his marriage,[139] he embarked on his extraordinary programme of expansion.

In 1716, at the age of 30, he insured Turkey mill[140] indicating (and supported by other evidence given below) that he had embarked on his papermaking career and had taken over the lease from his father, a lease that had another 20 years to run. In the same year he insured his own paper mill in Sandling, which from the name of the tenant mentioned in the succeeding policy of 1722, John Hillier, may be identified as Forstal mill.[141] By 1716 James

---

[136]   Shorter, A. Bib.9 50, a 1971 view : Harris, J. Bib.4 191, a comment of ca.1716.
[137]   Boxley PRs. KAO P 40 1/2.
[138]   There are 2 burial entries for George s.o. George Gill in the All SS' Maidstone PR in 1684, 1685; and a Thomas buried Boxley 1693 (corr:date).
[139]   Married at St Olave's, Southwark in 1713.
[140]   SFIP 7379 29:Oct:1716.
[141]   SFIP 7477 21:Nov:1716 and SFIP 25877 14:July:1722. In the first policy William Gill is described as a papermaker of Maidstone and in the second as a brewer of Maidstone.

The Estate of William Gill A Bankrupt

The Estate referred to was advertised in the Kentish
Post and Canterbury News Letter for the
8:March:1731 and is reproduced below with the kind
permission of the Kent County Library at
Canterbury.

A D V E R T I S E M E N T S.

To be Lett, forthwith, by the Affignees of the
Eftate of William Gill a Bankrupt,

A Large Brew-houfe with feveral Ale-houfes
thereunto belonging; and alfo a large Paper
Mill with the Appurtenances, together with feveral Acres of Hop-Ground, Orchard and Paf-
ture Land.

Likewife to be Sold, Some Utenfils belonging
to the Brew-houfe, with a large Stock of Beer
and empty Casks, and alfo the Furniture belonging to the Bankrupt's Dwelling Houfe; and a
large Stock of Hop-Poles, with a Parcel of Oats,
Beans, Barley, and Hay, with other Things.
Enquire at the Star-Inn at Maidftone.

There are several points to note about this advertisement:–

(i)   It provides evidence of his Brewing and Agricultural
interest over and above his papermaking involvement in a
"large paper mill", clearly Turkey Mill.

(ii)   The above items were "to be Lett" by the Assignees as
distinct from the others listed in the lower half of the
advertisement, items that were for sale including "the
furniture belonging to the Bankrupt's Dwelling House",
presumably Turkey Court ?

(iii) There is no reference to Forstal mill (discussed in the text)
nor his paper and rag stock in either mill.

(iv)   There is no specific reference to the six cottages that he
had built in 1718, known as "the Square" (a site known to
this day as "Square Hill"). Can these be considered as
among the appurtenances of Turkey mill ?

Brooke, the London Stationer and uncle of his wife,[142] already figures in William's affairs[143] and, indeed, must be seen as a powerful influence behind William's expanding interests. In addition to his extensive investment in papermaking[144] he owned a large brewhouse, pubs, hop gardens, orchards and pasture (see notice of sale opp.), all of which led to overstrained finances and, ultimately, bankruptcy for life.[145]

After his final bankruptcy in 1731 almost nothing is known about William snr. The Boxley ratebook shows that he was assessed for some small property in the Weavering Ward of Boxley between 1728–51, valued at £6 (a not insignificant sum) and later transferred to a John Day.[146] He was still a certified bankrupt as late as 1753, a year before his death and burial at All Saints', Maidstone, in 1754;[147] hardly a bankrupt's tomb. Through Brooke's direct intervention Forstal paper mill (and maybe through indirect influence Cobtree as well) was retained for the family, an additional benevolent act on his part.

William and Elizabeth were married in 1713 and it is reckoned that their son, William, was born ca.1723–25 and that he was probably the youngest of the family of four.[148] One would imagine that after George Gill had moved to Ditton in 1721 William and Elizabeth would occupy Turkey Court; but this is uncertain. Their domicile prior to 1721 has not been discovered nor has it for the period of William's bankruptcy; and moreover it is not known where any of his children were baptised.[149] William jnr., for instance, could have been baptised in Boxley, if his parents had been living in Turkey Court; but his baptism has not been recorded there.

---

[142] James Brooke(s) apart from being a resident of Lewisham and High Sheriff of Kent in 1731 is also known as a Stationer (Stationers' Company Apprentices entry 1158 : Ed. *D.F. McKenzie*, Oxford, 1978); also, the same man, as a Bookseller, Anchor & Crown, London Bridge, 1702–50; died 1750 (Dict. Booksellers and Printers, Bibliog.Soc.). See also Brooke's Will, end of this section on William snr.

[143] In 1716 James Brooke made 4 quarterly payments to William Gill for insuring the goods and merchandise in Turkey mill. (For details see Shorter, A. Bib.12 187).

[144] Subject already covered in (i) Chap.III pp.120–121; and (ii) The Papermaking Empire of William Gill Chap. IV.

[145] The first bankruptcy 14:May:1729; the second 15:Jan:1731.

[146] What was this property in the Weavering Ward ? It cannot have been Turkey mill or Court, but might have been the Square which though conveyed to Harris in 1738 and subsequently to Whatman in 1740 may have been let to Gill accounting for these assessments and might at the same time provide a solution to his domicile following his bankruptcy. After 1743 the ratebook entry shows Gill as an Outdweller, possibly having moved to London to live with his son ? (see also next footnote).

[147] A description of William's tomb and inscription is given in Dunkin, J. Bib.46 306. From this it is learnt that his wife, Elizabeth, died 23:Apr:1750 aged 60. This might explain the cessation of the Boxley assessments in 1751 or contribute to this.
      John Day, as the next occupant, is mentioned merely because his name appears in 1750 in the Apprentices' List as a Maidstone paper maker (Shorter, A. Bib.12 185) and since he was assessed in the Weavering Ward he may have been working at Turkey mill ? (A John Day was assessed in Ditton between 1700–24, possibly an employee at the paper mill and father of the apprentice ?)

[148] This assumption is based on his apprenticeship dates, the date of his indenture being 5:July:1737 (see further under William jnr.)

[149] The Boxley Ratebook entries (see App.III) showed that William snr. replaced his father in 1723, but the baptisms of his family have not been found either in Boxley or Maidstone, Ditton, Aylesford, parishes where one might have expected them to be; the search was extended to Dartford, Darenth (where a William and Mary Gill were having a family in the 1740's/50's), Sutton–at–Hone, Lewisham (Brooke's domicile) and a casual search in many other parishes. It is possible that the children were baptised at his wife's family home (she was co–heir of John Laurence Esq.), but this has not been located.

The rest of the family consisted of two daughters one of whom, Anne, married Thomas Wright in 1746; he was to become her brother's partner as a Wholesale Stationer (see below under William jnr.). William snr. had another son named "Brooke", evidently after his father's benefactor; he died in 1744 without issue and because of this designation he is assumed to have been William jnr.'s elder brother.

The Will of William Gill snr., if ever there was one, has not been found.[150] As mentioned in the comment on the Notice of the Sale published in the Kentish Post Forstal mill does not appear as one of the items on offer. Probably because only the lease of Turkey mill was for sale, Brooke decided that it was in his power to salvage the freehold of Forstal leaving it to William jnr. in his Will (see below) in 1750. In view of William snr.'s lifelong bankruptcy it could scarcely have been passed back into his hands and was thus held in trust for his son.[151] It may or may not be a matter of any significance but in William jnr.'s indenture (1737) his father was still described as "of Maidstone, Papermaker"; this was at a time when critical events were taking place at Turkey mill with Richard Harris having just taken over the tenancy of it and Brooke as chief mortgagee trying to secure the freehold.[152]

ABSTRACT OF THE WILL OF JAMES BROOKE OF LONDON BRIDGE (31:0ct:1750)*

He bequeathed (1)   to his nephews THOMAS WRIGHT and WILLIAM GILL Estate Interests and years to come of my Paper Mill near Maidstone, Kent.
(2)   to WILLIAM GILL for ever my Paper Mill near Boxley, Kent.

(a)   Wright (through marriage) and Gill were Brooke's great–nephews.
(b)   The second item above clearly refers to Forstal mill which William jnr. insured as his own in 1754 (see next Section William jnr.).
(c)   The first item is puzzling; it clearly refers to a lease. The mill was almost certainly Cobtree which William snr. had set up himself in 1717; but why did not Brooke specify near Boxley also? The alternative, but less likely, could be Upper and Leg o' Mutton mills in Loose; William snr. may have had a hand in setting these up and the two young Stationers insured it as their own in 1770.

* Prerogative Court of Canterbury Ref: PROB 11–784. Search carried out on behalf of author by Mr. Simeon Clarke (Member of A.G.& R.A.)

[150]   The Will is not indexed under the Prerogative Court of Canterbury Wills at the PRO (1754–56); nor under PRC in the Kent Archives Office. There is a reference to a William Gill of Maidstone in the Consistory Court Act Books (KAO PRC/22 Book 25), but this refers clearly to another William Gill who was buried at All Saints', Maidstone, in 1738 (corr:date) and not the paper maker (possibly John and Mary's son baptised there in 1707).
[151]   The fate of the Cobtree mill after Edward Rowe's insurance of it in 1731 is discussed below under James Brooke's Will.
[152]   This subject and its implications for Harris and Whatman has been discussed fully in Chap.IV the "Aftermath".

WILLIAM GILL jnr. (ca.1725–1798)   (STATIONER)

Unlike his father William jnr. restored the family fortunes in no uncertain way.

As mentioned it is not known where he was born (possibly in the parish of his mother's home) nor exactly when, but to judge from the date of his indenture ca.1723–25.

He was apprenticed to James Brooke, the London Bookseller and Stationer, his great–uncle, on 5:July:1737 as "William, son of William Gill, Maidstone, Kent Papermaker".[153] He was made free and took up the Livery of the Stationers' Company in 1747; and in 1753 he became the partner of Thomas Wright, his brother–in–law, as a wholesale Stationer at London Bridge (1753–59) and then at 30 Abchurch Lane (1761–98). Up to 1777 they were styled Wright & Gill; from 1778–80 as Wright, Gill and Pettiward; and, later, as Wright, Gill and Dalton.[154] He was Lord Mayor in 1788 (3 years after his partner had been Lord Mayor) and Boyd's "Citizens of London" (p.225) described him as an "illustrious Lord Mayor". He died a very wealthy man (said to be worth £300,000).[155] He married Elizabeth Hassel of Wyrardisbury, Buckinghamshire in 1751 and settled there.[156]

In the context of this work he and/or his partner is of interest on several counts. The elder Whatman must have been acquainted with George Gill, albeit indirectly, and in much closer contact with William snr., and equally James Brooke, during Harris' tenancy of Turkey mill and, later, its acquisition. Though no positive evidence has come to light of Brooke acting as an outlet for Whatman, this seems highly probable; nor is there any direct evidence that when Wright and Gill began trading that the elder Whatman had dealings with them, although once again through their various connections this seems highly probable too. What is beyond question is the fact that the younger Whatman had established a trusting relationship with Wright, Gill & Co. by 1780 as may be seen from Whatman's Ledger (1780–87).[157] Thomas Balston pointed out that not only did Whatman trade with them but used them as bankers as well and paid many of his London accounts through them.[158] This evidence of "trust" suggests that the association had begun much earlier and the younger Whatman had inherited this goodwill from his father's contact with them.[159]

William jnr. as a Stationer had a reciprocal interest in papermaking. This is known from the fact that he insured Forstal mill, originally his father's property, within a few months of his father's death in 1754.[160] It had been held in trust for him by James Brooke who clearly had a considerable protective feeling for both father and son possibly through a special regard for his niece, Elizabeth. The question as to whether there was a similar operation carried out by

[153] McKenzie, D.F. "Stationers' Company Apprentices 1701–1800" Oxford 1978 entry No.1158.
[154] Maxted, Ian, "The London Book Trades 1775–1800" Dawson 1977.
[155] William Gill jnr.'s Will is held at the PRO ref. 1798 PCC 257 Walpole.
[156] Burke's "Landed Gentry" 1853 Edition 522.
[157] PSM.
[158] Balston, T. Bib.1 53.
[159] See Chap.IV p.170 and Chap.VII p.254 to see this subject in a wider context.
[160] Shorter, A. Bib.12 186. William Gill, Stationer of London Bridge, insured his paper mill in the parish of Boxley and his house in the tenure of Edward Russell, paper maker, SFIP 142250 (13:Aug:1754).

Edward Rowe, the Stationer, who insured Cobtree mill, has also been discussed earlier.[161] Eventually it must have passed into William jnr.'s sole possession, like Forstal, since both mills became the property of another branch of the Gill family at some point in the 18th C., but just when and how we do not know.[162] In contrast to what might be regarded as a passive interest in papermaking Wright and Gill took a positive step to acquire further paper mills for themselves in Loose and the first documentary evidence of these acquisitions is an insurance policy of 1770 for the "Old Paper Mill and the New Mill near",[163] identified by Spain as Upper and Leg o' Mutton mills just above the village of Loose.[164] John Farley had been the occupant of these mills since the 1740's and to judge from an insurance policy of 1750 he owned at least one of them.[165] In 1770 Wright and Gill acquired the property for themselves, but whether the Gill family had any prior interest in the "Old" paper mill viz. Upper mill, Loose, is a subject that has been discussed in the text already.[166]

It is not known for how long Wright and Gill retained an interest in this project. Certainly by 1774/5 the younger Whatman took over Upper mill, Loose, and rebuilt it.[167] But the Farley family continued to occupy Leg o' Mutton mill until 1778, it thereafter passing into the hands of the Golden (Golding) family. It might be legitimate to assume that Wright and Gill's interest in this mill ceased somewhere between 1775–78, possibly because they no longer felt able to compete with Whatman's renovation of Upper mill and the consequent increase in his output together with the new advances in the technology of waterwheels and transmission systems.

Just why Wright and Gill undertook this venture is not clear. No doubt the rapid growth of the Domestic White paper industry after 1760 was responsible for this. Perhaps they saw more potential in these two Loose mills, as Whatman was to also, than in Forstal and Cobtree.[168] Whatman may even have dissuaded them from continuing with this in view of their close relationship with him and his far greater expertise in White paper manufacture.

The only other documented evidence of a connection between Whatman and Wright in this context occurred when they were both amongst the signatories to a Petition to the Treasury concerning the collection of Duty by Excise Officers where they were acting on the behalf of the Paper Makers of Great Britain. The trade was finding this Collection a burden increasingly intolerable to bear.[169] This was in 1765 and it is perhaps a measure of the esteem in which a Stationer like Wright was held at that time (though by then he may still have had an interest in a Maidstone mill and could thus be regarded as a paper maker Stationer as were two other signatories, Robert Herbert and John Vowell). The other signatories included two notable paper makers Joseph Portal and Richard Ware.[170]

---

[161] References to Edward Rowe, Stationer, and his insurance of Cobtree mill in 1731, 6 months after William Gill snr.'s bankruptcy, will be found under Cobtree in App.III. Brooke's Will (q.v.) implies that the lease of this mill had passed later into his hands from Rowe.

[162] The documentary references to "Mr. & Mrs. Gill" in the Boxley Land Tax Returns see App.III.

[163] SFIP 292237 19:Dec:1770.

[164] Spain, R.J. Bib.10 47–49. See also Chap.VII; and earlier references to these mills in Chap.IV.

[165] Shorter, A. Bib.12 194 citing SFIP 124147 (28:Dec:1750), "*his* paper mill, drying lofts and storerooms adjoining..."

[166] See Chap.II p.87 and Chap.IV p.154.

[167] Balston, T. Bib.1 Chap.VIII.

[168] See Chap.VI note 96 and Chap.VII p.314.

[169] Coleman, D.C. Bib.15 131/2.

[170] Ibid. 132 note 1.

THE PETER MUSGROVES

The Musgrove family figure in the early papermaking history of Maidstone, engaged in it for nearly 100 years but never progressed beyond the status of minor paper makers. It is an unusual name and Reaney places it mainly in Westmorland[171] and not found locally in the Maidstone district until Peter snr. arrived in Boxley ca.1678 when the spelling changed from Musgrave to Musgrove. The author's searches suggest, however, an origin in the vicinity of Goudhurst or Marden and in view of his later career in papermaking and age (m. before 1678) he may have trained at Goudhurst or, alternatively, he could have been a millwright and learnt his papermaking after his arrival in Boxley. If a millwright, one could see him helping George Gill convert Turkey mill,[172] working for him and learning the manufacture of White paper before setting up on his own in 1686 at Lower Tovil mill (see below). This proposal has certain points in its favour, namely, Lower Tovil was much closer to Turkey mill than Forstal mill, the only other mill in Boxley (1.5 v. 4 miles); and, much later, his son Peter jnr. became the occupant of Turkey mill (1735/6) after William Gill jnr.'s bankruptcy indicating that his family perhaps had associations with this mill earlier.

As the precise date of Turkey mill's conversion is not known and the first documented reference to George Gill is 1680, then since Musgrove was domiciled in Boxley one can only postulate him working for Willard at Forstal as an alternative. This is a less satisfactory solution in that Peter Archer snr. must have taken over this mill only two years later and as a mill it was much smaller and with less scope for employment than Turkey mill.

The baptism of Peter Musgrove snr. estimated 1637–57 has not been found; nor the date and place of his marriage to Sarah. He could have arrived in Boxley with children,[173] but the only records found (Boxley PRs. KAO P 40 1/1, 1/2 and 1/9 covering BMB from 1558–1809) show that by his first marriage 2 sons were baptised there, only Peter jnr. surviving (b.1680). He remarried, at Wrotham, Mary Smith in 1682 and had a William and Mary in Boxley before moving to Lower Tovil mill in 1686. Peter snr. was buried in All SS', Maidstone in Aug. 1701.

There is no direct evidence that Peter snr. was a fully trained paper maker[174] when he arrived in Boxley, but it is clear from his later history that he occupied a paper mill and took on apprentices including his son, Peter, (see below). It is uncertain whether he or Henry Lanes, a fuller, undertook the conversion of Lower Tovil. The ratable value of the premises increased from £4–£12 (a paper mill rating) between 1684–5 while Peter Musgrove only appears as occupant in 1686.[175] If Musgrove had been millwright one would have expected him to have made the conversion, but there is just the hint that Lanes might have tried to make paper himself and failed and that this provided an opportunity for Peter snr. to take over.

PETER MUSGROVE jnr.

The last entry in the Loose RB. against Lower Tovil mill for 1701 was in the name of Peter snr.'s widow.[176] Peter jnr. was apprenticed to his father in 1692 at the age of 12 and he took over the mill and was assessed for it in 1702, remaining there until 1721 when he was succeeded by William Gill, who had an interest in it from 1722–28.[177]

---

[171] Reaney, P.H. "Dictionary of Surnames" (Routledge, Kegan and Paul 1976); he points out that it may have been connected with the Norman Mussegros. It is, however, one of the better known names of the "Border Rievers" and in theory Musgrove could conceivably have come south from one of the very early Lancashire paper mills. But a Kent origin for Peter is more probable (All SS'.Maidstone PRs. show a m. of John Musgrove, a clothier of Goudhurst and Anne Hills of Marden, 1656); the main families appear to have been in Goudhurst and Marden, a few scattered references elsewhere in SW Kent, but no Peter bapt. only an Abraham in Goudhurst (see note 173 below).

[172] Spain, R.J. Bib.8 85 considers that 2 additional waterwheels were added during this conversion, making a total of 3 observed by Harris ca.1716.

[173] Abraham Musgrove m. Elizabeth Post (both of Yalding) in West Farleigh in 1721; could have been a s.o. Peter snr. before arrival in Boxley. An Abraham was also apprenticed to William Quelch at Gurney's mill in 1743 and occupied this mill from 1751–58.

[174] Shorter, A. Bib.12 57. The validity of his reference to Archer, Musgrove and Radford has been questioned earlier in App.III "The Occupants".

[175] See Chap.II note 137.

[176] Shorter, A. Bib.12 190.

[177] Spain, R.J. Bib.11 178 observed that from 1723–5 the paper maker was John Robbins; 1726 Robbins now Gill; 1727/8 Gill; 1728 Mr. Pine late Gill.

Peter jnr. in the meantime had moved to Upper Tovil mill (replacing Stephen Manktellow, see earlier in this Appendix) where the ratable value was restored from £8 to £20 shortly after. Although the entries continued to be made against him for this mill up to 1745, he insured it in the tenure of others e.g. Gifford and Pilchard.[178]

The Ratebook Extracts for Turkey mill (see Appendix III) show that from June 1735 a "Mr. Musgrove" occupied Turkey mill, almost certainly Peter jnr. It is thought that William Gill jnr. having some acquaintance of the young Musgrove may have suggested him to the James Brooke as a suitable paper maker to help them out of their current difficulties. It also lends support to the notion that Peter's father had worked there earlier and knew the Gills and the mill. Whatever the motive the occupation only lasted for one year and either Musgrove considered it too big an undertaking, or more probably, Brooke and Gill found him incapable of meeting their needs, rumours of Harris' and Whatman's activities at Hollingbourne Old mill may have reached their ears. In the event it was Richard Harris who displaced Musgrove (See Chap.IV the "Aftermath").

THE WALTERS

In the context of this work the name "Walter" is associated either directly or indirectly twice with papermaking at the beginning of the 18th C. Unfortunately the name occurs commonly in that district at that time and it has not been possible to confirm any relationship between these two occurrences.

Thomas Walter was George Gill's son-in-law and George, as a widower, went to live with him in Ditton ca.1722 (see earlier under George Gill) but apart from knowing that he was having a family there from 1717 nothing more about Thomas' origin is known or whether he was connected with papermaking or had come, for instance, from the Dartford area (see reference to John and Sarah Walter living at "Gill's Farm" in Darenth from 1717). The baptisms of 2 Thomas Walters are recorded in the Aylesford PRs., 1663 too early and 1696 more likely for our Thomas, the latter s.o. Lawrence and Mary, possibly related to other Walters in the area which might include A. Walter at Court Lodge Farm.

The John Walter, and there were several contemporary ones all marrying girls in Aylesford, with whom we are concerned here came from Bredhurst and married Elizabeth Archer in Aylesford in 1703. The conjunction of these two names in that area is either a pure coincidence or leads one to conjecture that Elizabeth may have been a daughter of Peter Archer snr., baptised during his "missing period" and that John may have been trained by or worked for her father (see under The Archers) and that this couple, John and Elizabeth Walter, moved to No.1 paper mill at River, nr. Dover, sometime before 1717.[179] (See Radford below).

THOMAS RADFORD

As mentioned above Shorter bracketed a Thomas Radford, resident in Boxley towards the end of the 17th C., with Archer and Musgrove as trainee paper makers and suggested that Radford or a descendant of his later became the paper maker at River No.1 paper mill from 1742.[180] Again, as in the case of the Walters, it has not been possible to substantiate this suggestion. The Boxley PRs. have 2 Radford entries (i) a 1683 bapt. Margarett d.o. Thomas & Jane Radford; and (ii) a Mar.1687 bur. Thomas Radford, householder. The name does appear in All SS., Maidstone, PRs. before 1670 and one might speculate that the "householder" had lived there and was buried in Boxley and that the 1683 baptism referred to a granddaughter. No other references have been found to Radfords in the Maidstone district nor in the PRs. of St. Mary's and St. James' in Dover (but a reference to it has been found at New Romney, East Kent, but with a Maidstone association).[181] The River paper mill Radford was born in 1710 (somewhere ?)

---

[178] Shorter, A. Bib.12 191 SFIP 41763 18:May:1727. He adds that William Musgrove may have been working this mill for his father in 1742; William turned up later at Eynsford mill (see Chap.VII p.303). For Abraham Musgrove see under Peter snr. above (footnote 173).

[179] Inf. from Mr. Douglas Welby of River. John Walter's first wife was an Elizabeth and was buried at River before 1740; he married a Mildred Bowers there in 1740, his second wife.

[180] Inf. from Mr. Douglas Welby of River. Thomas Radford, first mention, m. Sarah Lamber at River 1735; became lessee of the mill in 1742 following John Walter there; and was buried there in 1760 aged 50.

[181] All SS'. Marriages, Maidstone, record marriage in 1656 of Thomas Hunt to Mary Grigsby d.o. late Robert Radford of New Romney.

and, if connected, would have been a grandson of the Boxley Thomas and Jane, following the Walters (see above) to River, perhaps as a journeyman.

THE RUSSELLS

The relationship between the papermaking Russells, or the absence of it, represents another unsolved enigma; the solution to this may lie in the fact that, if they all came from one family, some of them may have followed a Nonconformist persuasion and the others not. Numerous 17th C. and early 18th C. entries are to found in the registers of the parishes in which they lived; in other cases they are not and in the case of John Russel (sic) the baptisms of his family are all recorded in the registers of the Presbyterian Chapel that existed in Earl Street, Maidstone, in the 1770's/80's (see App.II note 72). These registers start in 1732 but other, earlier, Russell entries were not seen there. The Russell genealogy, more particularly the Boxley Russells, has not been pursued further mainly because none of the Russells, including the two Ancient Paper Makers, appear to have distinguished themselves in papermaking (see especially Chap.IV and the reference to Andrew Johannot as undertenant at Shoreham mill).

The reason for including them in this Appendix is due partly to the possibility that Alexander Russell might have originated from the Goudhurst area, though there are alternatives to this suggestion; and partly due to the fact that through Constant Russell, who may have been his sister married to John Hillyer, he might have been linked to the early paper makers in the Maidstone area (see under "John Hillyer" in this Appendix) as indeed he was in a general sense through his association as an Ancient Paper Maker; and finally, because of a possible link with "Robert Russell of Ailsford" (see below).

ALEXANDER RUSSELL (Ancient Paper Maker) (bur.Shoreham 1732)

The first information of him is as a signatory of the Petition presented to Parliament by the Ancient Paper Makers in 1690 (see Chap.I). Questions arise from this event. How did one qualify for a place amongst these paper makers? and where was Alexander operating a paper mill in 1690 ? and was he related in any way to William Russell another of the signatories?

Taking these questions in the same order it has been remarked that the Ancient Paper Makers should not be regarded as a band of rustics but rather as established domestic manufacturers of White paper. The first Petition made by the paper makers (1689) in their opposition to "The CASE of the company of WHITE-PAPER-MAKERS" made it clear that the Ancient Paper Makers had 20 years experience in making White paper to the COMPANY's 3 or 4. How do we reconcile this claim with the fact that the first known position occupied by Alexander is not until the year 1692, two years after the Petition was signed?[182] It was also pointed out that William Harris only made his first appearance at Gurney's mill in 1689 and the possibility that James West might have begun his papermaking in 1688? There at least two ways of answering this question of qualification. Either one accepts that a given paper maker had in fact been making paper for 20 years or more e.g. like George Gill whose apprenticeship dates *at the latest* were ca.1667–74 with William Blackwell even earlier; or one assumes that others came from long established papermaking families and provisionally William Harris has been assigned to this group (see App.II note 20). In Alexander's case one could postulate either of these explanations (a) taking his marriage date of 1683 as a marker, then, like George Gill, one could point to an apprenticeship period starting early in the 1670's possibly at Goudhurst (see below); or (b) that he was related to William Russell, another Ancient Paper Maker, who has been recorded at Tredway mill, Loudwater, Buckinghamshire, between 1693– ca.1726(?).

As to the second question where was Alexander in 1690? In the absence of a RB for Shoreham, the only evidence of his appearance there, is the baptism of a daughter (1692) and the burial of a son *William*, in 1693, the latter described as a "child" and *not* baptised there, in fact the only indication of Russells in the PRs before these dates is the baptism of a Marie Russell in 1640. These facts do not rule out the possibility that Alexander may have been at a paper mill there in 1690, but there is no evidence to prove that he was (or indeed that there was a paper mill there in 1690, although the fact that he remained at Shoreham until his burial there in 1732 together with insurance policies

---

[182] See Chap.II note 22; and Chap.VII pp.297ff., Shoreham mill.

etc. proves that there must have been a mill for him to go to, whether a paper mill or a mill to convert is not known). Accepting that he was at a paper mill in Shoreham in 1690 does not solve the question of his origin or training.

Before considering this last problem one must needs consider the question as to whether he was related to William Russell A.P.M. in Buckinghamshire. Little has been discovered about William except that he was married there in 1682 (a year before Alexander) and that he appears to have been replaced at Tredway by 1726.[183] An origin that would accommodate these facts is the baptism of a William s.o. Thomas and Elizabeth Russell of Horsmonden in 1651. There was a large colony of Russells in the Goudhurst/Horsmonden district. Russell is not an uncommon name but Thomas and Elizabeth were contemporaries in the parish with a John and *Constant* Russell and although no daughter (Constant) is found there, a Constant Russell of almost the same age as Alexander married John Hillyer, the future paper maker of Forstal, in Boxley (1686), and believed to share the Russell's anti–COMPANY sentiments. Alexander was married to Elizabeth Bound in East Malling, close by, in 1683. Hillier was succeeded at Forstal by a William Russell (1725–32) and a long succession of Russells ending in a Hillier Russell. The fact that Alexander had a son, William (albeit deceased), could also be seen as a pointer. It is also conceivable, if the William Russell A.P.M. had left Tredway in 1625,[184] that he (or a son of the same name) had transferred to Forstal. The most one can say in these circumstances is that there are several indications that Alexander and William were related and that both, perhaps, had served apprenticeships at Goudhurst. To cover his "missing period" (see Chap.II) he too may have followed Peter Archer to an unidentified mill through lack of opportunity at Goudhurst.

That there are undeniable blocks to these suggestions is demonstrated further by (a) in spite of searching 25 parishes in Kent (1°, 2°, & 3° search areas) Alexander's baptism has not been found;[185] and (b) the relationship of the Forstal Russells has not been ascertained either, possibly because they might have adopted another religious persuasion (Presbyterian).

In view of the frequency of the occurrence of the name Russell it has not been possible to pursue some of the other "tail–ends" raised in the Text and in App.III (under Forstal). Was there any link, for instance, between Jane Russell (second wife of Sir Edward Duke) and the later Russells ? Jane came from "Ailsford". Or, again, did the other Shoreham family of Russells connect with any of those who appeared later in the Maidstone area, linked in partnerships with Edmeads (with a Wilmott ancestry) and the Clement Taylors, all from the mid–West Kent district?

[183]   Shorter, A. Bib.1 135 139/140.

[184]   Shorter, A. Bib.12 140 SFIP 39048 24:June:1726. A later policy of 1762 shows Turner as a millwright insuring his paper and corn mill called Tredway mills in the tenure of John Spicer SFIP 193117 11:Oct:1762.

[185]   In the Maidstone (1° search) apart from a marriage of 1638 (Robertus Ellis to Iana Russell) no Russells in Boxley before Constant; nor in Aylesford except for a William (a Wharfinger of Maidstone) marrying there in 1656; none in Ditton, Burham, Snodland, Loose, Barming (West Malling & Wateringbury start too late), but there was a baptism of Sarah d.o. Francis Russell in E.Malling in 1669. In Maidstone itself there was William the Wharfinger who occupied a house in King's Meadow and was a member of the Common Council in 1682, but only 1 baptism, a daughter, recorded; a John Russell also with a daughter bapt.1680. (There were clearly earlier Russells in Aylesford but no records).
In the West Kent Region (2° search) 10 parishes covered including Dartford with Russells but no Alexander and no Russell family 1683–91; Darenth, Sutton–at–Hone n/a; Eynsford no 17th C. Russells; Otford a Robert & Jane 1681; Shoreham (before Alexander) only bapt. Marie 1640; Wrotham one 17th C.; Ightham & Plaxtol 18th C. only; West Peckham nil.
In South Kent (2° & 3°) a very large colony at Goudhurst from 1599, but no Alexander; Horsmonden 3 families from 1638; early Russells at Benenden & Penshurst n/a; Chiddingstone nil; Russells at Brenchley but no Alexander or Constant.

OTHER EARLY MISCELLANEOUS FIGURES IN THE MAIDSTONE PAPERMAKING DISTRICT

THOMAS NEWMAN

Thomas Newman, paper maker, (referred to in Chap.II) was buried in Boxley in 1702. Newmans had been present in the parish as far back as 1633 but no Thomas baptism or marriage found. It is remotely conceivable that he might have been a Master Paper Maker at some point earlier in the history of Forstal mill e.g. in the early 1680's. On the other hand as suggested in the case of Peter Musgrove (see previous section) he too could have worked for George Gill at Turkey mill, a large mill requiring certainly more than one paper maker; or under one of the Master Paper Makers at Forstal or even at one of the other early paper mills; but on account of his Boxley burial the most likely place would have been Turkey mill on account of its size and the status of its occupant.

JAMES SMITH

James smith, paper maker, was buried in Snodland in 1705. No evidence has come to light of a paper mill operating in Snodland before 1755. Smith may have retired and gone to live with relations in Snodland and died there: (Smiths were present in Snodland). If, however, he worked somewhere in that area, since Snodland lies on the left bank of the R. Medway, one might see him employed at Millhall mill, Ditton; assessments are made against a John Smith there between 1677–94 in the Church Book (KAO P 118/5); and buried there in 1703. Or he may have been working at East Malling rather than, say, in any of the Boxley mills.

RICHARD BURNHAM (Great Ivy Mill)

Burnham is a name traditionally associated with Buckinghamshire and East Anglia and one solution to his origin would be to see him coming from the former. But there was a family of Burnhams in Maidstone itself in the 1650's, though no sign of a Richard; entries in the All SS.' PRs are very scarce at this time (William s.o. Henry Burnham 1657). The name has not been found in 25 parishes surrounding Maidstone, Goudhurst and in West Kent. No marriage or burial in Maidstone either (there is the possibility that he might have become Nonconformist ?). An alternative is that his name could have been misspelt. Brumans, sometime spelt Brooman, had been former occupants of Great Ivy mill from 1675 (Spain, R.J. Bib.10 70) and Henry Bruman was Richard's immediate predecessor. Whatever the truth, Richard would have required training which he could have had at Forstal mill. This seems to be a more reasonable suggestion than seeing him come from the West of London and set up a mill in Maidstone without previous knowledge of the area. He disappeared from the Maidstone scene before 1703, being succeeded at Lower Tovil by Thomas Manktellow, Stephen's son.

Two points may be added to the above (i) he took an apprentice, John How, in 1693 (Shorter, A. Bib.12 192); and (ii) apropos of the potential of this mill for making White paper, (Spain Bib.10 73) noted that three springs rise within 150 m. upstream of the mill so that initially, and as indeed happened later in its history (see Chap.VII TABLE XII), it had the facilities for making White paper, though this potential does not seem to have been realised until late in the 1780's.

NOTES ON SOME OF THE CONTEMPORARY VICARS

BOXLEY

1646–1678 (decd.)	Thomas Heymes	Described as "a creature of the Puritan Party" but who managed to conform in the shifting winds.
1678–1687 (decd.)	Humphrey Lynde	Lynde's father was a bitter Puritan; but his son though Puritan was conciliatory. Humphry also had the Curacy of Maidstone from 1677.
1690–	John Wyvall	

MAIDSTONE

1618–ejected ca. 1643	Robert Barrell	Formerly a Puritan but somewhat lax and changed to High church later.
1643–1653	Thomas Wilson	Strongly Puritan (managed 2 wives and 11 children); Cambridge graduate; numerous appointments ending with Otham prior to succeeding Barrell. The Maidstone PRs. are especially low in the number of entries after 1644 for a number of years.
1653–1662 (decd.)	Crump	Mild Puritan; ejected 1662 because he would not comply with the Act of Uniformity; allowed to preach occasionally at Boxley by Lynde; (d. 1677).
1662–1677 (decd.)	John Davis	Had to revert hurriedly to conformity.
1677–1687 (decd.)	Hunphrey Lynde	(see above).

CRANBROOK[186]   (*GOUDHURST* a similar situation existed there)

Before 1662 1662 (act of Uniformity)	Mr. Goodrich	Not an Episcopalian nor a strict Presbyterian or Baptist. Dr. Calamy reported "10 Ministers cast out of this town and places adjacent". The act of Uniformity was only one of the 4 contemporary Acts passed by the Anglicans to repress all forms of Dissent. (1661) The Corporation Act denying municipal office to dissenters; (1662) The Act of Uniformity compelled the use of the Elizabethan Prayer Book, which led to widespread ejection of Ministers (including those referred to above); (1664) The Conventicle Act penalizing attendance at Nonconformist Services; (1665) The Five Mile Act banishing ejected clergy from their places of ministry, from teaching or taking in boarders. Severe persecution followed.
After 1662	John Cooper	Episcopalian. It was noted that the more well-to-do conformed, the artisans nonconformist. Among the more important churchgoers the same people occupied the same pews as they had before the Restoration. The change did not seem to make much difference.

---

[186] Notes taken from "The Annals of Cranbrook Church", Tarbutt, William (1870).

APPENDIX V

Unlike the other Appendices this is devoted to certain special features connected with the papermaking of the period, too detailed to be included in the text without a danger of losing track of the main issues but which, nevertheless, are necessary for a proper understanding of the subjects covered there.

WHITE PAPER AND ITS CELLULOSIC CONTENT

PART I

THE NATURE OF THE RAGS USED FOR MAKING WHITE PAPER IN THE 17TH AND 18TH C. PAPER MILLS OF THE BRITISH ISLES

In order to appreciate the impact of the introduction of the Hollander Breaker and Beater on the papermaking industry fully it is now necessary to define the "Rag" more precisely than is usually the case in histories of papermaking, especially so for those readers not already familiar with the raw materials in use at the time of this change in the process. At the end of the 20th C., when rags have almost entirely disappeared from papermaking furnishes, these remarks could well apply to more people than one might imagine at first. It also provides an opportunity to consider the cellulosic nature of the rag and its relationship with water during the process, the full significance of this in the beating and sheet–making operations, and on the subsequent behaviour of the paper itself (PART II).

In terms of White paper manufactured in the past the reader should be aware by now that the rags used for this were not just any piece of waste material. In modern times when rags were still the principal raw material employed in making high quality papers, technical or otherwise, success in their application depended not only on the skill and experience of the rag sorter but on the knowledge accumulated over the years by the technician in selecting and evaluating the physical and chemical properties of the different raw materials available to him and appropriate for the kind of paper that they would eventually give rise to. This is easier read than comprehended. It has been said that *communication* is most effective between persons having an equal knowledge of a subject and that even then it rarely exceeds seventy per cent. For persons who have not had the benefit of experience in the use of rags to make the more difficult classes of paper, it is all the more important for them to grasp the significance of the different factors that so easily escape notice and which can affect the underlying characteristics of a rag, stemming as they do from the previous history of the fibres, their extraction and the textiles from which they are derived.

Although the rag was raw material to the paper maker, it has to be seen here in a deeper perspective, namely that it was derived from a textile which in turn was manufactured from more basic materials. PART I of this section is intended to give an insight into this aspect of papermaking. However, it has to be remembered, as the frequent references in the literature demonstrate, that searches were made during the 17th and 18th Cs. for alternative raw materials,[1] a fact indicative of the chronic shortage of the staple elements that beset the industry up till the latter part of the 19th C. In many cases, had these been successful, they would have bypassed the textile stage; indeed this is true of papermaking from its very beginnings. One might say that the use of rags was peculiar to European papermaking until recent times. Although some of these efforts resulted in varying degrees of success,[2] the

---

[1] Summarised by Voorn, H. Bib.52.

[2] Shorter, (Bib.9 42) lists a number of attempts made in England during the 18th C. to use alternative materials, such as nettles and willow bark, but none of these ever came to anything; whereas Matthias Koops was highly successful in making significant quantities of paper that have lasted to the present day without deterioration from straw, flax waste, wood and waste paper at his Neckinger and Thames Bank (Millbank) paper mills in London in 1800. (See also Index Paper Makers, Koops; and this App. note 146).

earlier European paper makers were effectively restricted to the use of rags derived from Linen, Hemp and Cotton materials; with the addition of "Woollen Rags or such other Offal as happens amongst Rags" to the furnish of inferior papers.[3] Finally, abundant though the supply of suitable textiles may appear to have been during this period, it is important to remember that only a very small fraction of these ever reached the paper maker for conversion into paper, a point that has already been made in Volume I.[4]

COTTON

Cotton, present in rags, had been available to paper makers on the Continent from early times;[5] although it may have been minimally present in British papers from the beginning of the 17th C.,[6] it is most unlikely that paper makers in this country would have had access to significant quantities of cotton textiles until late in the 18th C.[7] For our purpose then it may be eliminated from this discussion. We will revert, however, to Cotton in Part II of this Section as an example of cellulose in its relationship to water in the papermaking process.

That there was a growing interest in the possibility of using cotton as a raw material is perhaps demonstrated by the Society of Arts' action in sending a sample of Wild Cotton to the younger Whatman in 1774 to examine and experiment with. His comments, which are most instructive in more ways than one, have already been discussed in the main text.[8]

As mentioned, cotton derived from textiles is not likely to be identifiable in British papers much before the 19th C. Even then it was regarded as an inferior material, possibly because the bulk of it came either from printed cloth, which would have required chemical bleaching, a process that had not been fully mastered at that time, or in the form of floor sweepings. This view was not shared by everybody as is shown by Renouard's admiration of early 19th C. French papers made from cotton.[9] Nevertheless a certain prejudice against its use in the top qualities of paper lingered on into the 20th C., although cotton would have been

---

[3]   Extract from "Additional Instructions to be observed by Officers in the Country employed in charging the Duties on Paper" (1722).

[4]   See Chap.V p.185 + n.50.

[5]   For an account of Cotton in Europe see Wescher, H. Bib.53. Cotton is not mentioned here as a papermaking raw material; the reference is to its use in Textiles and the implication is that these would have found their way into early European papers.

[6]   Lowe, Norman, Bib.71 Chap.VII (1972) and also Wadsworth & Mann, Bib.55 (1931) both refer to the use of real cotton in Fustians, Bombast or Downe and another cloth called "Checks" appearing in Lancashire on a significant scale during the second half of the 16th C. (These contained true cotton and should not be confused with the earlier usage of this word, where "Cotton" referred to woollen cloth soft and fluffy with a raised nap.) True cotton had been used in England since 1200 for candle wicks, stuffing and quilting, in a form that would have been impractical to use in papermaking; but it is mentioned in Bladen's Beater Patent of 1682.

[7]   Though references to cotton are made in the 17th & 18th Cs. by Houghton, J. Bib.20 I.343 (17th C.) and Wadsworth & Mann (loc.cit) again in the form of cotton wicks, inkhorns and quilting and, later, to imported woven textiles like calicoes and muslins, it was not until the new inventions in spinning and weaving machinery became effective in the latter part of the 18th C. that cotton textiles were mass produced in this country. Because of their relative cheapness (and fashion) these found a ready market amongst the poorer classes especially in clothing and furnishing fabrics. For a fascinating account of this see Edwards, M. Bib.54 (1967).

[8]   See Chap.V p.185. Whatman's reply is printed in its entirety in Balston, T. Bib.1 36/37.

[9]   Renouard, A.A. Bib.56 (3rd Ed.) 383 footnote. Freely translated he wrote, "the brilliance, strength and durability of the cotton papers, now invading our printing works, are entirely consistent with the durability of the cotton textiles and garments from which they originate". (1834).

almost universally present in the Fine rags used in the manufacture of these papers by the end of the 19th C.

For the sake of completeness, cotton had come into its own as a furnish for the highest qualities of paper by the mid–20th C. in the form of pure cotton textile cuttings and cotton linters, a raw material with a 2–6 mm. length fibre recovered from the cotton seed hull.[10]

In the hands of competent paper makers cotton linters can be converted into papers that are, for most purposes, much the same in quality as those made from linen though, typically, inferior in tear strength but superior in other properties. The linen fibre, other factors being equal, is much stronger than cotton; and, it has been claimed, is more resistant to degradation caused by the action of sunlight (though at the same time more easily disintegrated by certain chemical action especially during bleaching). These properties, together with its lustrous quality and the fact that it is a better conductor of heat than cotton, contribute to producing a superior textile. But, by the time it has reached the rag stage in its history, passed through a papermaking process (a process which can be adjusted to meet the particular characteristics of different fibres), it is doubtful whether many people could distinguish a modern linen-based paper from one produced from a purified Cotton Linter furnish, especially if the papers concerned have been sized subsequently with gelatine size.

WOOL

Although woollen rags may also be eliminated from the discussion, it is perhaps worthwhile mentioning them here and showing why this should be. Wool is not a papermaking fibre in the normal sense. It is a protein fibre in contrast to the cellulose fibre used in conventional papers. Where it was incorporated in the furnish of early papers, it would have enhanced the fermentation of other rags (a subject to be discussed later in this section); but otherwise it would only have acted as a filler, weakening rather than contributing to the strength of the paper. In fact to complete the quotation used earlier[11] wool would have been used mainly for very low grade brown paper, which "is commonly thinner and more uneven, and always more rotten.......".

References to "Woollen Rags" or sometimes "Linsey-Woolsey"[12] occur sporadically in the literature on papermaking; almost always these are references to their use in Brown paper;[13] or similar categories such as grey or scrap paper;[14] and, although not specifically instanced, probably for Wrappers, Littress and other non-descript papers.[15] They may also have been

[10]   For a detailed technical account of the Cotton Linter see Temming & Grunert, Bib.57.
[11]   This App. p.184 (opposite).
[12]   Houghton, J. Bib.20 Vol.II 410 (1699). Mr. Million in his account of the papermaking process mentions that "among the rags will be some linsey-woolsey (originally a wool/flax mixture) which the dirt makes indiscoverable till they are once washed".
[13]   Evelyn, J. Bib.30 338 (Aug.24 1678). "They cull the rags which are linen for white paper, woollen for brown...."
[14]   Keferstein, Georg, "Unterricht eines Papiermachers an Seine Sohne, diese Kunst betreffend" (1766).
[15]   Woollen rags were included with other materials such as "Blues" and "Bagging", obtained from two of his Rag Merchants, Neatby and Stower, in the younger James Whatman's Ledger for the years 1780-1787. It seems probable that these would have been combined with vat bottoms and dirty broke etc. to make wrappers, littress etc.

used in the 19th C. in Filter papers[16] (of a kind) and this has led to the suggestion that they were also added to the furnish of Blotting papers[17] though there seems to be no documentary evidence for this. Finally, wool had a rather different function when it was incorporated into 19th C. Cartridge paper by John Dickinson in order to prevent the retention of sparks after a detonation.[18]

In following the fortunes of White paper manufacture, woollen rags unless they found their way into the furnish by accident are of no further interest.

HEMP (Cannabis sativa)[19]

For our purpose here raw materials for papermaking based on hemp can also be eliminated from this discussion, although one cannot rule out the possibility of small quantities of the fibre finding its way at times into the furnish for White papers. In the normal way hemp was used for canvas (especially sailcloth) and cordage; and for reasons which will be given below these materials would only have been used in the lower qualities of paper. However, the cultivation of hemp (see below) and the subsequent processing it underwent to produce fibre is in many ways similar to that of flax, considered in greater detail next. Like flax the nature of the fibre extracted for spinning is very dependent on the conditions under which the plant is grown and harvested, as well as on the treatment it receives later. Thus under certain conditions fine hemp fibres can be produced and woven into a cloth that is roughly equivalent to coarse linen. It is conceivable, therefore, that some of this material might have reached the paper mill and passed as linen rag; if so, the likelihood is that this would have applied to consignments of rag received from the Continent rather than from domestic sources. The British Paper Industry was very dependent on continental supplies of rag and especially from northern Europe during the 18th C.

The use of hemp fibres for canvas and ropes is very ancient and dates back several centuries B.C.[20] The plant from which the fibre is extracted grows almost anywhere, but prefers a temperate climate and a rich (manured) soil. In more recent times it was quite widely

---

[16]   Shorter, A. Bib.12 24 note 12 mentions that wool was used in limited quantities for some Filter papers and Blotting papers as well. This, however, is a late 19th C. reference. If one looks back, William Lewis describes in his "A Course of Practical Chemistry" (London, 1746, 10 footnote k) filter paper as follows..."there are several sorts of paper employed for this purpose. Some use coloured papers, which are in many cases extremely improper, as they are apt to tinge the liquor which passes through them. The best filtering paper, is of a spongy and rare texture, which does not easily break when wetted; such is the grey paper which covers the pill boxes, as they come from abroad". It is conceivable that these pill box covers may have been one and the same as the "Papiers Fluants" referred to by both Lalande and Desmarest (Bibs 23 & 21), but neither mention wool as part of their furnish. Desmarest, however, (Bib.21 559 under "Brouillard") does refer to "une pate bulle pour filtrer les liqueurs" known under the name of Papier Joseph; although he does not mention wool specifically, he describes the "pate bulle" as stuff furnished from weak rags.

[17]   Blotting paper had been used in this country from very early times (Horman, William, "Vulgaria" 1519 "Blottynge Papyr serveth to drye weete wryttynge lest there be made blottis or blurris") and seems to have been something of an English speciality according to Desmarest; although here again there is no positive evidence that wool formed part of the furnish.

[18]   Dickinson, John, (Paper Maker) Patent No.3080 (1807).

[19]   True Hemp (*Cannabis sativa*) should not be confused with Manila Hemp or Abaca (*Musa textilis*), more commonly used today, which was exported to America ca.1818 and first used there as a papermaking fibre in 1843. See also note 25 below.

[20]   The leaves of the plant were also from early times dried and then smoked or chewed as Hashish in the middle and far East.

cultivated in this country, especially in Lincolnshire and Suffolk,[21] but the best qualities of hemp came from Russia. Before the 17th C. most of our ropes and canvas were imported, but during the 17th C. ropehouses were established in the Royal Dockyards,[22] which made the Navy independent of foreign sources; and other coastal towns, like Bridport, specialised in cordage for merchant vessels. Similarly, by the mid–17th C. the better qualities of canvas were manufactured in this country, mainly in the West Country and, later, in Lancashire (Warrington).[23] From this it will be clear that the chief sources for hemp as a material for papermaking would have been naval dockyards and ports and this is confirmed by advertisements.[24]

Hemp is grown (like flax when grown for the fibre rather than for Linseed oil) as a high density crop; it is pulled up unripe; allowed to ripen in stooks; the better qualities are water retted (the lower, dew retted); it is broken, scutched, heckled and spun in the same way as flax. The hemp fibre is enormously long, less flexible and less easily bleached than flax and had been used from early times for maritime purposes because, it has been claimed, it was more resistant than most other fibres to the rotting action of salt water.[25] In spite of this beneficial property canvas sails (the Royal Navy apart)[26] were usually tanned with horse grease, tar and ochre; and cordage invariably tarred with Stockholm tar with the result that these materials "after they are unfit for the purpose for which they were made" were also unsuitable for the manufacture of White paper.

FLAX (Linum usitatissimum)

The flax plant is the source of the linen textile which, as we have seen, is likely to have been the principal source of the rags used to make White paper for most of the period that we are concerned with here.

Flax is the strongest and oldest of the vegetable fibres used by man; it has been used by him since paleolithic times. The weaving of linen cloth had reached an advanced state in Ancient Egypt; it was in use in Roman Britain, but the spinning and weaving of it declined after the fall of the Empire. By 1400 its cultivation and use had been firmly re–established, but the quality of the domestic cloth was said to have been coarse. The industry, in spite of encouragement later from Acts of Parliament and Protection, never seems to have flourished

[21] Warden, A.J. Bib.59 (1864).
[22] Oppenheim, M. "The Administration of the Royal Navy" (1896). (Inf. kindly supplied by Miss R. Prentice, Nat.Maritime Museum).
[23] Rees, Abraham, Bib.50.
[24] Shorter, A. Bib.12 App.H p.404 gives a number of examples of maritime materials advertised for sale at ports ranging from Bristol and Barnstaple in the West to Deal, Deptford and Rotherhithe in the East.
[25] The phenolic element in lignin is thought to give the fibre its resistance to salt water action. Dr. Gascoigne (Messrs. Cross & Bevan letter 7:Dec:1977) kindly supplied me with the following information on the lignin content of certain fibres in the *raw* state:–

   Flax   2–4%
   Hemp   5–7%
   Manila Hemp   9–12%

He pointed out that hemp, and especially Manila Hemp, is thus more lignified than flax, but stressed that these figures refer to natural fibres and that the derived cordage would not necessarily show the same relative lignin contents.
[26] Steel, Bib.32 85 (1794). "The tanning of sails in the Royal Navy has been tried, but is not approved of".

in England, although there appears to have been no shortage of linen textiles in the country during the 17th and 18th Cs., a subject already discussed in the main text.[27] Increasingly the industry felt the competition from cotton. By the mid–18th C. the domestic manufacture of cotton goods benefited from the protection afforded to the growing linen industry, partly because Fustian (linen warp/cotton weft) was an important outlet for the linen yarn. The earlier Calico Acts (1701 and 1721), which had prohibited the wearing of coloured Indian calicoes and which were primarily designed to help the wool trade, ironically stimulated the growth of the English calico industry instead. Cotton in the end helped destroy the *English* linen industry, which had mothered its early growth and had lived with it for a considerable period.[28]

For the domestic production of textiles (and, subsequently, the supply of linen rags for the expanding paper industry) flax had considerable disadvantages when trying to compete with cotton, which ultimately superseded it in meeting the greatly increased demand for clothing and furnishing fabrics at the end of the 18th C.,[29] at a time when the early difficulties of maintaining an adequate and regular supply of raw cotton were being overcome as well as achieving an improvement in its quality.[30]

The cultivation of flax, compared to that of cotton, was a very slow and demanding business; it lacked the flexibility for increasing production that its rival possessed. It has been said that flax is the most costly, troublesome and precarious of all crops to cultivate;[31] Moreover, there were processing difficulties, the mechanisation of spinning linen thread lagged well behind that of cotton and even to-day, apparently, some hand feeding to machines is still necessary.[32]

Though the linen rag was virtually the only material available to the maker of the best qualities of White paper in Britain for the period under consideration here, in terms of its cellulose content flax in its dry raw state is not as pure as cotton;[33] these impurities had an important bearing on the method used to process the rags. There are many factors which also affect the quality of the linen and which have to be taken into account. These are examined next.

[27] See Chap.V p.185 + n.50.

[28] Harte, N.B. Bib.65(a) 110.

[29] In case this statement should create a mistaken impression, it has been pointed out to the author that "Cotton did not oust flax, which continued to be produced in large quantities up to the start of the Second World War. It simply filled the requirements of a rapidly expanding market" (letter J.T. Mitchell, Snr.Res.Officer, Shirley Institute, 8:Sept:1976). We have to remember also when considering this developing situation that although the linen industry may never have flourished in England, quite the opposite to this was taking place in the north of Ireland where the industry expanded dramatically during the first half of the 18th C., but its output could not match cotton's (Chap.V, n.52).

[30] Edwards, Michael M. Bib.54 Chap.III.

[31] Gill, C. Bib.38.

[32] Mitchell, J.T. Snr.Res.Officer, Shirley Institute, letter 8:Sept:1976.

[33] For comparison the cellulose content in the dry raw state (all cases) of cotton 91%; flax 82%; wood 60%; and straw 40–50%.

The Quality of the Linen

It has been said that the texture of some early European papers was coarse and that they had inferior felting properties (this means poor consolidation during sheet formation, a subject covered in Part II). Some of these defects may have been due to the inability of the beating equipment to cope with the materials properly; but evidently these papers also contained varying proportions of imperfectly retted flax fibres.[34] The same source states that even as late as the 18th C. some papers, all watermarked, were quite brown from the use of unbleached "flax" (sic).[35] The use of imperfectly retted flax fibres or of unbleached linen would almost certainly account for the presence of shive in inferior "white" papers of this period, papers that were otherwise of reasonable strength. This sort of paper probably found its way mostly into printing rather than writing uses. It can be seen then that the use of strong linen rags did not automatically produce the top qualities of White paper. In the passages that follow we shall be looking briefly at some of the possible reasons that may have led to variations in the quality of the linen rags that reached the paper mill.

There were, for instance, political, geographical and cultural factors that could have affected the quality of the linen thread; and there were, in addition, process differences that would have caused variations in the quality of the finished textile. Over and above all this by the time the linen rags reached the paper mill and came to be used for making paper, the cloth would have passed through a whole series of random as well as organised selection processes which effectively prevent positive identification of any single factor as the cause of the defect that we may observe. All the same any one of these influences could have had an overriding effect in determining the general level of the quality of the linen produced and thus the rag reaching the paper mill.

One can reasonably assume that in the course of the long history of the linen industry the skills needed for producing thread from the flax plant would have been perfected as far as the empirical procedures for control at the disposal of the producers would allow. This assumption is important in that the quality of the fibre is greatly dependent on both the method of cultivating the flax plant and on the processes used subsequently to isolate and purify the fibre, methods involving stages of great technological complexity. Though the skills may have been there, it is necessary to qualify this assumption by saying that this does not imply that the quality of the production of linen from flax was universally high at this time.

It will be recalled (see Chapter II) that during the early part of the 17th C. the British paper industry was very small and, in fact, as Professor Coleman has pointed out,[36] such rags as were available in this country were being *exported* to France. But by ca.1670 the situation was changing and an expansion in the domestic paper industry was taking place with a

[34] Beadle & Stevens, Bib.58 (1909).
[35] Ibid. The term "flax" is applied loosely here; the distinction between flax and linen is blurred. To avoid confusion the definition adopted in this book is that flax and flax fibres refer to the plant and its constituents in the raw and pre–retted state; and during subsequent processing as flax straw and remains as such for scutching (beating) and heckling (dressing the flax with a steel flax–comb or hackle); but after spinning it becomes linen yarn woven ultimately into linen cloth. In this usage Beadle and Stevens "unbleached flax" would be interpreted as "unbleached linen". In a more ancient papermaking process it is possible that natural flax fibres, of great length and strength, might have been used to make paper and account for poor felting. (See also under flax waste; Chap.V p.184 + n.48.)
[36] Coleman, D.C. Bib.15 53.

consequent demand for raw materials to support this growth and the supply of these depended, though not necessarily proportionately, on the level of textiles available in this country. But the conditions that determined this supply were complicated at this time and remained so for the next century. Both domestic and foreign sources have to be taken into consideration and these changed radically during this period.

It has been estimated that in the late 16th and early 17th Cs. 14% of all English agricultural labourers were employed in producing flax (a further 15% hemp); the industry continued to grow rather than diminish, so it could be said that at the beginning of the 18th C. the English linen industry was an important one; but by international standards it was small and completely overshadowed by imports of textiles, chiefly those from France. However, several events conspired to complicate and change this situation, these stemming principally from the effect of raising the Duties on these imports (to provide revenue to pay for the current wars rather than to encourage the growth of the domestic industry).[37] The consequences of the various rates of Duty imposed between 1660–1760 (too complicated to cover here)[38] were far reaching and concern us at this point only to demonstrate that the sources of supply for the linen found in this country were very varied and altered frequently. Thus, besides having a domestic source, we were importing initially large quantities of linen from France; other suppliers included Flanders, Holland, Germany and, later, Russia.

The effect of higher tariffs from 1690 onwards was to hit French imports the hardest; as the duties increased Flemish and Dutch imports were also affected, though the bias still remained anti–French. The only source favoured by these actions was Germany. The rates of Duty were related to the width and not the quality of their cloth. The Germans made narrow width textiles and were thus able to improve the quality of their cloth at the expense of the French and at the same time pay much lower Duty.[39] Harte maintains that it was the Duty of this period (1697–1712) that had the biggest impact on the realignment of the sources of imported linen. Imports of French linen declined sharply towards the end of the 17th C.; Flemish and Dutch by the early 18th; but German imports continued to grow until the 1730's, only declining after 1750. Some cloth continued to find its way into this country by the practice of fraudulently misrepresenting its country of origin.[40] Also, as mentioned in the main text, it has been reckoned that about 33% of the linen entering this country by–passed the Excise and was smuggled in mainly from France and Holland.

At the same time as all this was taking place, an Act of 1696 allowed hemp and flax to be imported Duty free from Ireland,[41] ostensibly to encourage foreign Protestants to settle there; but, be that as it may, it also had a galvanizing effect on the local linen industry of the north. Although earlier attempts had been made to foster a linen industry in Ireland,[42] it was not until the early 18th C. that it really got under way and then in the northern provinces. This resulted from a combination of different events and conditions. Gill has identified the

---

[37]   Harte, N.B. Bib.65(a) 76.
[38]   Ibid. Generally, Sections II & III; in detail, Section IV especially pp.97–99.
[39]   Ibid. pp.76–79.
[40]   The Dutch got some of their linen through as "Silesian"' and some as "narrow Germany".
[41]   Scotland was not included under the terms of this Act until after the Act of Union in 1707. As a consequence their industry tended to lag behind the Irish.
[42]   See Chap.V, note 52.

following factors that led to this rapid and unexpected expansion; the expulsion of the Huguenots from France brought capital and technical know-how to the north of Ireland (amongst other places); as seen above an Act of William III's permitted flax, linen yarn and cloth to be imported into England from Ireland Duty free; Antwerp, the main market for linen, was cut off by war; to these may be added the contribution made by Scottish, Dutch and Lancastrian Settlers together with the very important effects deriving from the local system of land tenure; the soil, the climate, the abundance of water and the fact that flax could be grown remuneratively on small-holdings; untapped skills in weaving and a ready market for the cloth in England.[43] According to Harte[44] by 1750 imports from Ireland and Scotland exceeded continental supplies (excluding the smuggled linen); by the 1760's imports from Ireland alone exceeded those from the Continent. Likewise by the 1730's imports of Irish linen were already driving out German cloth. By this time also the Northern Ireland bleacheries were eclipsing, though not completely, the famous ones at Haarlem.

Throughout this period the English linen industry had remained an important but small one; nevertheless it can be seen from the foregoing that the bulk of the cloth in this country (from which the papermaking rags would ultimately be derived) had come from many different sources and that these had chopped and changed dramatically between the latter half of the 17th C. and the first half of the 18th and continued to alter. The paper maker's interest in this altering situation would have lain more in the way it might have affected the quality of his raw material rather than the quantity. The industry itself was not confronted by any shortage of textiles but the scarcity of rags emanating from them. It was this, coupled with the inefficient methods employed in collecting them, that created the difficulties. To complicate this picture even further, as the 18th C. progressed because of the shortages it became increasingly necessary for the rapidly growing paper industry to import rags, the quantity of rag imports rising some twenty-fold between 1725 and 1800, a large proportion of them coming from Germany.[45] It will be obvious from this state of affairs that the quality of the linen, whether in prime cloth or rag form, could have varied considerably. The question may be asked, did it ? Though one cannot identify today specific examples of this, it is nevertheless indisputable that this was the case.

The quality of the fibres in the linen rags would have depended to a degree on the state of the art employed in cultivating the plant and extracting the flax from it and this would have varied from locality to locality as indeed the special skills required in each case. For example, retting was carried out in ponds or flax dams in Ireland and in flowing water at Courtrai in Flanders.[46] This would have produced important differences in the conditions required for this microbiological process and its control. Similarly, the density of the cultivation and the degree of ripeness when the crop was pulled up from the ground both affected the quality of the flax fibre,[47] these leading to fine or coarse yarn as the case may be. In addition, the effects of other variables in the extraction process, could lead to variations in the quantities of non-cellulosic residues remaining in the fibre bundles. To all of these must be added the effects of the various processes that led to the finished textile, like bleaching, as well as to

[43]   Gill, C. Bib.38.
[44]   Harte, N.B. Bib.65(a) 94/95.
[45]   Coleman, D.C. Bib.15 90.
[46]   Warden, A.J. Bib.59.
[47]   Ibid.

substances that may have become associated with the cloth later in its life such as applied starch etc.

Not all of these variables would have necessarily affected the manufacture of White paper, because the Rag Sorting process would probably have taken care of the more obvious differences between the textiles and the rags that they produced. But there were variables, such as those arising from the retting process, which would have been less obvious to the paper maker and which would have had an important influence on the effect of the treatment applied to the rags during the papermaking process. To-day, with the greater understanding that we have of the chemistry of the process, we can look back with hindsight and see the kind of dangers that lay in the path of a weaver who grew his own flax, as was the practice in both Ireland and Flanders. One wonders just how much control the producer had had over his process remembering that right from the outset he would have been saddled with all the other variables referred to above, the climate, the soil, the density of the crop etc. So at this point it would be instructive to examine the retting process at the centre of this issue.

The Flax Retting Process

In order to loosen the fibre bundles in the stem of the flax plant, the ripened straw is subjected to a fermentation process known as "retting". The retting process is of twofold interest. First, apropos variations in quality, the properties of the linen yarn and cloth are very dependent on the way this process is regulated. Second, a grasp of the process itself is absolutely central to understanding the fundamental changes that took place, mainly during the 18th C., in the method of preparing rags for papermaking.

According to the season and the methods used in growing the flax, the straw may vary in its degree of ripeness when pulled from the ground; and this in turn may affect the treatment and the interval that elapses before it proceeds to the next stage. The aim of the next process is to rot the straw in water to a point where, after drying it, it may be crushed; scutched with a swingling knife to separate and remove the chaff and shive from the fibre bundles, which are then drawn through metal combs (hackling) to tease the material open and separate out the tow (broken and short flax fibres). Hackling may be repeated using finer and finer combs. Warden[48] noted that in 1800 a stripe of flax a foot long could be drawn out to a length of 17 miles before being formed into thread; later this extension was increased to several hundred miles. After hackling the fibre bundles were ready for spinning into yarn. Warden also remarked that it was *necessary to have fine flax in the first place to produce fine yarn*; coarse flax would not break down into fine flax; and this fineness stems from the overall conditions of cultivation and harvesting, which the plants have been subjected to.

In the plant the flax fibre bundles are embedded in soft tissues that lie between the outer cuticle and the woody core (see Plate 18). The purpose of retting is to loosen the bonds between these soft tissues and the fibre bundles and the more effectively this can be done, the easier it is to separate the fibre bundles from the wood in the beating and scutching processes that follow. The loosening action is achieved by fermenting bundles of the straw immersed in special ponds (stagnant water) or in partially submerged crates placed at the edges of rivers as, for instance, at Courtrai on the River Lys (in slowly flowing water Eyre and Nodder[49] estimate a complete change of water over 3 or 4 days). The process is judged to be sufficient, when tests show that the fibre bundles may be separated from the woody part of the stem. The tests, however, are highly empirical in nature and, although in experienced hands they can provide a useful guide, control of the process is still open to a considerable margin of error.[50]

---

[48]   Warden, A.J. Bib. 59
[49]   Eyre & Nodder, Bib.60 T 262.
[50]   Eyre & Nodder, Bib.60 T 238. Two of these traditional tests are described, one of these being the "break" test, where the sample is bent over the finger and the degree of separation is noted. It is pointed out that, provided sampling of the mass is adequate, in *experienced* hands this test can provide a useful guide, but that it suffers from an important objection. Namely, the examination is carried out on *wet* straw, which may appear brittle when wet but which on drying (the next process to follow) will cause unfermented material to bind the remnants of the matrix in which the fibre bundles are embedded to the woody stem again, thereby affecting the efficiency of the later processes to a greater or lesser degree. This test was used mainly in Ireland and Russia. The Belgians used a similar sort of test known as the "pull out" or "loose core" test.

Plate 18

CROSS–SECTION OF THE FLAX STEM

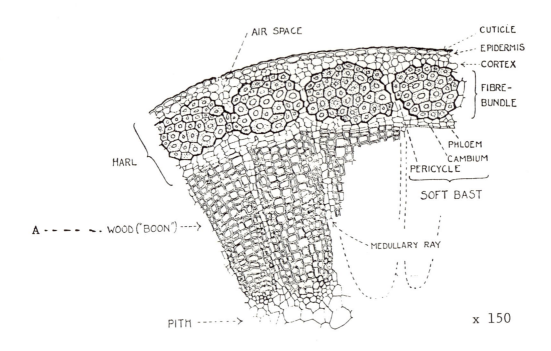

x 150

A Also known as the " BUN " in the 17th C.

(Reproduced here by kind permission of the Textile Institute, Manchester)

Eyre and Nodder,[51] under carefully controlled experimental conditions, identified 4 stages in the retting process. Very briefly, the first stage embraced the wetting and swelling of the straw and the solution and fermentation of soluble materials in it producing a great deal of temporary acidity due to carbon dioxide in this solution; during the second stage conditions become totally anaerobic and there is a rapid development of a different *permanent* acidity combined with further temporary acidity; and by the end of this stage the majority of the water soluble materials have been removed and the loosening of the fibre bundles is just becoming apparent. During the third stage permanent acidity continues to increase and fermentation of the soft tissue takes place resulting in further loosening of the fibre bundles. They reported that at the end of this stage the dried straw could be cleaned reasonably well in the scutching process.

It is *the fourth stage* that is the critical stage, because a balance has to be struck between continued loosening of the fibre bundles and the acid attack on the materials that *join the fibres to each other in these* bundles, affecting their strength adversely. It is at this critical point also that the effect of other variables, such as the ripeness of the straw when it was pulled up, becomes noticeable, the weakening of the bundles occurring much earlier for unripe than for fully ripened straw (a condition that is very much dependent on the weather).[52]

In these experiments Eyre and Nodder also examined the effects of other variables on the process, such as preliminary leaching; the character of the water and the differences arising from the use of stagnant and of "flowing" water; included also was the condition of the straw, fresh straw retting much more rapidly than dried and stored material. They also compared the Flandrian method, developed at Courtrai, of double retting in "flowing" water against a single ret, showing that the twice–retted material produced the maximum division of fibre strands without sacrifice of strength.

It can be seen from the above that with the aid of modern methods of analysis and control the effect of many of the variables in cultivation, harvesting and retting can be identified and to some extent regulated. But how many of these would have been known to the earlier flax grower ? and, if some of them were known, would he have had any control over them ? Clearly, with long experience certain factors must have been identified such as, for example, that the water or the soil or some other condition in one district produced better flax than in another; and to this extent some of the difficulties may have been avoidable. But it is really the critical end–point of the retting process that would have been the most difficult to judge, that is the conditions related to the degree of over or under retting that may have taken place; and, in turn, to the quantities of non–cellulosic residues that may have remained in the fibre bundles. Even with the most experienced judgement it cannot have been an easy matter to

[51]   Eyre & Nodder, Bib.60 T 244 fig.4.

[52]   The variations that occurred in the cultivation of flax have only been covered very cursorily here. Apart from the obvious effects of the climate on the crop, Warden cites, for instance, factors such as the density of the crop. Flax was grown either for its seed (linseed oil) or for its fibre, but not for both. In the case of fibre the crop density was high, especially for the finest fibres used in Lawn and Cambric. Again, when grown for fibre the flax is pulled before it is ripe, after the fall of the flower and when the stem was two thirds yellow. If pulled too early, then the fibre tended to be weak; if too late, it produced coarse fibre. A nicety of judgement was required. After pulling and being rippled to remove the bolls, the sheaves could have been stood for periods of varying length and these could have affected, as seen from Eyre and Nodder's work, the rate at which the straw retted in the pond.

determine the state of the flax, which is unlikely to have been uniform throughout the mass, in something as large as a flax pond. That the fibres of the linen rags, eventually used by the paper maker, have endured successfully in paper form for so long is undoubtedly a tribute to the skills of the people who cultivated and extracted the flax in the 17th and 18th Cs., though it must be added that the permanence and durability of the papers concerned are indebted in no small measure to the gelatine used to size them.

When one takes into consideration the difficulties likely to have been encountered in producing a uniform quality of flax in any one given area (now that we have had an opportunity of looking at the retting process) and then multiplies this by all the different countries, localities and climates from which the linen cloth in this country originated, it will be obvious that the quality of the fibres would have varied very considerably, variations moreover that would have been difficult to distinguish in the average paper mill's rag sorting process. All the same "acceptable" White paper was manufactured formerly, but as we have seen the quality improved considerably during the 18th C. largely as a result of a new found ability to by–pass some of the difficulties formerly encountered with the flax fibre.

The Rag Fermentation Process

It was said earlier that we have a twofold interest in the retting process. In the preceding pages the provenance of the linen textiles and the linen rags used by the paper maker has been examined and shown to be very varied, from which it has been inferred that the quality of the flax extracted for these would have been equally variable and, in particular, with respect to the quantities of non–cellulosic residues that may have remained in the fibre bundles.

It is the effect of and the quantities of these residues that may have given rise to the early papermaking practice of fermenting the rags before beating them, a contributory cause at least, if not the main one, because it has been mentioned before that other materials, such as starch applied to the textile in the course of its life, would also have had to be considered in the same context and be removed. The author is indebted to Dr. John Gascoigne (of Messrs. Cross and Bevan)[53] for most helpful suggestions concerning an explanation of the function of this process in the preparation of rags for papermaking.

Dr. Gascoigne pointed out that linen yarn is not easy to separate into its ultimate fibres by mechanical action *with reasonable energy* and the inference from this statement is that the hammers and mortars of the traditional papermaking process lacked the necessary energy to achieve this. Nowadays this separation may be facilitated by preceding the mechanical action, which in any case after the introduction of the Hollander Beater utilised far more powerful sources of energy, with a light alkali boil. Before such means became available to the paper maker it is not unreasonable to assume that he had observed, or been informed of, the effects of the over–retting of bast fibres, such as flax and hemp, which as we have seen from Eyre and Nodder's experiments leads to the almost complete separation of the ultimate fibres; and, later, if proceeded with, to embrittlement. Dr. Gascoigne has suggested that it is in this fourth

---

[53]   Letter 23:Apr:1976.

Plate 19 – Rag Fermentation Process

Illustration of "Le Pourrissoir", *une chambre voutée ou les chiffons, ayant été mouillés, subissent la fermentation.*   (1698).

After sorting into various grades, the rags (in those mills that used this process) were taken to the Pourrissoir, where they were either heaped in piles on the floor, wetted and turned periodically; or placed in false-bottomed tanks, where water was poured over them and allowed to drain away (a treatment also repeated at intervals). The dampened rags were left to ferment for periods varying from a few days to several weeks depending on the season, quantities and quality. After this treatment, which, it was claimed, prevented the paper from being coarse and harsh and not felting properly, the rags were cut into small pieces and rinsed before "breaking" in the troughs.

stage of the retting process that the clue to the origins of the papermaking fermentation process may lie.

It will be recalled (as Plate 19 shows) that the rags in the Pourrissoir, after wetting, were left to ferment for periods varying in length and Lalande went so far as to make the comment that when mushrooms grow on a heap of rags, the moisture is considered to be right.[54] To use his own words, Dr. Gascoigne has proposed that "the development of an active microflora under the largely anaerobic conditions of the wet rag piles would lead to acidity and this combined with attack on the hemicelluloses catalyzed by bacterial enzymes would lead to the hemicelluloses (with pectin and applied starch fragments) being rendered water soluble. The consequent loosening of the binding material from the fibre bundles would then lead to a rapid separation of ultimate fibres from the bundles during stamping (in the mortars). On the exposed surfaces of the rag piles fungi would grow and their hemicellulose–hydrolyzing enzymes would withstand the acid conditions (e.g. the *hemicellulase* of Aspergillus foetidus has an optimum pH value around 3.6) and fungal pigment would be produced".[55]

Dr. Gascoigne has also pointed out that the main fermentation described above would not be the same as with the plant bast fibres since in their case the soluble materials from these would have been removed already (Stages I, II and III). The early period of microbial growth would thus have coincided with the entry into Eyre and Nodder's fourth stage.

The explanation given above seems to fit in well with Lalande's description of the rags in the Fermenting House (Le Pourrissoir) and their behaviour. For example, he makes the following observation "le linge le plus fin se pourrit moins promptement que le grossier, et le linge usé plus difficilement que le linge neuf" because of what he calls "*l'humidité interne* qui dispose les fibres à la fermentation est plus considérable dans le linge neuf ou grossier que dans le linge fin ou usé".[56] In other words he is saying that the non–cellulosic residues, that feed the fermentation, were lower in the fine linen and in the well worn rags than in the case of the coarse linens and in new rags. These conclusions are backed up in a general sense by what we have learnt from Eyre and Nodder's experiments including those concerned with the effects of preliminary leaching. One has to remember also that the rags were not steeped in stagnant ponds, but wetted frequently in piles, so that the duration of the fermentation period could have varied very considerably according to the prevailing conditions. Other factors such as the quality of the water could also have contributed to these variations masking the true nature of the process.

In contrast to the above the Hollander Rag–breaker and Beater, at the time of its introduction (and more so at a later date) clearly had the necessary energy to break down some, if not all, of the linen rags into their ultimate fibres without the aid of the customary rag fermentation process. As seen in Chapter VI, the wind–powered paper mills of the Zaanland (the district where the "Maalbak" was developed) did not ferment rags; there is no evidence that they ever

[54]   Le Francois de Lalande, J.J. Bib.23 Art.16 para 6.   "Lorsqu'il croît des champignons sur le monceau des chiffons, on estime que c'est la marque d'une bonne mouillée".
[55]   Dr. J. Gascoigne's letter 23:Apr:1976.
[56]   Le Francois de Lalande, J.J. Bib.23 Art.16 para 6.

possessed "pourrissoirs".[57] Elsewhere the practice of fermenting rags was discontinued, more rapidly perhaps in Britain than in other continental countries, after sufficient experience of using the new Beater had been gained.

Though not strictly comparable, it is of interest to note that, in modern times, Clapperton stated that *New* White Linen (Damask) cuttings could, after sorting, be "put straight into the breaker, or even the beater, without any boiling or bleaching"; whereas *New* unbleached linen needed *very* drastic treatment, e.g. boiling under pressure caustic soda for several hours, before breaking and beating could be started.[58] This illustrates that the Hollander, albeit a modern version, was at least capable of tackling (in this case materials for very strong and durable bank−note paper) some classes of new linen textiles (scoured and bleached during the manufacture of the textile) without any of the usual papermaking process pre−treatment. If one allows then for the fact that the 18th C. beater at the disposal of the Whatmans was much less effective than the modern one and takes into consideration as well the fact that the Whatmans would not countenance the use of linen rags until "after they are unfit for the purpose for which they were made", one must assume that they would have had no difficulty in coping with the unfermented "used" materials available to them.

It can be seen that unlike many raw materials used for papermaking to−day, where pulps are tailor−made for the paper manufacturer, the flax grower and linen weaver in the 17th and 18th Cs. would have had the textile in their sights. The paper maker, whose interest would have been focused on the ultimate fibres as opposed to the fibre bundles of the yarn, would not have come into this picture at all. By sorting his rags many of the superficial differences occurring in the parent textiles, differences due to various regional and climatic factors and imposed on them in subsequent usage, would have been reduced. But, underlying all of this would have been more fundamental variables, that is seen from a papermaking point−of−view, arising from the imperfect control of the retting process. The paper maker with the Hollander could have taken these in his stride; but had he not had this equipment he would have had to face another hurdle, the random effects of yet another acid fermentation process.

---

[57] No yardsticks have survived by which this performance might be assessed. The Dutch wind−powered mills, it has been claimed, had very considerable reserves of energy at their disposal; and yet, later, the lower powered water mills also succeeded in achieving the required results.

There may have been other reasons than those that might be inferred here for the absence earlier of "pourrissoirs" in the Zaanland mills. For example, this may have been because the materials that they had used there formerly were in a sufficiently weak state so as not to require fermentation. It will be recalled that the first recorded advances in the development of the Hollander or Maalbak, during the 17th C. was synchronous with the intention to make White paper in these mills.

[58] Clapperton, R.H. Bib.61 34/35.

VARIATIONS IN OTHER CHARACTERISTICS OF THE RAG

Up to this point we have been considering the different fibres used in the textiles from which the rags were derived and, in the case of linen textiles, the variations that stemmed from the process used to extract the flax fibre and how this affected the paper maker and his process. But before the rag ever reached the paper mill other agencies, both chemical and physical, would have affected the parent textile and these would have in turn influenced the performance of the rag during papermaking regardless as to whether the mill employed the new or the old technology.

The Whiteness of the Rag

In the manufacture of White paper of any period one obviously has to consider the degree of whiteness that can be achieved and this clearly is related to the raw materials and any subsequent treatment to which they may have been subjected to make them white. The chemical bleaching of textiles, the use of chlorine gas in this case, was not practised industrially until 1787, in Aberdeen. This was followed very shortly afterwards by patents and experiments concerned with the chemical bleaching of papermaking rags and half–stuff.[59] But as earlier paper makers had no such bleaching process to resort to, they had to rely on the inherent whiteness of their rags to get the best results. To some extent they were able to heighten this themselves, as instanced by the younger Whatman's use of Stone Blue in 1765 which he added to his pulp to "whiten it".[60] It is not known how old this practice was, and whether indeed it was accidental or intentional, because it has been claimed that Smalts (a pale glue glass powder made by fusing cobalt oxide and silica) had been added to paper from as early as 1550.[61] In the main, however, the earlier paper makers would have had to rely on the highest degree of "bleaching" that the cloth manufacturer had managed to achieve.

The early bleaching of cloth was essentially a laundering process, alternate "Bucking" (or "Ashing"); and "Whiting" (or "Crofting"), processes that will be described below. Bleaching is a very ancient Art, but it was not until the latter part of the 18th C. that more scientific

---

[59] For a comprehensive account of this development and the younger Whatman's interest in it see Balston, T. Bib.1 103–108.

[60] Stone Blue was a mixture of Indigo (not a fast colour), starch and/or powdered chalk. The reference to the younger Whatman's use of it is given in Balston, T. Bib.1, pp.147/48; see also note 61 below; and notes 78 and 158.

[61] Dard Hunter (Bib.57) claims in his chronology that Smalts had been used in paper since 1550. Blue, however, may have been used intentionally as a whitener from an early period. The OED, for instance, cites under the substantive use of *Blue* the special application of the word for material used by laundresses, giving an example dated 1618. But the first documentary evidence known to the author of it being added to papermaking stuff is that of Whatman's in April 1765. "Formerly our papers were of a very yellow cast. We have improved the manufactory by throwing blue into it, as people do, in washing, into linen." The inference from this is that, at least to Whatman, this was a contemporary innovation in Britain and suggests that the earlier use of Smalts in paper was for colouring rather than for its whitening effect. However Le Francois de Lalande's reference to the Dutch practice of using blue must obviously predate Whatman's (Bib.23 Art.130). The Dutch had also noticed the yellow cast of their papers. There seem to be only two possible causes of this, common to both countries, the use of unfermented rags or the improved washing performance of the Engine. Gelatine size tends to impart a slight yellowish colour to paper, but this would have been apparent long before the use of the Engine or unfermented rags. Whatman's reference to a *very yellow cast* must relate to some other factor where the background colour of the waterleaf paper had either been lessened as a result of improved washing, thus placing greater emphasis on the colour of the gelatine film; or heightened by the use of unfermented rags; the colour of the latter is a debated issue (see note 78).

foundations for this process were laid by Home and Black, by 1771 to be more precise.[62] By the beginning of the 18th C. the Dutch had established a reputation for being the best Bleachers,[63] but their process was based on a highly empirical procedure and even Lalande was aware, rather later, of the fact that the Dutch did not have the right kind of woods to produce good quality lye (the alkaline extract of wood ash) so vital to the success of the process. Edelstein[64] gives the proportions of different ashes used in making up a typical lye for use in bucking; expressed here as percentages of the total weight of ash:−

3.5%	Blue Ash
3.5%	Pearl Ash
23.0%	Marcost Ash
35.0%	Cashub Ash
35.0%	Muscovy Ash

Home discovered that it was the presence of the muscovy ash from Russia that made this mixture effective. This was because the muscovy ash contained sufficient lime in it to produce potassium hydroxide, a much more powerful alkali than the normal potassium carbonate found in most wood ashes (lime, calcium oxide, reacts with potassium carbonate in the lye solution to form potassium hydroxide and insoluble calcium carbonate). Home demonstrated as early as 1756 that it was much cheaper and just as effective to mix quicklime with the common pearl ash than it was to buy and use the more expensive muscovy ash. It was by trial and error, however, that the earlier bleachers had arrived at some sort of optimum for the composition of their ash mixture.

It had been known for a very long time that the addition of quicklime to ashes increased the effectiveness of the lye and accelerated the "bleaching" process. But it was not known why until Francis Home quantified this addition, empirically, by methodical analysis. Hitherto there had been no means of knowing how much quicklime had to be added without producing an excess of it, because lime, either by itself or in excess, was known to affect the strength of the cloth adversely. Its use, therefore, had been prohibited by law, a law not always adhered to in practice and unwittingly broken every time muscovy ash was included in the liquor.

The bleaching process itself varied from place to place; but it consisted of the following sequence of treatments. Typically, the Dutch (according to Warden) steeped their cloth first in *hot* waste lye, followed by fresh lye, the steeping lasting up to 8 days. Edelstein gives another recipe in which the linen was first fermented for 36−48 hours in water containing a little bran or rye before treating it with the lye. This process of steeping the cloth in the lye was known as bucking and in Holland it might have been repeated 5 or 6 times, the strength of the lye being reduced each time. Sometimes (Edelstein) the liquor contained small amounts of soap. In other cases (Warden) the cloth was washed after bucking with (soft) Black Soap (the impure, dark coloured settlings found in the soap kettle after removal of excess lye and allowing it to stand); the cloth was then wrung out.

[62] Edelstein, S.M. Bib.62.

[63] It has been suggested that the early location of the bleaching industry in Holland and Germany may have been determined by plentiful supplies of buttermilk being available there. (Sykes, Sir A.J. "Concerning the Bleaching Industry" Manchester, 1925).

[64] Edelstein, S.M. Bib.62.

The accounts tend to differ somewhat in their description of the processes that followed. In Warden's account the next process was to sour the cloth in a vat of buttermilk (lactic and other acids) for a period of 1–3 weeks. This was followed by "whiting" (crofting or grassing), which consisted of laying the cloth out on the grass, where it was exposed to sunlight for long periods. This is where bleaching proper took place.[65] Other accounts relate that the cloth was whited immediately after the bucking operation, being watered periodically during grassing. The whole bleaching operation could take up to 5–6 months to complete.

The Irish used much the same process, but the farmers only undertook the weaving there; the bleaching was too long and expensive a process for them to carry out. It was left to the Clothiers to contract this work out.[66]   After 1756 the process time in England was considerably reduced by the use of sulphuric acid for souring the cloth after bucking.[67]

The bleaching process, prior to the advent of specific chemical bleaching agents, was in reality predominantly a washing process assisted by the oxidation effects of sunlight. This is in contrast to the later methods which relied on chemical oxidation with alkali treatment as an auxiliary. The earlier process was dominated by the quality of the lye whose active component was potassium carbonate (closely allied to sodium carbonate or "washing soda" but not produced in quantity until after 1787). It had been known for a long time that the effectiveness of these alkaline carbonates could be increased by the addition of quicklime, which converted the carbonates into strong alkalis. These powerful alkalis are capable of swelling the cellulose to a much greater extent than the carbonates, liberating impurities from the more inaccessible regions of the fibre. Home had shown in 1756 that controlled addition of lime to lye accelerated the bleaching process without affecting the strength of the cloth adversely; as a by-product of this discovery he also demonstrated that muscovy ashes used on their own, though bleaching satisfactorily, reduced the strength of the cloth. It was to be another 15 years before his work was explained in chemical terms by his colleague Joseph Black.[68]

It could be said then that it was not until after 1771 that there was any scientific basis for controlling the composition of the liquor used for steeping the cloth. This is only part of the story because Gascoigne has pointed out that "in the 18th and 19th Cs. little control would

---

[65] Dr Gascoigne (Messrs. Cross & Bevan) in a letter of 7:Dec:1977 pointed out that cellulose is tendered by exposure to sunlight, mainly as a result of photochemical oxidation, which includes the oxidation of certain other impurities present, such as grease, natural colouring matters etc. converting them into compounds that are soluble in alkali. The effectiveness of the ultra-violet rays in the sunlight is tempered by the temperature and humidity conditions prevailing at the time of exposure. Over long periods this would have led to some weathering action on the cloth as well.
    In Ireland Gill (Bib.38) states that there was a close season for bleaching; no bleaching took place between August and February.

[66] Gill, C. Bib.38.

[67] A contemporary comment on the introduction of sulphuric acid for use in souring is of interest (O'Brien "The British Manufacturing Companion", 1795). "Bleaching with Milk or Vegetable Acid, not being in use now among Callico-Printers, the mineral is here chiefly spoken of; of which mode it may be observed, that it unquestionably took some time to get into general practice. Vitriol being of so corrosive a nature it might naturally be expected to injure the cloth very materially."
    "But experience has shown, that by being properly weakened, generally so much that it is not stronger than Vinegar, and may even be drank, that consideration has therefore vanished." Like the chemical bleaching processes that followed later, it is clear that the introduction of sulphuric acid as a souring agent must have passed through a damaging phase in its history; the last traces of sulphuric acid are *very difficult* to remove from a material like cellulose.

[68] Edelstein, S.M. Bib.62.

have been exercisable because the necessary chemical criteria were not available, only physical strength and whiteness were assessed. The heating in alkali can lead to oxidative damage and this can be added to by hydrolytic and oxidative damage in bleaching". The latter can be particularly severe under certain conditions. It was to be a another 100–150 years before analytical methods were developed for quantifying the damage inflicted by the action of these chemicals on the cellulose.[69]

It can be seen from the foregoing that the traditional bleaching process of textiles, which the paper maker relied on indirectly for the whiteness of his rags, was of an empirical nature founded on trial and error and confirmed by experience; and to that extent unpredictable. One has no means of telling to–day just how much material, through lack of proper control over the process, turned to dust in the course of its life between the plant from which it was extracted and the rag stage ready for the sorters in the paper mill.

From Textile to Rag in the 17th and 18th Centuries

What is a rag? In an age accustomed to using pulps processed to a tight specification the reply to this question is usually expressed with a shrug of the shoulders. If one reverts, yet again, to the younger Whatman's definition that "Linnen and Ropes, after they are unfit for the purpose for which they were made, are the best Materials that can be procured for making paper", it is obvious that apart from the acquisition of superficial dirt, stains, grease, applied starch, soap residues etc. more fundamental changes must have taken place between the former and the latter states.

Setting aside for the moment the element of accumulated dirt, this passage from cloth to rag may be conveniently divided into the effects of chemical and physical agencies on the original material. If one assumes that any injurious chemical treatment in the processing of the textile, or during its later history (see later), had manifested itself by the time the rag stage had been reached, then one might also assume that, after sorting out the affected material, the best qualities of linen rags, those used for Writing paper, would have been composed of cellulose in a reasonably undamaged state.[70] The reasons for thinking this are twofold.

Firstly, modern experience has shown that the best qualities of *used* rags (known as No.1 linens and No.1 Fines, a mixture of cotton and linen) had on average an acceptably high alpha–cellulose content and low copper number, values indicative of their good chemical state.[71] After a light alkali boil under pressure and controlled chemical bleaching, the quality

---

[69] These include methods such as Cross & Bevan's Alpha–cellulose test; Schwalbe's Copper Number test, later modified by Clibbens & Geake; and Farrow & Neale's Viscosity tests, also modified and standardised by Clibbens & Geake in work carried out at the Shirley Institute, Manchester.

[70] For those readers unfamiliar with the characteristics of the cellulose chain molecule (not yet discussed) the significance of this comment may not be appreciated at this point, but will become apparent later.

[71] It must be understood that the term "acceptable" is applied here to the rags as *raw materials*. In this state their alpha–cellulose values may have been 88–89%, when one would be looking for a value > 97% in purified cellulose; likewise their copper numbers would have been 0.6–0.7 compared to 0.2 or less for the latter. But these rags would inevitably have contained an element of weak material that the boiling, bleaching and rag–breaking of the present day process would have removed, raising the values in question.

of these rags could be improved so as to give papers of the highest standards. Sometimes, but not always, it was necessary to add a small proportion of new rags (New Pieces) as half-stuff to stiffen the furnish a little. In the 17th and 18th Cs. the paper maker would not have had these improving processes available to him, nor would his beating equipment have had the power to handle the new pieces, and yet he produced papers that have endured centuries of handling and storage, often under unfavourable environmental conditions, and despite this remained in a state comparable to that of good papers made to-day (making some allowance for the dirt and discolouration acquired with the passage of time). The conclusion to be drawn from this is that the best qualities of rags in those times were in as good a state chemically as the modern equivalent of *used* rags, rags that may have undergone more chemical treatment in the course of their existence than some, at least, of the earlier linens (see comments below on Poggi's "northern and southern rags").

Secondly, the survival of the best qualities of rag is also related to wear. The principal physical agents that convert a textile into a "rag" can be divided into those concerned with Flexing and those with Abrasion; of these two abrasion is undoubtedly the more damaging. The author consulted the Shirley Institute on these points[72] and they were of the opinion that flexing arising from movement of the body, for instance, would result in very little damage; with regard to breakage of the fibre, the cotton fibre is an entity that will not break down into finer fibres; but the fibre bundles of the linen yarn are susceptible to this and will break down into finer fibres. All the same the effects of flexing are negligible in comparison to the effects of abrasion. So here again, one can only assume that in earlier times, after rag sorting, the best quality of rags would have been those that had suffered the least abrasion. If they had been abraded severely, then it is likely that the yield of the fibre after the rag-breaking process had taken place would have been low and that much material had been lost down the drain. In other words it would not have been incorporated in the paper; only the paper maker would have suffered.

It is self-evident that the chemical and physical agencies described above would have affected a textile to a greater or lesser degree dependent on the manner in which it had been used, for instance, as clothing, table or bed linen, or a lining to other, more costly materials such as vestments, curtains, bed-hangings etc. and, if it was considered to have been re-usable, whether it was used again; and in addition to all these possible uses and re-uses, one has to consider how often the article in question may have been laundered in its life and under what conditions and whether starch was added. The life span of a textile may have been very long in some cases. Finally, regarding their provenance, as noted in the main text, the younger Whatman commented on the fact that the largest quantity of the best rags was likely to come from "Families of opulence" to which he added the cynical comment "where the fewest are saved".

---

[72] Wiseman, L.A., late Director of Research at the Shirley Institute, 25:Aug:1976. In a general comparison of this kind one has to bear in mind that in the case of the 17/18th C. papers the protective effect of the gelatine size might mask to some extent the true quality of the underlying cellulose (see Poggi's comments on the quality of contemporary rags this App. p.205).

Extracts from Anthony Poggi's letter to the younger Whatman[73] show that contemporary laundering practice was another important feature in the history of a rag's antecedents. This appears to have differed very considerably between the northern and southern countries in Europe and paper makers in the United Kingdom received rags from many different sources during the 18th C. ranging from Italy and "The Straights" in the south to Germany, Poland and Russia in the north though mainly from Germany and to a lesser extent Flanders.[74] By the sound of Poggi's comments the rags from the northern countries would have received little damage chemically from repeated laundering "by dint of soap alone". The deposition of lime soaps in the textiles, which one would expect from this treatment, would lead to harshness or, as Poggi put it, to "an inflexible quality and substance, which render it unfit for *our* purpose".[75] In this context one wonders how often the cloth was washed, probably more often in the latter part of the 18th C. than earlier. Cotton materials which were introduced at the end of this century, though less hard wearing than linen, were easier to wash, dry and iron, a factor of some importance when soap was not very effective.[76] Also at this time men of fashion were much more conscious of hygiene than they were earlier and hence a preference for materials that would wash easily.[77]

The southern rags, that Poggi refers to, were evidently severely affected by the "hot strong lees" that were poured repeatedly over the laundry. In fact he adds the comment that "the corrosive nature of the Lees ennervates the fibres of the Flax, and in time reduces it to a mere Caput Mortuum". He goes on to suggest that similar treatment of the two kinds of rags in the Mortars (Beaters by this time in England) "at the end of a certain time" might reduce French rags to dust "while the English would be found far from being so". This is a revealing comment and the difference in laundering procedure might have been one of the reasons why Germany became the principal source of rags for Britain during the latter half of the 18th C.

Before leaving this subject it should be said that it is a mistake to consider "rags" *solely* as a sort of débris or litter left behind by mankind. It is perfectly true to say that by definition rags were a waste material and in the 17th and 18th Cs. they were indeed collected from dunghills and laystalls. All the same rags had a certain character of their own and this manifests itself in quite different ways when one compares their use in the old and the modern processes. The ultimate demise of the rag as a raw material took place with the

---

[73] Bib.39, written in the 1790's and quoted again in this App. note 143.

[74] Coleman, D.C. Bib.15 107 Table VI "Import of Rags, 1725-1800". See also this App. note 157.

[75] Poggi's comments were directed at the paper's performance in the printing of engravings and he is referring here to the more receptive character of French papers in this respect compared to those made from the soap-laundered northern rags.

[76] It is interesting to compare these comments on the advantages of cotton (Edwards, M.M. Bib.54 Chap.III) with Houghton's comments (Bib.20 I.348 : 15:Feb:1694) on the manufacture of soap and its use in laundering in the 17th C. Of the effect of soap, he wrote, "Alkali-salt, Oil, and Tallow, intimately mixed by boiling by help whereof, the unctuous parts of the tallow and oil do incorporate with the grease that is on foul linen, clothes etc. and the alkali-salts do so far divide those greasy particles, that they are capable of being diluted [mixed with water] which of themselves could not be." On the making of soap it is interesting to note (in view of Francis Home's work later) that Houghton's prescription for good soap was to take 2 parts of olive oil, 1 part tallow boiled with Soap-lees, extracted from Salts of Wood (Pot-ashes mostly imported) and *"slak'd with Lime to make them yield more Lee"*.
With regard to the quantities of soap used in London, he said that "I find upon enquiry that in good citizens houses, they wash once a month, and they use, if they wash all their clothes at home, about as many pounds of Soap as there are heads in the family, and the higher people be, the oftner they change, the less pains the washers are willing to take, and the more soap used; but on the other side the poorer people are (and the poor are the most numerous) the seldomer they shift, and so use less; and 'tis probable we may allow a pound a head once a month for every soul in the bills of mortality."

[77] Edwards, M.M. Bib.54 Chap.III.

increasing use of man–made fibres combined with vegetable fibres in post–world war II textiles. But before this there was a certain air of romance about the origin of the modern "rag". After a preliminary grading by the rag merchant the rags reached the paper mill in every sort of form and from every sort of place ranging from ship's linen on the one hand to collars and cuffs on the other (with memos often scribbled on the cuffs). But once these rags had been dusted, sorted, cut into small pieces in the rag–cutter, carefully and skillfully blended to suit the paper they were destined to become part of, they were boiled and bleached and by this stage had acquired an anonymity that may be contrasted with the fate of some rags in the earlier paper mills. Since the paper makers of the 17th and 18th Cs. would not have had any of these processes at their disposal the result would have differed in two respects.

First, although in the lower grades of paper the kind of rags used would have quickly assumed the anonymity of their modern counterparts, in the manufacture of the best qualities of White paper the paper maker would have chosen the best and the whitest rags at his disposal, because there was no other option open to him, so that with nothing but an intervening cold water wash the rags in this case would have retained an element of their individuality in their passage from textile to paper.

Second, in the pre–chemical era a variable proportion of the non–cellulosic components of the flax fibre would have persisted through the textile stage to the rag and have affected the quality and performance of the papermaking stuff. Before the arrival of the Hollander beater the fibres in the rags were in many cases too stiff for the traditional hammers and mortars to handle adequately to produce uniform stuff and it was necessary, therefore, to resort to a fermentation process to loosen the fibre bundles. In the modern papermaking process it would be fair to say that the boiling and bleaching processes would have enabled the paper maker to have used the traditional stamping process, had he wished to do so. But there was a transition period, namely the 18th C. when the paper maker had the use of the newer beating equipment, the Hollander, on the one hand, but not the chemical treatment introduced in the 19th C. on the other. This difference appears to have left its mark on the colour of the paper. As to whether a change in colour was due to the paper maker's use of unfermented rags or to some other innovation is a debatable point (cf. Note 61). But the fact is that, apart from the "very yellow cast" that the younger Whatman had observed in his papers, a noticeable distinction can be drawn between cream coloured papers made by the elder Whatman and the grey cast of another maker's paper bound into the same volume of a printed book; a grey cast is common in papers made under the conditions employed in the Old Technology whereas both of the Whatmans used the New.[78]

---

[78] The colour differences referred to could have been due to many different factors and it would be difficult, if not impossible, to identify the causes in any particular case with certainty. Some paper makers might have used a higher proportion of rags derived from unbleached as distinct from bleached linens; others may have added pigments. Another reason might have stemmed from more efficient rag washing in one case than in another. A distinction also has to be made between papers made from fermented and unfermented rags. In general the causes can be divided into extrinsic and intrinsic. The fortuitous use of this or that rag or this or that washing procedure could have occurred in any age. But the 17th/e.18th C. rag in Britain (and much later in France) differed fundamentally from those used later in the 18th C. in that one was fermented and the other not. (Some of these differences are dealt with in more detail later in this appendix. See also Desmarest Bib.21 486/487 who points out that where there were seasonal shortages in water, with a consequent diminution of water power, a certain degree of rag fermentation persisted in France even after the Engine had

(continued...)

RAGS TO PAPER

In the previous pages an account has been given of the materials from which rags for use in 17/18th C. White papers were derived; as well as the different treatments they encounted in their journey to the paper mill. In Chapter VI the mechanics of the operations required to convert them into papermaking stuff have also been examined. At this point these two aspects have to be married in order to see what actually happens when this conversion takes place.

It might be thought that to do this was just a matter of sub–dividing the rags by attrition thereby forming a pulp of dispersed fibres suspended in water from which a sheet of paper could be made and where the physical properties of this sheet were dependent on some kind of fibre "entanglement". This was probably the view commonly held until the second and third decade of the 20th C. and there may still be people who believe this to be the case. Dismemberment of the textile and the dispersion of the fibres in water, of course, take place; and it is also true to say that some kinds of paper can be made simply by dispersing fibres in water and forming a sheet where fibre "entanglement" may play a significant role in holding the sheet together,[79] but with rags and the making of conventional papers the process is much more complicated than this.

First, one has the plain mechanics of reducing the textile to papermaking stuff. It will be recalled that broadly this is achieved by stages in order (a) to conserve material; and (b) at the same time produce a pulp that will suit a desired end–product. Second, in the course of reducing the rags in this process the surface area of the fibres is developed in various ways to produce a pliable matrix which, during deposition from being suspended in water and subsequently being dried, has the property of being conducive to the formation of inter-cellulose hydrogen bonds. It is these bonds that are responsible for holding the sheet of paper

---

[78](...continued)

been introduced, unlike the Zaanland where it had never been practised).

There seems to be some difference of opinion as to the inherent colour of the fermented versus the unfermented rag. Peter Bower (letter to author 19:Apr:91) refers to his experiments in making papers from both unfermented and fermented rags and this is reflected in his observations on papers used by J.M.W. Turner (*Turner's Papers*, Tate Gallery 1990 pp.35, 44 & 46). This is not the place to comment on these except to say that English White paper by this time had long abandoned the practice of using fermented rags. But one point in his findings that emerges clearly confirms Le Francois de Lalande's observation that rags that had been overfermented bore a yellow colour that was difficult, if not impossible, to remove by washing. As seen earlier the fermentation process embraced many other non–cellulosic residues and acquired impurities which would not be present in 20th C. rags, making comparisons difficult.

In fact both Le Francois de Lalande and Desmarest noted that contemporary Dutch papers, made in the Zaanland from *unfermented* rags, had a "teinte jaunâtre" that required the addition of blue to compensate for this (cf. note 61). Certainly, the colour of "White" English papers made during the pre-chemical era from unfermented rags had a much better creamier colour than those made in earlier times, a general observation to which no doubt there may be exceptions. It seems from this that one cannot draw any firm conclusions as to the colour that may have been imparted to papers using or not using this process. The fact is that many early papers are characterised by a greyish cast compared to the warmer colour of those made in the post–fermentation era.

[79] Papers, for instance, can be made from various non–cellulosic fibres. Edward Lloyd made paper from asbestos fibres as early as 1684 (although Voorn, H. Bib.52 3 reported that a Jesuit priest, Athanasius Kirchner, is alleged to have made some as early as 1646). Lloyd noted that the "Paper made of it proved very coarse and apt to tear". Today, many other non–cellulosic fibres are in use, such as glass micro–fibres etc., for making special papers. But, apart from the fact that these fibres are very expensive and scarce compared to the abundance of cellulose, *without special aids* papers made from them have a low dry strength and are very easily dispersed again in water. In contrast very considerable dry strength properties can be achieved in waterleaf cellulose papers (that is without the use of any additives) and not insignificant wet strength properties as well.

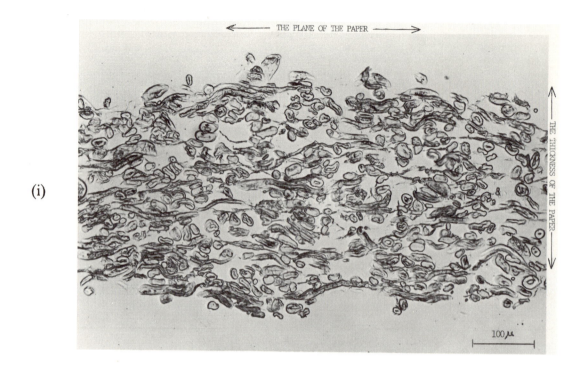

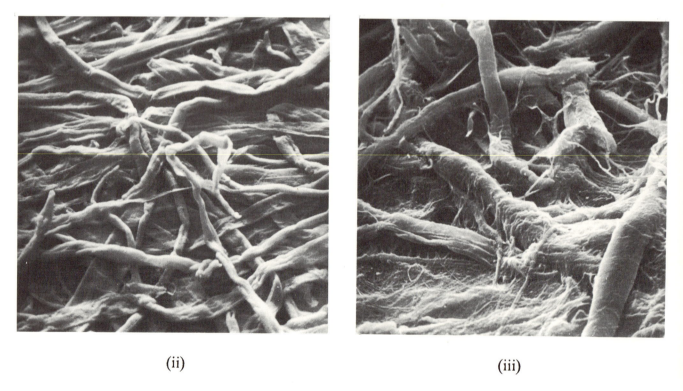

(ii)   (iii)

(i)   Cross−section of a Whatman Filter paper (free beaten cotton)
(ii)   Stereoscan Electron Micrograph of Filter Paper [240x] (free beaten cotton).
(iii) Stereoscan Electron Micrograph of highly beaten cotton fibre [600x]

Note the contrast in fibrillation between (ii) and (ii), free v wet beating.

(i) Original photograph by kind permission of the Shirley Institute. (ii) and (iii) Original photographs by kind permission of Mr. Jouko Laamanen and the Finnish Pulp & Paper Research Laboratory.

together. The underlying principle of this treatment is the enhancement of the special relationship that exists between cellulose and water.

No attempt will be made here, apart from a very brief mention of the subject, to explain this relationship in terms of *a sheet of paper*; it is an extremely complicated one and quite outside the scope of a book of this kind. Indeed, it is doubtful whether anyone has yet succeeded in producing anything like a comprehensive explanation. One of the reasons for this is the very intricate nature of the structure of the fibre itself and its behaviour under various conditions, a point referred to later. Another difficulty arises from the enormous differences in scale between the matrix of fibres that make up a sheet of paper and the cellulose molecules of which these fibres are composed and through which they form the hydrogen bonds mentioned.

Temming and Grunert illustrate this difference in scale diagramatically[80] and show that if the dimensions of the β–glucose unit (the smallest in the cellulose chain) are considered as having a value of 1 (~7.5Å)[81] then by comparison the *diameter* of a cotton fibre is ~30,000 – 47,000 times greater than this; add to this the fact that the *lengths* of the shortest cotton fibres are ~1,000 times greater again, one can see that the differences in scale between the smallest unit of cellulose and the fibre dimensions are very considerable. When one considers this in relation to the distance over which the inter–cellulose hydrogen bonds, referred to earlier, are effective, namely, only a few Ångström units (in other words within the same order of magnitude as the basic units of cellulose), one begins to realise that the dimensions within the matrices, which make up a sheet of paper, are of an almost astronomically different order to the dimensions within which the bonds operate. In spite of these enormous spaces bonding does take place, but the difficulty lies in trying to relate this to the behaviour of a sheet of paper when, say, its structure contracts on drying and expands again on wetting with bonds forming and some of them breaking again and perhaps forming again at new sites on re–drying.

Many attempts have been made to analyse the structure of paper, some of these being based on the fact that the majority of the fibres lie in the same plane as the sheet itself and that the sheet may, therefore, be regarded as an accumulation of mono–layers superimposed on one another (see Plate 20). By studying the properties of a mono–layer, it has been possible to construct a mathematical model of a multi–layered sheet based on this data, which showed some similarity to a sheet made in the normal manner.[82]

(*Note*. The reader may be a little puzzled here by the switch from flax/linen to cotton cellulose in the last page or two of the text. This switch has been made because, when we come to consider the cellulose/water relationship in Part II, it is probably true to say that more facts and statistics are available and accessible for cotton than there are for flax. Since we are now considering the fine structure of cotton and the ultimate fibres of flax down to the molecules themselves, which after all are both cellulose, the differences between the two on a macro–scale need not concern us here. The moisture regain of the flax fibre is generally regarded as being rather higher than that of cotton (12% compared to 8.5%). In any case by the time these fibres have reached the rag stage, it is likely that they will have undergone modifications that cause them to behave rather differently to the carefully isolated and purified fibres on which many of the statistics are based. In the Section that follows this, for the sake of simplicity, pure cotton will be used solely as a model for presenting this relationship between cellulose and water).

[80] Temming & Grunert Bib.57 21.
[81] Å = Ångström unit = 10^{-8} cm.
[82] Kalmes, O., Corte, J. & Bernier, G. A series of papers published in TAPPI 1961–3.

PART II

THE CELLULOSE / WATER RELATIONSHIP

For the purpose of this book it will be enough to give a brief and simplified account of this relationship between cellulose and water merely to demonstrate that it has a vitally important connection with the beating of the rags and that it plays a key role in the processes of sheet formation and drying.

Up to this point attention has been focused mainly on linen rags and thus the flax fibre, from which they were woven. The flax fibre is a much longer and stronger fibre than the cotton hair, but it is less pure than cotton and more easily disintegrated by chemical means; both flax (see Plate 18) and cotton have a complex macro-structure. In the case of cotton the fibre is made up of many layers of "fibrils" embedded originally in non-cellulosic substances. These fibrils are laid down during growth in concentric rings (like the growth rings of a tree) and wind spirally round the longitudinal axis of the fibre.

The fibrils are themselves made up of much smaller units known as micro-fibrils;[83] and these, in turn, of micellar strands, both of which are close in scale to the basic β-glucose unit (diameters ~40 times and ~20 greater respectively). It is with units of this kind that the behaviour of the cellulose will be considered here. But first it is necessary to account for the forces that hold these units together, because it is the relative strength of these that determines the effect that water and other chemicals have on the cellulose.

Reverting once again to the Ångström unit scale, cellulose molecules take the form of *very* long chains made up from what are known as β-glucose units, the smallest units in the chain; Temming and Grunert estimate anything from 10,000–12,000 units in the molecule of *raw* cotton linters. These units are linked together (see Figure 8) by *strong* glycosidic bonds that confer on the chain its great strength. Projecting from the sides of these units are numbers of hydroxyl groups (--OH) and these groups, when placed in a favourable position in relation to other hydroxyl groups, that is favourable in terms of distance and orientation, will form bonds (hydrogen bonds) between each other. Due to inevitable variations in the distance and orientation occurring between opposing hydroxyl groups, the hydrogen bonds formed between them will tend to be variable in strength and are much weaker than the glycosidic bonds holding the β-glucose units to one another.[84] It may be noted at this point that this hydrogen bonding may take place between the hydroxyl groups of either other cellulose chains or, and *this is most important*, water (which may be written as H--OH).

---

[83] Temming & Grunert, Bib.57 20. The structure is illustrated diagrammatically.

[84] Apropos of this reference to the "favourable" position of hydroxyl groups *vis-à-vis* other hydroxyl groups, it has been shown that the hydrogen bonds in the cellulose matrix cover a wide range of bond strengths. Formed under the most favourable conditions they are not easily broken; but in the weaker examples, due to resonance or thermal vibration, the bonds may be broken much more easily and the hydroxyl groups thus freed will compete for other hydroxyls as, for instance, those of water.

Figure 8

CELLULOSE

[structural formula]

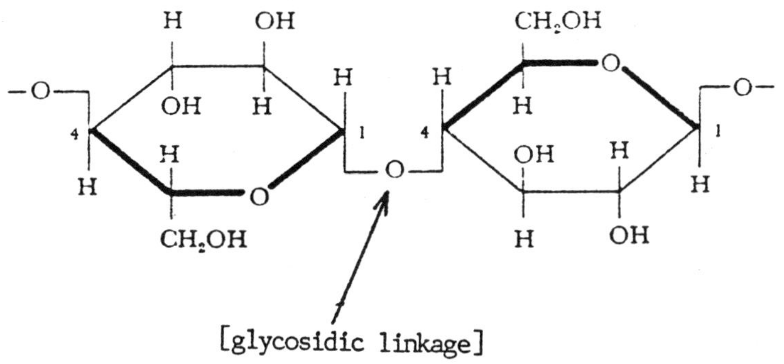

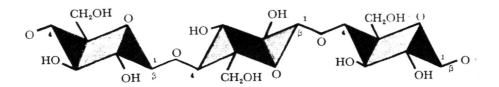

[glycosidic linkage]

Illustration of two β–glucose units joined by a glycosidic linkage

Illustration showing differing orientation of alternate β–glucose units

During growth these long cellulose chains become organised into "bundles" which are held together by hydrogen bonding. These bundles are, in fact, the micellar strands and multiples of these become the microfibrils.[85] Because of the movements of the cellulose chains during growth and dehydration and compounded by the fact that the β–glucose units are oriented alternately in different planes along the axis of these chains (rather like the links of a common metal chain, where one link lies in a plane roughly at right angles to the next) a complex 3–dimensional pattern results, which prevents the micellar strand from being absolutely regular throughout its length. Consequently, the bonding in these strands will also vary in degree along their length.

The term "Crystallite" has been given to those regions in these strands where the degree of bonding is regular and extensive, and therefore strong. The less well ordered regions in these strands are known as the "Amorphous" zones. *This is an important distinction* particularly in the context of this discussion.

Unlike the amorphous zones, the crystallites are highly resistant to the penetration of water (in fact the reason why cellulose does not dissolve in water has been attributed to their impenetrable nature); and their disruption can only be effected by the action of powerful chemicals, such as concentrated solutions of alkali. In the purer sources of cellulose, such as cotton, the degree of crystallinity is high (~90%); among the impurer sources, such as wood, the porportion of non–cellulosic substances (such as lignin) acquired during growth is much higher and whereas this may interfere with the attainment of a high degree of crystallinity (though there is some evidence that in the raw state the degree of crystallinity is high too), the much more drastic treatment needed to extract the cellulose from the raw material reduces it significantly. In passing it may be noted that although very pure cellulose can be obtained from wood, it is obtained at the expense of its chemical stability, which cotton and linen possess to a much higher degree.

Having described above this propensity of the hydroxyls of one cellulose chain for hydrogen bonding with either the hydroxyls of another *or* with water, depending on the prevailing circumstances and the strength of the bonds involved; and having described also how the micellar strands, formed by the bonding together of several cellulose chains, are composed over their length of highly organised and regular zones (crystallites) interspersed with less organised regions (amorphous zones), one is now in a position to examine certain of the more significant characteristics of cellulose behaviour that stem from these properties. In the context of Rags and of Paper they may be divided conveniently into two:–

(a)   Chemical degradation that leads to chain shortening.

---

[85] Dr. A.R. Urquhart, (Bib.63) has described the formation of cellulose molecules as being laid down from the beginning in the presence of aqueous solutions and hence most, if not all, of the hydroxyl groups will have water molecules attached to them. In the case of cotton, for example, when the boll opens and is exposed to the sun, *the fibres* start to dry and much of this water will be removed, with a corresponding shrinkage of the fibre. In the early stages of this drying, movement of cellulose chains will take place and be relatively easy, because of the highly swollen state of the fibre. This movement will result in the cellulose chains moving into positions that favour the formation of the strongest possible hydrogen bonds with neighbouring chains. As drying proceeeds, this movement will become progressively more restricted by the strong bonds already formed, which in turn will lead to less favourable bonding conditions and hence to an increasing number of weak bonds.

(b)   The various states in which water may become associated with the cellulose matrix and their effect on the formation of a sheet of paper.

Chemical Degradation

This subject will only be touched on. It has been seen in Part I how in the extraction of the flax fibre from the plant permanent acidity is produced in the liquor during the retting process[86] and, if the process is allowed to proceed too far in Stage IV, the strength of the fibre bundles is affected adversely by this acidity. The same will apply to any kind of hydrolytic acid[87] attack on the cellulose whether it originates, for instance, from prolonged exposure to the acid conditions produced during *rag fermentation* (where this was used); incorrect use of sulphuric acid in the early history of the new method of souring cloth in the bleaching process; or in more modern times the indirect effects of chemical bleaching that has not been properly controlled.

The solutions containing the acids will penetrate the amorphous zones of the cellulose strands and attack the glycosidic linkages, leading to chain scission which (as can be seen from Fig. 8) will progressively shorten and weaken the micellar strands and thus, ultimately, undermine fibre strength. Even under the most modern and carefully controlled conditions used for preparing cotton linters on a commercial scale, the cellulose chains are shortened from ~10,000 plus β–glucose units in the raw state to ~1000–7000 in the purified bleached state.[88] If the degree of polymerisation, as it is known, falls below a value of ~600, then serious degradation has taken place, which will lead eventually to the cellulose crumbling to dust.

Oxidative damage, another form of degradation, can arise from the effect of hot alkalis, particularly in the presence of oxygen, on cellulose. In some cases oxidative damage leads to nothing more serious than discolouration of the cellulose (yellowing); but in other cases it too can lead to progressive chain shortening (see, for example, Poggi's comments on the repeated effect of "strong lees" on French rags). The reason for chain shortening in this case is not due to the scission of the glycosidic bonds, but to the rupture of the β–glucose units themselves.

It is also probable that this repeated laundering treatment with hot strong alkalis (as described by Poggi) led to the increasing penetration and opening up of the crystallites themselves, making them more accessible to the acid solutions produced later during rag fermentation (which was practised in France to a much later date than in England) and hence to more severe chain scission at the glycosidic links.

Translated into 17th and 18th C. papermaking experience, it would seem probable that rags that had been chemically degraded seriously during their history would, because of the effect of chain shortening, have lost cellulose as "fines" down the drain during the process. The net

---

[86] See this App. p.195.

[87] Hydrolysis (here hyrolytic) refers in this case to the decomposition of organic compounds by interaction with water in the presence of acids.

[88] During the extraction and purification of cellulose some of the hydroxyl groups are chemically modified into aldehyde and carboxyl groups, the latter exhibiting significant ion–exchange properties. (Temming & Grunert, Bib.57 27.)

result would have been a very low yield for the paper maker from his rags rather than degraded paper. Once chemical processing had been introduced (late 18th/early 19th Cs.) the story could have been quite different.

The Relationship with Water

Depending on the circumstances in which the cellulose exists at the time, for example, whether it is fully immersed in water (papermaking stuff) or has been converted into a dried sheet of paper, water will be associated with it in various ways, all of which affect the sheet–making process and the subsequent behaviour of the sheet itself.

If one takes the case of the dried sheet of paper first, then under normal atmospheric conditions, the air that surrounds it will contain water vapour. On a dry day the air will contain less water than on a humid one, given the same temperature conditions in both cases. The air, of course, will not only surround the fibres in the paper, but it will also penetrate the amorphous zones of the micellar strands; and the water vapour held in it will thus in the first place attach itself to any *free* hydroxyls there may be within these zones, including those on the external surfaces of the crystallites; and then, as shown previously, any water molecules not already satisfied will compete for the hydroxyls wherever the weakest inter–cellulose hydrogen bonds exist. Because of the constant making and breaking of these weak inter-cellulose bonds, due to thermal vibration, water will become bonded there too (and, indeed, may even proceed further than this by hydrogen bonding through water bridges[89]).

In other words, in these circumstances, there is *chemical combination* of water to individual sites both along the exterior surfaces of and within the cellulose strands. This state is known as one of *"Adsorption"* or the chemical sorption of water by the cellulose; and, for a given set of conditions, is a fixed quantity.

By decreasing the humidity of the ambient air (in other words a drying process), the quantity of sorbed water will be reduced (the drier air will encourage the cellulose to give up water to it during the constant making and breaking of the weaker bonds). Conversely, if the air is humidified, the uptake of water by the cellulose will increase. It will be obvious from this that wherever paper is stored, this process of sorption and de–sorption will be going on all the time due to fluctuations in the atmospheric humidity; and this leads not only to constant movement of the fibres in the paper, but at high humidities it can provide conditions for various forms of attack resulting from, say, residual acidity or from micro–biological agents.

The real process of sorption and de–sorption is a great deal more complicated than the description given above may lead one to believe.[90] But the example given has been used

---

[89] Emerton, H.W. "Fundamentals of the Beating Process", B.P.B.I.R.A. 1957, 108.

[90] When the water content of cellulose is plotted (graphically) against the Relative Humidity % of the atmosphere for a given temperature, one obtains a curve of sigmoidal shape (see fig.9). Dr. Urquhart, (Bib.63) has pointed out that the shape of this curve is probably the result of two curves combined, one representing the hydrogen bonding of water molecules to hydroxyl groups in the cellulose (possibly augmented by some combination of water and other groups as well) and a curve representing in part weaker water/cellulose bonds to less active sites in the substrate; and, in part, the addition of further molecules of water to the sites already occupied, that is what might be termed "water bridges". He goes on to point out that at very high humidities the sorption curve may represent actual condensation of liquid water

(continued...)

Figure 9

SORPTION ISOTHERM (25 °C) FOR A WELL–SCOURED COTTON
(see footnote 90 on opposite page)

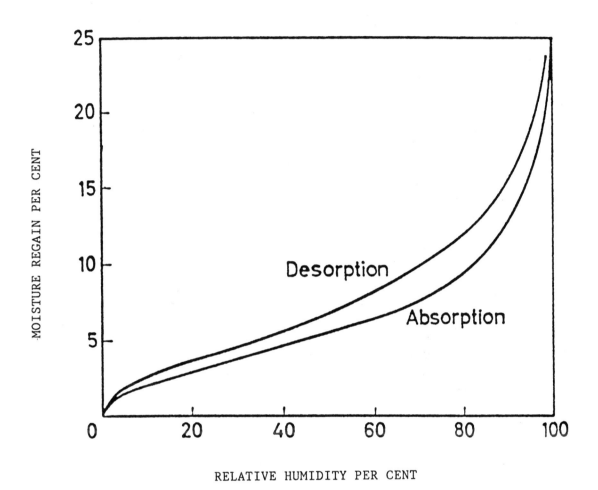

RELATIVE HUMIDITY PER CENT

(Reproduced here by kind permission of the Shirley Institute, Didsbury,
Manchester and The Textile Institute)

to show that in the final stages of drying a sheet of paper the residual water is combined chemically with the cellulose and cannot, therefore, be removed by, say, squeezing or pressing.

It is not intended to take this subject of chemical sorption any further here for two reasons, Firstly, in this book and in this particular section, we are really considering the progress of the cellulose from the rag state, through the pulping process, to sheet-making and drying – in short, one long sequence of saturating and de-watering operations; and, secondly, having achieved the dry sheet of paper, in the majority of cases that we are likely to encounter for the historical periods in question, the paper has been sized subsequently with gelatine size to the tune of some 6–12% by weight. Gelatine is itself a hygroscopic (water adsorbing) substance with quite different properties to cellulose. Consequently, one has a product which may contain variable amounts of either constituent and whose condition at any given point would be impossible to resolve into the expected behaviour of its component parts. An added complication may arise from the fact that gelatine also has *intra*molecular bonds and it is conceivable therefore that, when the sized paper is dried, *inter*molecular hydrogen bonds may be formed between the cellulose and gelatine.

If one now goes to the other extreme and considers, say, cellulose "stuff" immersed in water, then one would find that there was a surplus quantity of this water surrounding the cellulose and filling the cavities between fibres. This is sometimes called "free" water in the sense that when, for example, the vatman dips his mould into his pulp and lifts it out again, the surplus or "free" water will drain away rapidly through the wire cover; it is not bound to the cellulose in any way.

This surplus water is necessary in the first place in order to beat the rags; and an even greater quantity is needed to dilute the stuff to enable the vatman to manipulate it and form a good sheet of paper on his mould. Hence, during the conversion of the rags into papermaking stock and even more so when this stuff is formed into paper, a very large excess of water is necessary. The author has known conditions on a certain kind of paper machine, where sheet formation takes place from stuff in the ratio of 1 part cellulose to 2000 of water. A ratio of ~1 : 200, however, is more typical; and in the making of paper by hand the consistency of the stuff may be much higher than this, even as much as 1 : 50.

One has then "free" water at one end of the process and "chemically sorbed" water remaining in the cellulose at the other (the latter is present all the time, that is to say whether the cellulose is saturated with water or in its so-called dry state). There is, however, a stage intermediate between these, in which water has an important effect on the early stages of

---

[90](...continued)
    in the finer capillary structure of the cellulose matrix.
    It has been found, however, that the sorption and de-sorption curves follow different paths (see fig.9 : there is a wealth of literature on this subject stemming from the classic work done by Urquhart, A.R. and Williams, A.M. published in the J.Text.Inst. 1924, 15 T433). The difference is largely attributable to the fact that during de-sorption, at any given point, due to the constant breaking of hydrogen bonds, there are more sites available on the cellulose for the competing water than in the case of cellulose that is in the process of absorbing water under exactly similar conditions.

consolidating the sheet of paper as it is made. The water which brings this about is known as "capillary" water.

This water is essentially the same as the "free" water referred to above, in that it is *not* chemically bound to the cellulose; but it differs merely in its physical relationship with the cellulose matrix. After the "free" water has drained away (and, in the case of handmade paper, after further free water has been expressed from the matrix by the Press), there is a certain residue of water left in the sheet, which is made up of water held *physically* in the finer capillary structure of the cellulose *plus* that proportion of the water that is satisfying the potential bonding sites within the cellulose matrix (in other words the water that has been chemically sorbed *under that particular set of conditions*).

This so-called "capillary" water affects what has been described earlier as the consolidation of the sheet of wet pulp through forces known as *Surface Tension*. At the surface of water (e.g. at the water/air interface) the cohesive forces of the water molecules are not balanced as they are, for instance, in the main body of the liquid. The net result is that the outer layer of molecules behaves like a thin elastic film always trying to contract (it is these forces that makes rain contract into drops). It is well-known that, if one dips a fine glass capillary tube into water, the water will rise up the tube until the weight of this column of water reaches a state of equilibrium with the surface tension forces that caused the water to rise in the first place.

The capillaries in the cellulose matrix, however, are:-

(a)   not rigid like the glass tube

(b)   irregular; they may, for instance, be two-ended, taper or just be the minute crevices formed at the crossing points of fibrils.

(i) two ended   (ii) tapering   (iii) crossing point

(Barkas, W.W. Bib.64 471)

Under saturated humidity conditions all such capillaries will be filled to the brim with water. As the humidity falls, the largest capillaries will empty first. The tensions in the capillaries above a Relative Humidity of 99.9% (i.e. near saturation) are small; below 99% RH the tensions within the capillaries increase rapidly and reach *very high* values as the relative humidity conditions approach 60–70%, by which time the remaining capillaries will have diameters of molecular dimensions, in other words the surface films will become non-existent and the water will have evaporated.

Plate 21

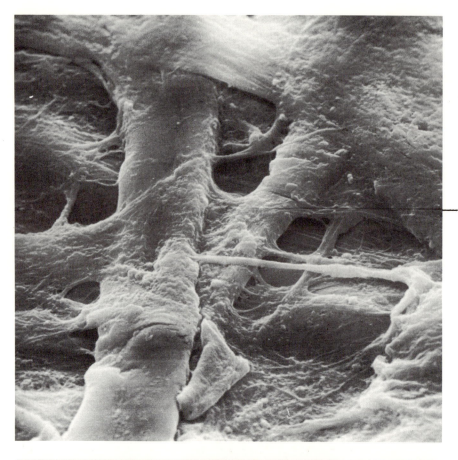

Stereoscan Electron Micrograph of Banknote paper [1200 x]

Highly beaten fibre

—— (indicator line)

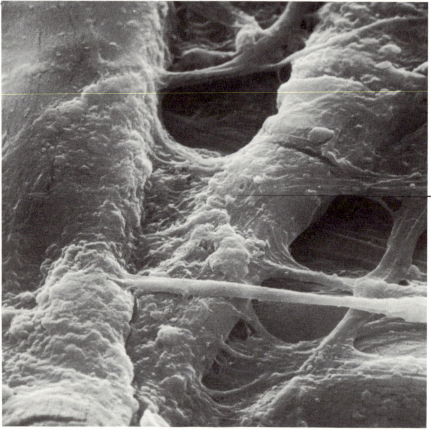

Stereoscan Electron Micrograph of Banknote paper [2400 x]

Highly beaten fibre

—— (indicator line)

Original photographs by kind permission of Mr. Jouko Laamanen and the Finnish Pulp and Paper Research Institute.

Although the fibres in these two illustrations are still identifiable entities it will be noticed that their highly fibrillated and plasticized surfaces extend to and drape themselves over other fibres and where the dimensions are appropriate for this interfibre bonding will take place when the paper is dried.

It can be seen then that after the "free" water has drained away from the sheet a new set of conditions takes over. (It must be remembered that this is an average effect, since the water in the paper is not necessarily uniformly distributed; some parts will be drier than others and so reach a more advanced stage earlier than other parts). At first the cellulose mass is saturated with water; but as drying conditions set in and water progressively evaporates to the atmosphere from the capillary tubes, the surface tension forces will increase drawing the pliable surfaces of the capillaries or the fibrils closer together. Illustration (iii) (on previous page) is purely diagrammatic. The fibrils, shown there, far from being simple rods are, as electron microscopy has shown (Plate 21), a mass of micro–fibrils and micellar strands, which when drawn together by these powerful surface tension forces will progressively increase the number of inter–fibre chemical bonding sites in the cellulose matrix.

One of the objects of the Beating process is, therefore, to develop this fibrillated surface and thus promote inter–fibre bonding, which gives a dry sheet of paper its cohesion and strength; and there is experimental evidence to show that beating does in fact increase the specific surface of the fibres.

Another object of Beating is to develop the plasticity of the fibre in order to assist the surface tension forces, described above, in bringing the fibrillated surfaces closer to one another in the first place. From this it can be understood why the paper maker prefers the used rag to the unused textile (to quote Whatman once again "Linnen and Ropes, after they are unfit for the purpose for which they were made, are the best materials that can be procured for making paper" 1774). Good quality used material would have had sufficient, but not too much, abrasion and flexing to make it more responsive to the beaterman's action. A stiff rigid fibre would in Whatman's day have needed so much beating to get it into a malleable state that it would have been too "wet" to handle effectively at the vat and, if made into paper, would have produced a sheet with a hard, hornified surface.[91]

In the case of cotton linters (and it is assumed here that the same would apply to linen) Dr. Urquhart shows[92] that the *adsorptive* capacity is increased by beating (unlike wood pulp). He attributes this to the fact that the vigorous processes used in the extraction of the cellulose from the wood have already opened up its internal structure, whereas in the case of cotton the extraction process is less drastic and thus leaves scope for the beating process to produce this increase.

---

[91] In order to produce a well–formed sheet the vatman must have stuff that responds to his actions. If the stuff parts with its surplus or "free" water too rapidly, then he may not have sufficient time to form his sheet properly; conversely, if the stuff fails to part with its water, then equally he will have difficulty in manipulating it. Stuff in these states is described as "free" and "wet" respectively and will result from the conditions under which beating takes place.

If the furnish is made up of fibres that are new and stiff, in other words intractable, then it will require prolonged treatment in the beater to increase fibre plasticity; and, almost inevitably, this will result in a high degree of fibrillation. Taken to the extreme this will produce highly swollen pulp (due to the penetration of large quantities of water into the amorphous zones) with reduced void space. This in turn will restrict the drainage of the "free" water and the stuff would, therefore, be described as "wet" (and, indeed, may be recognized by taking a handful of stuff from the beater and squeezing it. It will have gel–like characteristics, stuff tending to escape between the fingers rather than water alone). Of course, if the vatman can manage to manipulate this "wet" stuff, then the subsequent removal of the larger quantities of retained water will result in powerful consolidating forces and a strong paper.

[92] Urquhart, A.R. Bib.63 quoting unpublished work carried out by Shaw, J.T.B. at the Shirley Institute, Manchester).

Finally, although its has proved possible to "beat" cellulose rags in other liquids like toluene, for example, and in spite of the fact that some of these liquids will generate significantly high surface tension forces, no inter–cellulose hydrogen bonding takes place and the resultant paper has no strength. So, apart from its very important function in acting as a vehicle in which rags may be washed and fibres dispersed, water has other *essential* roles to play in the sheetmaking process that other solvents are incapable of performing.

Summarizing, it can be seen that:–

(a)   With its capacity for hydrogen bonding with the cellulose, it makes the action of the beater effective by penetrating and swelling the fibres, thereby providing both the medium and the conditions needed for the formation of paper.

(b)   By creating powerful surface tension forces within the capillary structure of the cellulose matrix its removal consolidates and gives cohesion to the wet sheet.

(c)   As a result of chemical de–sorption during the final stages of drying the loss of water liberates sites for inter–cellulose hydrogen bonding, which gives the dry sheet of paper its strength.

Water is thus an essential ingredient of the papermaking process from the very outset.

The Effects of the Cellulose/Water Relationship in Practice

At the beginning of this discussion it was pointed out that the actual beating and sheet-making operations were much more complicated than the mere maceration of rags and entanglement of fibres that some people might imagine constituted the papermaking process.

The rag, as it is received then, is a piece of woven cloth made up of *threads* (it may contain other features like hems, for instance, or rags made up from a different fibre, all of which would have been sorted into different categories as a first step). The first task of the beaterman is to dismember the woven textile into its warp and weft threads with the minimum damage; and during this process some of the threads will be brushed out into fibre bundles and fibres. Even the first two or three minutes of this treatment will produce a significant "beating" effect. This process is known as "breaking the rag" and the skill required for it is second only to that of the vatman making his sheet of paper (this referring to the practice employed in the manufacture of high quality paper; in the 17th and 18th Cs. the furnish of lower quality papers using rotten materials to start with would have had a much more casual treatment).

Having completed this process, the beaterman can then (after making suitable adjustments to the consistency of his stuff) "beat" the rags. It is during this process that the plasticity of the fibre is increased and its surface developed; and, later, knots and strings are brushed out so that the stuff when ready is homogeneous.

By now the stuff is saturated with water and is ready for dilution and making into paper. As described earlier, once the vatman has the requisite amount of stuff on his mould to make a sheet (the deckle acting as a rim for retaining a small pond of stuff), the "free" water drains away rapidly through the wire cover. The sheet left on the face of the mould still contains a lot of "free" water and at this stage it is much too wet to handle because of this. By couching it onto a felt and sandwiching it between this and another felt it can be pressed to remove further free water. It is only after the de-watering of the pressing stage that the surface tension forces created by the capillary water become effective to the extent that the Layer (the third member of the vat crew) can now separate the wet sheet from the felt and handle it without it falling to pieces. Further traces of "free" water are removed during the pack-pressing process. A certain amount of "free" water will still remain in the sheet after this, because there is a limit to the pressure that can be applied beyond which the sheet would become crushed.

It then becomes a case of drying the paper which begins with the emptying of the larger capillaries, the water diffusing from the interior of the sheet to the exterior. This is followed by the emptying of the smaller ones and the creation of powerful consolidating forces bringing the fibrillated surfaces together; this process proceeds at a slower pace as the capillaries collapse and diffusion is impeded. Ultimately, this leads to the removal of *part* of the chemically sorbed water depending on how far the drying process is taken. In the 17th and 18th Cs. this stage would really not have proceeded further than an "air-dry" condition (say, ~65% RH which with cotton would produce a moisture content ~9% – see Fig. 9; for linen ~12%). On a good drying day one might lower the moisture content further; but it was not until drying aided by steam heating had been introduced that lower moisture contents than this could be achieved (and the papermakers' troubles began).

Shrinkage

During this dewatering process two shrinkage effects will manifest themselves. First there is a shrinkage of the whole sheet structure, a macro-effect produced largely by the surface tension forces and by the draining of the void spaces of free water. It results in a contraction of the *thickness* of the sheet and is for the most part reversible on re-wetting the paper. There is, however, an element of area contraction also caused by surface tension forces, which tends to be irreversible.

Second, there is the actual shrinkage of the fibres themselves and this effect takes place on a micro-scale. It will be remembered that at the beginning of this discussion it was said that the effects of this shrinkage are so complicated that they lie quite outside the limits of an account of this nature; part of these complications arise from the fact that the shrinkage of a fibre takes place in the plane of its diameter and only marginally in its length. Since the majority of fibres lie in the plane of the sheet of paper, one has to seek an explanation for a *considerable area contraction* (of a macro-order) resulting from fibre shrinkage which is on a micro-scale. The general theory advanced to explain this is based on *the communication of fibre contraction* through the hydrogen bonded areas at fibre crossings; but it is by no means a straightforward relationship.[93]

---

[93] Page, D.H. & Tydeman, P.S. "Formation and Structure of Paper", Proc.Tech.Sec. B.P.B.M.A. 1962 397.

Figure 10

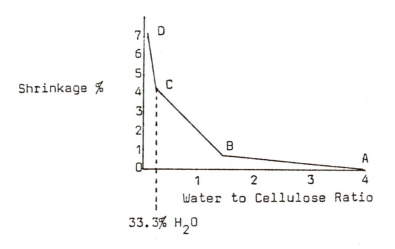

A to B large drop in water content: little shrinkage
                                         initial surface tension
                                         thickness contraction

B to C lesser drop in water content: considerable shrinkage
                                         further surface tension
                                         thickness and area contraction

C to D small loss of water: considerable shrinkage
                                         fibre shrinkage

(Reproduced here by kind permission of Dr. H.F. Rance)

Originally published in TAPPI, Vol.37, No.12
December 1954. Copyright TAPPI 1954.

To give one an idea of the order of shrinkage that takes place during the sheet forming process, Figure 10 (which is an example of shrinkage in a paper made on a paper machine) illustrates three distinct phases.[94]

CONCLUSION

From the foregoing account of this special relationship between cellulose and water it should not be difficult for the reader to appreciate the far-reaching effects that a fundamental change in the method of breaking and beating rags, particularly linen rags, might have had on the sheetmaking process and the achievement of desirable properties in the paper.

To-day, with a greater understanding of the physical and chemical nature of this relationship, one can see some of the pitfalls to be avoided and the advantages to be exploited, no small consideration when one takes into account the enormous widths of the web and the high speeds of modern paper machines. In contrast to this, the 18th C. paper maker would have had to discover these things the hard way, by trial and error.

---

[94]   Rance, H.F. "Water Removal" Fundamentals Symposium TAPPI. 1954 640.

TWO DRAWINGS BY JAMES WHATMAN, THE YOUNGER, OF THE RAG PREPARATION FACILITY AT OLD MILL, HOLLINGBOURNE IN 1771–1772

A "NEW ENGINE AT HOLLINGBOURN PUT UP JULY 1771"

The discovery of Engineering Drawings (of which there are two) relating to the installation of a new rag-washing and beating facility at Old Mill, Hollingbourne, and dated 1771–1772 not only provide us with valuable information regarding the state of the design and operation of the Hollander at this time, but since the subject concerns Old Mill and since the engineering drawings appear to have been made by the younger Whatman himself, this find is particularly appropriate to the subject of this book.

They are quite possibly the first engineering drawings of a beating engine that have come to light.[95] Up till now one has had to rely on the diagrammatic illustrations of van Zyl (1734) and Natrus, Polly and van Wuuren (1734), followed by those specially engraved for Le Francois de Lalande's "Art de Faire le Papier" (Bib.23) in 1761 and Desmarest's Plates (Bib.21) which appear to have been taken from the "Encyclopédie" and thus dated 1767. As information on this subject is so scarce, anything new that throws light on it deserves close study; interpretation, however, will inevitably be somewhat speculative in character because of the fragmentary nature of this new information.

Unfortunately, there is no definitive scientific account of the Rag Beating operation; and there is never likely to be one now[96] because of the almost universal replacement of the cotton and linen rag by wood and other pulps, a substitution which began in the 19th C. and which resulted in more attention being paid to improving methods of preparing the new materials than on the rag and its treatment, which was something quite different.

In spite of this a great deal of attention was devoted to the subject of beating during the 1920's–1930's, but the facilities for measuring the changes that took place in the engine were not adequate for the purpose at the time. (To place things in perspective it was not until 1928 that Meyer and Mark produced a model of the structure of cellulose that fitted experimental data and it took a great many more years before a reasonable picture of the cellulose/water relationship found in papermaking began to emerge). The early accounts of beating tended to be highly subjective and based on very empirical procedures. Far more sophisticated methods were being developed during the 1950's–1960's, but these were applied to the much more homogeneous and tailor-made new materials including, for instance, cotton linters. The demise of the rag, for all practical papermaking purposes, took place in the 1950's, when more and more textiles were made containing a proportion of man-made fibres rendering the rags derived from them useless for conventional papers. So, for an assessment of the Hollingbourne facility one has to fall back on recollected experience and opinion.

---

[95] This supposition raises quite a number of questions. For example, who made these Engines and did they not have working drawings? What prompted Whatman to makes these drawings? It is conceivable that contemporary working drawings of Beating Engines may have survived but have not been recognized as such. One imagines that by this date the demand for this type of equipment was such that some firm somewhere was engaged in constructing them, but the whereabouts of such a firm has not been discovered yet.

[96] In case this statement is misunderstood, it should be said that several serious attempts (like Sigurd Smith's work) were made to reduce the rag-breaking and beating operations to something approaching a science with results, both theoretical and in terms of improved beater design, that led unquestionably to more efficient practices and to better control. All the same one cannot point today to any standard work on the beating of rags in the papermaking process.

As described earlier there are two principal stages in the preparation of Stuff from Rags, namely, (a) the washing and breaking stage; and (b) the development of the fibres to increase the potential for intercellulose bonding. In the case of the former the objective is to dismember the textile and reduce it to its more manageable threads and fibres; thus one is less concerned but not altogether unconcerned, at this stage with the sharpness or bluntness of the fly–bars than one is when developing the fibre during the beating stage; the aim is to homogenize the heterogeneous furnish made up of irregular pieces, in size and strength, of rag and at the same time to keep damage of the fibres themselves to a minimum. Inevitably some development of the fibre will occur, but in good hands this can be kept under control; incorrect treatment at this stage can be disastrous. In addition, to wash the rags efficiently in the Breaker they must still be in a discrete form; in a reduced state the dismembered textiles would quickly block the chess. Effective dismemberment of the textile is achieved (a) by using relatively low consistencies compared to those used in the beater; and (b) by careful control in lowering the roll as homogenizing proceeds.[97] The "carding" action of the edges of the fly–bars is improved by the use of low consistencies; and by careful roll lowering procedure the cutting effect of sharp bars, should these be used, is lessened. Nevertheless the use of blunter bars is preferable.

In the beating operation one is concerned with treating the fibre in accordance with the properties required in the final sheet of paper. Except where "freeness"[98] is specifically required, as in Blotting paper etc., blunt rather than sharp bars are used in beating; blunt bars will bruise and plasticize the fibres and develop highly fibrillated surfaces increasing both capillary and, to a much smaller extent, sorbed water held by the cellulose matrix and reflected, ultimately, in sheet strength. Where a high degree of absorptiveness is required in the paper, sharp bars were often employed in order to reduce the bruising effect; but in the great majority of cases the objective was to achieve strength and hence one was more likely to find blunt bars in use than sharp.

The situation is not always as straightforward as suggested here. For instance, in order to achieve satisfactory drainage conditions on the papermaking mould when making *thick* papers, the stuff must be freer than in the case of thin papers, other requirements being the same. This is usually achieved by shortening or cutting the fibre to a greater degree in the beater and can be accomplished with the blunter kind of bar using lower consistencies and/or a weaker furnish. Again, one might be confronted with an exceptionally refractory kind of fibre or textile, needing special treatment; as we shall see later, what Whatman has described as "coarse" rags may have been just such a one as these.

---

[97] In an electrically driven rag–washer (as indeed in a beater also) the process of homogenizing can be seen very clearly by plotting power consumption against operating time and noting (on a recording instrument) the effect of each change in the roll position. Two features will be observed. First, initially with each position the curve fluctuates violently as the roll copes with unevenly distributed clumps of rag. As these become more and more dispersed the curve smoothens and at the same time falls as homogenizing proceeds. Second, when the curve at this roll position starts to flatten out and the homogenizing process ceases to be efficient, the roll is lowered another stage and the form of the curve repeats itself to a lesser degree. By the time the roll has been let right down and the fluctuations cease altogether the rag furnish will have been reduced to a homogeneous state. It must be understood that this account has been simplified to illustrate the principle; in reality the sequence may be complicated by variations in washing procedure and (introduced at a later date) engine bleaching.

[98] See this App note 91

Finally, as described in Chapter VI, many other important features have to be taken into consideration to achieve an effective beating operation besides the type of bar used. Taken together these may be summarised as:–

(i)   The nature of the bars employed is not only a matter of sharpness versus bluntness, but of bar thickness which is more important than the edge; the beating effect depending on the total thickness of the bar. Their grouping and setting is equally important in providing adequate roll cell space.

(ii)   Overall beater design to improve performance and reduce energy consumption include the ability to adjust the roll position progressively and precisely; a sufficient cutting angle; effective contouring to produce good circulation and reduce dead spots.

(iii)   Consistency, much higher in the beater than in the rag breaker, can be varied together with roll pressure to produce different beating effects.
        Consistency also plays an important role in the type of fibrages formed.

(iv)   The nature of the rag furnish.

With these points in mind one can now examine the drawings and notes relating to the installation at Old Mill, Hollingbourne.

General Description of the Drawings

A description of the circumstances surrounding the discovery of these drawings has been relegated to the author's working papers. As their subject matter is quite different they are considered here separately and in date order. Both of them are "Engineering Drawings"; there may have been others but, if there were, they have been lost.[99]

No.1 Drawing of the new Engine at Old Mill shows details of part of the beater roll and bedplate with notes on its performance in what is almost certainly the younger Whatman's handwriting, an assertion supported by the personal style of the comments and the manner in which they are arranged to illustrate various points. It appears equally certain that he made this drawing himself, seemingly to scale though not indicated.[100]

No.2 Drawing of the Swimming Wheel is drawn to scale, but has no writing on the recto side.

[99] This is an interesting new sidelight on the younger Whatman's accomplishments. We have no knowledge of Whatman's contact with the rising generation of Engineers prior to his contact with John Smeaton and John Rennie in 1787, in both cases connected with waterwheel design and transmission.

[100] No.1 Drawing has been drawn on what appears to be Laid Medium Writing (ca.22.5 x 17+ in.) and is countermarked in the right half "J WHATMAN" (see below). The drawing, as it is now, is in 2 halves and appears as if it had been formerly bound into some sort of notebook. No.2 Drawing is on paper that bears no countermark and does not appear to have been bound into any sort of book since the drawing extends right across the sheet of paper. Both drawings may have been executed in 1772; there seems to have been no good reason for doubting the dates indicated in the drawings as differing from the dates when the drawings themselves were made.

The tail of the "J" in the countermark belonging to Drawing No.1 curls back on itself but does not quite cut through the "J" itself, ending instead with a slight upward flick. This point is mentioned because James Wardrop (Signature 1938 July Appendix) suggested that the tail or twist in the J" had diminished by 1781 and was vestigial by 1789. Though the form which the "J"s took undoubtedly changed from early ones with a tail to later ones with a loop at the bottom end, Wardrop's dates cannot be regarded as a reliable bracket. The tail of the "J"s of the late 1780's had a tail that turned back and crossed itself and these, albeit, diminished, are still to be found in papers of the early 1790's by which time they were being replaced by "J"s with loops. The type found in Drawing No.1 may possibly be characteristic of the 1770's? It may be noted in the context of these observations that the double line "W"s of this period were quite variable, sometimes with the left "V" in front of the right and sometimes the other way round; again, sometimes the junction is open at the top and in other cases closed, cf. forged countermarks Preface to App. I & II Plate 16 and other double line "W"s in TABLE XXVI.

No.1 DRAWING

(New Engine at Hollingbourn put up July 1771)
(Plate 22)

As will be seen from the reproduction of the drawing opposite, the upper left half shows a clump of 4 fly–bars passing over a bedplate with 17 bars set close together in it. This part of the drawing also illustrates that the surface of the Roll has been cut back to a new radius between the clumps (called Tiers in this drawing). The thickness of the tips of both beater and bedplate bars has been projected down to the lower half of the left–hand side of the drawing and appear as lines "A" and "B" respectively. Underneath these there are two paragraphs of comment on the performance of the above. A transcription of these comments, and the notes in the drawing, is given below. In the right half of the drawing a bedplate, slightly smaller in width (13 cm. as against 15 cm. for the other) is shown with only 8 bars as against 17 and well spaced to give proper roll cells between each bar. This bedplate may represent an improvement in the design of the beater or, on the other hand, it might illustrate the bedplate belonging to the Half–Stuff Engines. Against this drawing, to the right of it and faintly, the date "Aug. 31 1771" has been written. Whatman's punctuation has been retained.

TRANSCRIPTION OF THE NOTES IN THE DRAWING (working from the top downwards)

Top Line

New Engine at Hollingbourn put up July 1771.

Between the Clumps [called Tiers in this drawing] of Beater Roll Fly–bars

(left)   When the Roll was first made the Wood between the Tiers was up to this line

(right)   March 15th.1772 lowered the Wood between the Tiers down to this line which made the Stuff turn in the Engine Faster and cleared the Stuff better from Knotts. [Lowered clearly means here a reduction in the radius].

Underneath the drawing of the bedplate bars

This Plate is 3/4 of an Inch out of parallel with the Bars in the Roll

Two paragraphs of comment below the drawings

The above Roll having 15 Tiers in it of 4 Bars each and 17 Bars in the Plate makes it necessary to have the Bars kept at all times very sharp and particularly if you are working coarse stuff which it is but barely possible to beat in this Engine so as to work fast enough for a Thick coarse Paper. In 1771 I attempted to make some fine Litteris with all Seconds beat in this Engine. The Medium sharpness of the Bars in the Roll is shown above at A and the sharpness of the Plate at B. And notwithstanding the work appears so very sharp. the stuff worked so extremely wett that I was obliged to leave off making it and go upon thick Post made all of Fine Rags, which it beat very well.

Rolls and Plates with such a Number of Bars in them do very well for a Thin Paper made of coarse Rags. The Bars being so sharp cuts the Shive in the Rags into very fine pieces. And makes the Paper appear to be made of a finer Rag than it really is. And where coarse Rags, as at Mr. Willmotts are washed separate and afterwards about a Third part mixed with the fine and all work'd into a fine Paper. the coarse Rags could not be so well prepared for this

Plate 22 229

New Engine at Hollingbourn put up in 1771 (drawing by the younger James Whatman)

purpose as in an Engine where the Bars are extremely Sharp. it being absolutely necessary to clear the Shive out of the coarse Rags as much as possible which this Effects.

THE AUTHOR'S COMMENTS ON DRAWING NO.1

Dates

(a) Installed July 1771.
(b) The alternative bedplate design (shown on opposite page in drawing) Aug. 1771.
(c) Whatman states that "in 1771" he experimented with the beating of various types of rag using this installation.
(d) The alterations to the Roll, lowering the wood, were made March 15th 1772.
(e) Drawing No.2 is dated Sept. 19th 1772.

General Comment

It has been argued in the main text that Old Mill, Hollingbourne, must have been built with the aim of exploring and exploiting the capabilities of the Hollander rag–washing and beating engine, to be followed later with the conversion of Turkey mill to the new method of preparing rag stuff. Obviously the design of the Beating Engine must have passed through many different stages since Nicholas Dupin's model in 1690, improved forms being installed no doubt at mills like Chartham, Old Mill and Turkey mill in the 1730's. But one wonders in what respects the new Engine differed from the original installation (cf. comments Chap. VI p.244 [h]). Was there some new feature that Whatman wished to try out before making either a larger installation at Turkey mill or a smaller one for the Loose mill (acquired 1774), a mill that was no doubt already in his sights. Was this episode pre–planned or did it just happen that Old mill became vacant again in 1770?[101] Whether the initiative was his own or whether he was prompted into taking this action by advances made in other paper mills is something we can only guess at.[102] To support the former of these suggestions one could cite the fact that Clement Taylor jnr. leased the mill shortly after (1774), one might almost say a parallel action to that of Quelch and Terry in 1739; and it could also have been an object of interest to a new generation of paper makers like Lewis Munn who came to Hollingbourne shortly afterwards.[103]

---

[101] See Synopsis of Hollingbourne Old Mill App.I.

[102] It will be recalled that the younger Whatman and his first wife, Sarah, made two very brief visits to Holland, the first of these in 1765, the second in 1772 (Balston, T Bib.1 146). There is no evidence that he visited any of the Dutch paper mills or that the Engine had been developed there to the advanced stage that we find in these drawings. In fact G. Husslage's drawings (20th C.) of the traditional Engines used in the Schoolmeester (Zaanland) show a very much more primitive double grooved fly–bar than the one drawn by Whatman. If he had learnt anything relevant it could only have been during his first visit. On balance, to judge from Whatman's drawings and notes, it would seem that he was in the process of working out these innovations himself.

[103] See App.II, p.89.

Alterations to the Roll (March 15th 1772)

By "lowering the wood" between the Tiers, Whatman to some extent increased the roll cell space, but perhaps not as effectively as if he had lowered the wood between the bars themselves. Increasing roll cell space leads to improved "fibrages"; and in this case this would probably have only affected the leading fly–bar. Whatman notes, however, that his action improved the carrying power of the roll as a whole with "the Stuff turning in the Engine faster" and possibly increased turbulence generated by the limited improvement in roll cell space helped "clear the Stuff better from Knotts" by making the fibrages on the leading fly–bars more effective and thus dragging out and dispersing lumps in the stuff. These are keen observations by Whatman.

Angling of the Bedplate relative to the Flybars

Whatman noted the fact that the Bedplate was 3/4" out of parallel with the Flybars. The fact that he took the trouble to note and measure this suggests that angling of the bedplate was probably intentional and perhaps current practice. If one assumes a roll width of 24 inches (this is what Drawing No.2 indicates), then

$$\text{Tan } \theta = 0.75 \div 24$$

$$\theta = \text{ marginally } < 2°$$

20th C. opinion was that to be effective angling should be $> 3°$ and certainly not $> 43°$. Even though the angling here is on the low side, the principle seems to have been accepted.[104]

The Sharpness of the Bars

Except where a "cutting" effect was specifically required, as in Blotting papers etc., later practice favoured square edged rather than sharp beating tackle. Even as early as 1794 Johnson wrote "the duller these bars are, the stronger the paper will be, if they are too sharp, it will be rotten, for the rags will be cut too fine, which will likewise occasion a great number of broken sheets".[105] Whatman's bars do appear from these drawings to have been very sharp. How is this fact to be reconciled with what he has said in his notes about "the Stuff working so extremely wett" ? He goes on to say that by using *Fine* Rags he managed to overcome this difficulty.

Whatman makes a distinction in his notes between *Coarse* and *Fine* Rags; he also refers to the use of "Seconds" when he attempted to make some Littress and, if his meaning has been correctly interpreted, then it was during this attempt that the stuff worked extremely wet. The inference is that in this case the Coarse Rags and the Seconds were synonymous. Generally speaking, the grading of Rags at that time seemed to follow a sequence of "fine", "second" and "tres" with "offal" at the bottom of the list with the aim of distinguishing between the

---

[104] That is to say if the interpretation of Whatman's comment has not been misunderstood.

[105] Johnson jnr., B. "The Paper Maker's Guide or the Art of Paper Making". (Publ. by the author and sold at his Wholesale Paper & Rag Warehouse, Ludgate Hill, London 1794). Johnson's comment applied to linen rags; cotton, used later, was less demanding.

cleanest, whitest rags and the dirtiest and poorest in quality. However, from what Whatman has said, there seems to have been a further distinction made between strong rags i.e. those with a greater proportion of non-cellulosic residues due to a lower degree of retting the original flax, in short a coarse fibred material, and Fine Rags, evidently those derived from finer linens where either the retting of the flax was taken a stage further than in the coarse rags or that the flax itself was finer due to a different method of cultivation or to its having been harvested at a different stage.[106] (Other authors of the period, for example, have described cloth as "coarse, fine or middling, the rags coming out of it make paper of Different Degrees of Goodness").[107]

Following this interpretation coarse rags would inevitably have contained more "shive" in them than fine rags and hence Whatman's concern in cutting them into very fine pieces; but it is likely that the imperfectly retted would have been much stronger than the finer rags and thus to reduce them to uniform stuff they required longer beating causing the stuff to become "extremely wett" and unusable. The fact that later practice was to show that "duller bars" produced stronger paper and in Whatman's situation would have only made the treatment of his "coarse" rags even more intractable indicates that he may have lacked the means of controlling the beater roll position accurately; lowering the roll gradually was an important feature of treating rags of this kind correctly in both the Rag Breaker and Beater. This is one possible solution; alternatives to this are suggested below.

The solution to this problem may depend on a combination of different factors rather than on a single feature such as the strength of the fibres and the extra time and energy required to reduce them. For instance, another important factor in this situation could have stemmed from a failure to appreciate the marked effect that consistency of the stuff would have played in this situation. In general,

> *Thin consistency* coupled with heavy roll pressures produces CUT fibres and, usually, FREE stuff, but not always; very short fibres, though in essence relatively undeveloped and "free", can create stuff which is difficult to drain through a mould cover, and thus behave as "wet", due largely to the presence of "fines".

> *Thick consistency* coupled with lower roll pressures produces TORN and CRUSHED fibres i.e. when fully developed WET.

It could well have been the case here that with less power available than we have been accustomed to in more recent times, consistencies were generally lower then than became customary later. This, together with lighter rolls, less accurate control of roll position, unusually sharp bars and, in addition, a desire to maintain maximum through-put, all these could have led to excessive cutting and poor draining i.e. the stuff behaving as "wet". This view that "fines" may have been a more important factor than heavy beating as the cause of apparent wetness is leant support from Whatman's statement that "Rolls and Plates with such

---

[106] See this App. pp.191,195/196.

[107] Ref. prior to 1770. It has been assumed here, especially in view of Whatman's reference to the presence of Shive in his furnish, that "coarse" rags literally meant rags composed of coarse flax and containing a proportion of non-cellulosic residues and *not* rags that may have felt "harsh" due to the presence of excessive amounts of lime soaps, i.e. washed "by dint of soap alone" (this App. p.205).

a Number of Bars in them do very well for a *Thin* Paper made of *coarse* rags". He had failed, it will be recalled, to make a "*Thick* coarse Paper" because the stuff would not "work fast enough".

Cutting the fibre does seem to have been Whatman's objective with coarse rags, "cutting the Shive into very fine pieces". And to reduce drainage problems at the vat he suggests following Willmott's practice of blending only a proportion of these with finer stuff to produce a more workable mixture.

It was generally accepted from very early times that it was bad practice to process strong and weak rags together. Provided the condition of the coarse and the fine rags were of a comparable order, then much could be done to harmonize the effects of their disparity by giving the rags treatment appropriate to their differences in character in the rag–breaker and washer and blend them at a later stage. For instance, it was modern practice sometimes to add a proportion of "New Pieces" to "Fines" (i.e. a mixture of good quality linen and cotton *rags* and not "fines" as used in the sense above, fine particles) when filling in the Beater in order to give the furnish additional strength, both kinds of rag being by that time in the half–stuff stage. Whatman's comments are not absolutely clear on this point, but it has been interpreted here that Willmott[108] followed much the same procedure, "the coarse Rags.....are washed separate and afterwards about a Third part mixed with the fine and all work'd into a fine Paper".

Apropos of Whatman's comments, it will be remembered that one of the criticisms levelled at English papers made a few years earlier, in 1755,[109] was that "English Rags, being cut by Engines make the fibres so short, tho' coarse, that more Size is required to bind them together, to render them firm and serviceable, and makes the paper of a harder nature". To judge from the sharpness of Whatman's beater and bedplate bars in 1771 it would seem that, though advances in some directions had been made, and Whatman is seen here as clearly making experiments to improve the design of the beater (increasing roll–cell space and circulation) and studying the behaviour of different furnishes in them, the more modern treatment of rags and the equipment for this had yet to be developed. If one goes back to the illustrations of earlier beaters (e.g. Le Francois de Lalande's) and their description, it will be seen that they used grooved bars which appear to have been quite blunt and may have produced much more developed fibres than Whatman's sharp tackle.[110] However, although paper makers were to revert to "duller bars", the earlier Hollanders probably suffered from other serious defects such as poor circulation and dead spots and a low output. These drawings undoubtedly represent a significant attempt to improve the overall design of the Engine.

---

[108] The Wilmott referred to here might have been William snr. at Shoreham (act.1737–75), one of the elders of the Kent paper industry and a contemporary of the elder Whatman; or William jnr., a contemporary of his own, who had been operating Sundridge Corn and Paper mill since 1766 and no doubt equipped with the latest kind of Engine. Though both almost certainly manufactured White paper, no Wilmott countermarks have been discovered that can be attributed with certainty to either before the 1780's (see unidentified "W"s this Appendix pp.277 ff.).

[109] Royal Soc.Arts, Guard Book Vol.VIII ca.1755 (details of memo are reproduced in Bib.1 35; and Bib.47 362/3).

[110] cf. also comment on Dutch Engines (this App. note 102).

Some Comparative Figures

	Roll diam. cm.	No. of Fly–Bars	Clumping	No. of Bedplate Bars
Old Mill, Hollingbourne (Whatman's figures)	61	60	4	17
The alternative, possibly the Rag–washer bedplate				8
Figures from Rees' Cyclopaedia (John Farey 1813) (believed to be the original Springfield Mill installation).				
Rag Washers		40	3	12–14
Beaters		60	3	20–24
Springfield Mill 20th C. (Old mill beater floor)				
Rag washers	66	42	3	22
Beaters	66	42	3	22

In the last case the fly–bars were made of Steel; the bedplate bars of Gun–metal. Whether this had always been the case is not known.[111] The bars were certainly quite blunt in the roll, but relatively sharp in the bedplate. In neither of Farey's accounts (Rees' Cyclopaedia or under the name of Thomas Martin) is the material of the bars mentioned. Johnson, however, in his Paper Maker's Guide (1794) describes the fly–bars as being made of Steel "about 48 in number, one inch apart between every three". Earlier authors like Croker, T.H. "Complete Dictionary of Arts and Sciences" (1764–1766) and the 1st Edition of the Encyclopaedia Britannica (1771) also refer to Steel bars; the 3rd Edition differentiates between French and Dutch Engines, the latter having bars of brass and copper.[112] On the whole it seems likely that Whatman's roll had steel bars.

It should also be noted that from 1794 at least (Johnson's account) the clumping of the bars has been in threes as distinct from Whatman's four in the Hollingbourne installation.

[111] Cross & Bevan, "A Text–Book of Paper–Making" (Spon, London 1900) 170. "In making the finer kinds of paper, the roller bars or knives, instead of being made of steel, are made of bronze; thus any contamination with oxide of iron is avoided". (See also note 112 below).

[112] Between the 3rd Ed. (1797) and the 6th Ed. (1823) Encyclo.Brit. there is virtually no difference, the account being a slightly up–dated version of Desmarest (1788 Bib.21), hence the continued comparison of French and Dutch Beating tackle. According to these accounts the Dutch Rag Breakers had steel bars, but the Beaters had a mixture of brass and copper. The 7th (1842) and 8th (1858) both refer to steel bars clumped in twos in the Washer (37/38 bars) and in threes in the beater (54/57 bars), but how accurate and typical these accounts are is open to question.

No.2 DRAWING

Verso ("A Plan of the Swimming Wheel. The two Half Stuff and the Whole Stuff Engines drawn by a Scale of 1 Inch to the Foot. Sept.19.1772. The Fine End").

The drawing is made (as in the case of No.1 drawing) in pen and ink on uncountermarked laid paper. There is no writing on the recto side. (Plate 23 overleaf).

General Description

The drawing shows the layout of two ragwashers and one beater arranged radially around a central "Swimming" wheel with two sets of cogs viewed from above (i.e. showing a rectangular face with the cylindrical cog on the other, under side). The Engine in the 6 o'clock position is set above the other two engines at 3 and 9 o'clock.

Comment and Interpretation

It has been assumed here from various details of the layout of the engines and their dimensions that the engine in the 6 o'clock position was the beater or whole stuff engine and the other two the washers. Based on this the interpretation of the drawing is as follows, remembering that in the absence of any written notes this can only be speculative:–

(a)   *Dimensions*

The dimensions of the beater are much the same as those of the rag washers, marginally (2 inches) narrower; but the rolls are the same width and diameter. All of these engines are smaller than those referred to by B. Johnson jnr. (1794), who gives a length of 12–13 ft. but the same width, 4 ft. 6 in.; and the "Old Mill" Beaters at Springfield mill (which may have had the same overall framework as the beaters described by Farey in Rees "Cyclopaedia") were again slightly larger, ca.13 ft. long and 5 ft. 7 in. wide with a wider roll and larger roll diameter.

The walls of the engines, as shown in Whatman's drawing, appear to be 3/4 in. thick and might have been of cast–iron construction with a lead lining; they no longer have the massive appearance of the engines illustrated in Le Francois de Lalande (Bib.23 Pl.V & VI) and Desmarest (Bib.21 Pl.IX) unless one assumes that the "supporting" structure "J" in the sketch is solid up to the top of the engine.[113]

Finally, there is no indication of the depth of these engines though one might reckon from the diameter of the roll that they were ~18–20 inches deep compared to ~21 inches for the Springfield engines.

---

[113] This point has not been resolved. It is the opinion of some that this is too early a date for cast–iron construction of this sort. All the same none of the earlier illustrations of the Hollander (van Zyl, Lalande or Desmarest) show anything comparable to the wall of the trough as illustrated by the younger Whatman. It might, of course, represent a lead lining.

Plate 23

A plan of the Swimming Wheel : two half stuff and the whole stuff Engines :
the Fine End, Sept. 1772 (Drawing by the Younger James Whatman)

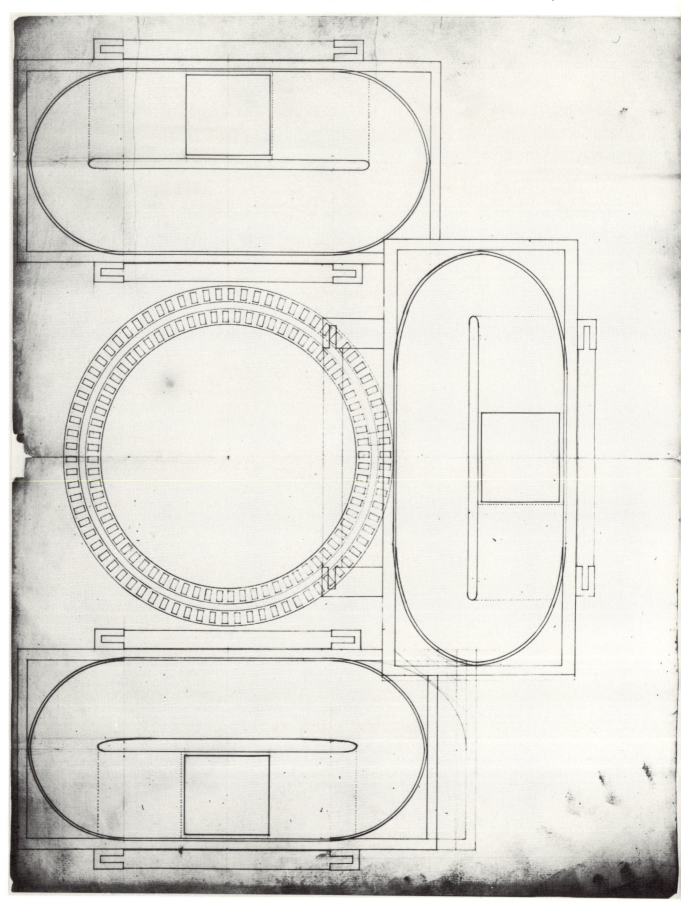

(*photograph B. Duffield*)

If the scale of the drawing is taken to be approximately accurate, the following observations may be made:-

Figure 11

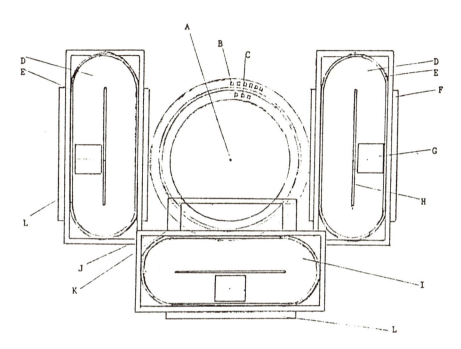

Rough Sketch (not to scale)

A.	The Swimming Wheel	Diameter	8' 7"	2.62m.
B.	Outer Ring of Cogs 78 in number	Radius to cog	4'	1.22m.
C.	Inner Ring of cogs 72 in number	Radius to cog	3' 6"	1.07m.
D.	2 half–stuff Engines (Rag Breakers/Washers)	Interior length	10' 3"	3.11m.
		Width	4' 6"	1.37m.
	(Shaft to outer row of cogs)	Length	5' 7"	1.70m.
	Contoured area same as length as mid–feather	Left Engine	6' 9"	2.06m.
	(see dotted line in drawing)	Right Engine	7' 3"	2.21m.
E.	Wall of the Engines (approx.)	Thickness	¾"	1.90cm.
F.	This projection represents the block carrying the outer Roll Shaft Bearing, which in the older type of Engine was raised/lowered by means of a jack, later by a lighter.			
G.	Plan of Engine Roll	Width	2'	0.61m.
		Diameter	2' 3"	0.69m.
H.	Mid–feather (length)	Left Engine	6' 9"	2.06m.
		Right Engine	7' 3"	2.21m.
		Whole Stuff Engine	7' 2"	2.18m.
I.	Whole–Stuff Engine	Length	10' 3"	3.14m.
		Width	4' 4"	1.31m.
	(Shaft to inner row of cogs)	Length	5' 2"	1.57m.
J.	The supporting structure for the Whole Stuff Engine or Beater lies above that of the 2 Half–Stuff Engines. This may be the base only that is shown overlapping.			
K.	The inner counterpart of L appears to project over the swimming wheel.			
L.	As in F above (see also Fig. 12)			

Figure 12

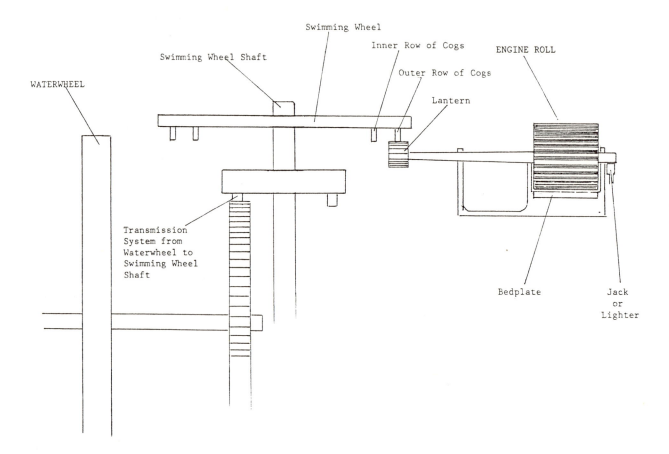

Diagram of suggested transmission system

(Drawing J.N. Balston)

(b)   *The Drive*

It has been assumed here that the transmission from the waterwheel follows the same
pattern as that illustrated in Le Francois de Lalande and Desmarest, a rough sketch
of which is given in Figure 12.

To judge from the distances of the Engine Rolls from the Swimming Wheel and
assuming that the roll shafts for the washers and the beater are of approximately the
same length, then it would appear that the washer lanterns engaged with the outer
row of cogs in the swimming wheel and the beater shaft lantern with the inner row.
The inner row of cogs would supply the beater with greater power. Whatman's
drawing does not show the transmission wheel and cogs connecting the swimming
wheel drive to the waterwheel shaft.

(c)   *Emptying*

No indication is given in this drawing as to how the engines were emptied.
According to earlier practice the contents of the washer would have been emptied
into some sort of container with a false bottom so that the half–stuff could be
drained and transferred later to the beater, where it was made up with additional
half–stuff until the consistency was judged to be correct. The illustrations of the
French Engines (Bib.21 and 23) show a "porte" or sliding panel at one end of the
trough with an outflow pipe on the outside, a method probably used in the Old Mill
installation. The notion of transferring the half–stuff to the beater by means of
gravity does not appear (certainly from this drawing) to have been devised yet.
Plate I in Rees' "Cyclopaedia" (Bib.50), though not illustrating a "porte", shows the
washers placed at a much higher level than the beaters and indicating in the text that
the contents of the former are 'let down through a plug hole in the bottom of the
latter. The valve position in one of the Old Mill Springfield *beaters* is illustrated in
a plan of this.[114]

(d)   *Starting and Stopping the Engines*

The procedure for carrying out these actions raises problems that have not been
resolved here although a number of solutions can be proposed. In van Zyl's
"Moolen–Boek"[115] there is another Plate (part of which is reproduced overleaf,
PLATE 24) illustrating the elevation of the exterior of the Engine as seen from the
driven end of the roll shaft. This shows that at the backfall end of the beater trough
there was another side lever which could be raised thus disengaging the lantern from
the drive; or, conversely, lowering it to put the roll shaft in gear again. There was
a special chock for keeping this lever in the raised position. The inference from this
is that the teeth of van Zyl's gearwheel must have pointed upwards. In Whatman's
drawing, however, the gear teeth of the "swimming wheel" point downwards, an

---

[114] Balston, T. Bib.74, figure facing p.153.
[115] For earlier reference to this work and his Plate IV, see Chap.VI Plate 13 and note 11.

Plate 24

TRACING OF INNER SIDE ELEVATION OF ENGINE ILLUSTRATED IN "THEATRUM MACHINARUM UNIVERSALE" (AMSTERDAM 1734)

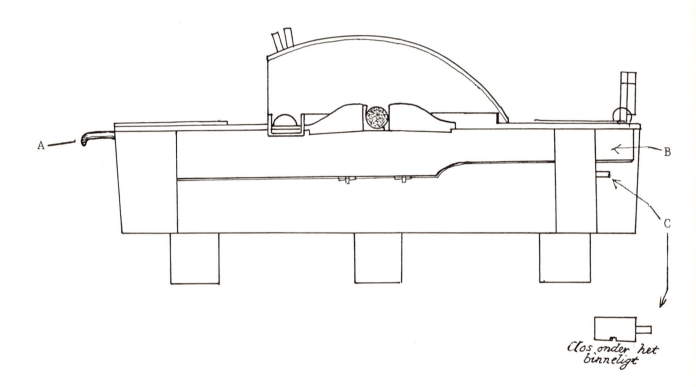

Clos onder het binneligt

A. Outer side lever for lifting roll

B. Inner side lever to disengage the roll shaft lantern from the drive

C. The chock (Clos) for keeping the lever in the elevated position

(Tracing made by the late Mr. E.G. Loeber)

arrangement that conforms to that shown in Le Francios de Lalande's Plate V and Desmarest's Plate VII.[116]

Apart from the obvious method of stopping the Engine by cutting off the water driving the waterwheel, a very cumbersome method of managing the operations on the beater floor, since the drive would have continued to work for quite some time after this, were there any other methods open to the beaterman to disengage his Engine?

The simplest method, if in fact possible, would have been to have had some means of raising the "swimming wheel", indeed, it may have been an ability to do this that gave rise to its name. But there seems to be nothing in contemporary literature to indicate that this was so.

No satisfactory alternatives have been forthcoming from modern authorities on watermills. The idea that a side lever, similar to van Zyl's, operating in the reverse direction, i.e. dropping it to disengage, is impracticable because one would be lowering the roll on the side already closest to the bed–plate which would lead to the bars being ripped out.

Spain suggested that another possible method of disengaging the roll shaft might be to have jacked it up at the roll end using the roll lifting gear on the outer side of the Engine to bring this about. Some rough calculations indicate that this does not seem to be a feasible answer to the problem either.[117] Mr. Jarvis, Chairman of the East Kent Mills Group, suggested that it might have been possible to immobilize the Engine removing 2 or 3 staves from the lantern, knocking them out of position. This would appear to be a somewhat hazardous operation, if the swimming wheel was rotating; indeed, if possible at all; but with the main drive stationary the Engine could be immobilized by this method for repairs whilst allowing operations to continue for the other Engines.

The problem discussed above would only have arisen if the beaterman had wanted to stop his Engine in an emergency, to remove an obstruction etc. Provided that there was a satisfactory method of emptying the the Engine, then it may not have been essential to stop the roll rotating during normal operations. An Engine can be filled in with water and rags while the roll is rotating in the raised position. The only difficulty that might have arisen would have been when either the half–stuff or final stuff was ready and the beaterman had nowhere for it to go. Though the roll could have been raised, with a full Engine further "development" of the stuff could have

---

[116] Le Francois de Lalande, Bib.23; Desmarest Bib.21.

[117] In a telephone conversation Mr. Spain had no other solution to offer; he suggested that it would perhaps only be necessary to depress the lantern 2 or 3 inches to disengage. However, if the bearing on the inner walls of the engines is regarded as the fulcrum, then based on the dimensions shown in the drawing to depress the rag washer lanterns 2 or 3 inches would require the outer end of the shaft to be raised 10 or 15 inches respectively; and in the case of the beater lantern 13 or 20 inches, which on the face of it does not sound feasible; a 1 inch depression, if that was effective, might have been more practical, requiring the outer end of the shaft to be raised 5 inches for the washer and 6.7 inches for the beater.

taken place if the waiting period was very long ("raised" in this context means raised off the bed–plate *not* out of the stuff).

(e) *The Imbalance within the Rag Preparation Operation*

There is a fundamental imbalance between the working times required for the rag washing and breaking stage and the beating stage, that is for a given furnish. There are so many unknown quantities in this situation and so many possible variations in the procedure that might have been practised in a paper mill that it is impossible to suggest a single solution here to this problem.

If one was to consider a single case, then according to the figures given in Table VII of Chapter VI one might assume that in a typical instance the rag washing operation would have been completed in 4 hours and the beating of the rags in 6 hours. Add to this the fact that the consistency of the stuff in the beater was in normal practice higher than in the washer; in fact, Whatman mentions that Wilmott added "about a Third part [of his separately prepared coarse rag half–stuff] mixed with the fine and all work'd into a fine Paper". Even if the assumption in this case has been incorrect, the furnish in the beater always had to be bulked up with an extra quantity of half–stuff. Thus there are two factors that favour an imbalance, operating time and consistency.

To illustrate this point and to demonstrate the options open to the paper maker let us assume, purely as an example, that:–

(a) the operating times given above are realistic.
(b) the washers are furnished with 100 lb. rags and the beaters contain approx. 100 + 30 = 130 lb. half–stuff.
(c) for this example any loss of material is ignored.
(d) emptying and drainage time for the half–stuff is similarly ignored.
(e) and, finally, that the example occurs on Day 2 ff., because on Day 1 the rag washers clearly have to have several hours start to prepare enough half–stuff for the beater to use.

With these conditions in mind (and knowing that the Old Mill rag washers and beaters had roughly the same cubic capacity; in actual fact a slight disparity favouring imbalance) it will be seen that, if both washers and beaters work a 12–hour day, by the end of this day the 2 rag washers will have produced 600 lb. of half–stuff and the beater will have consumed only 260 lb. In time this would have led progressively to an ever–increasing imbalance in output. One might ask the question, therefore, why did Whatman install 2 rag washers to 1 beater ? Did he have, for example, another older engine at Old Mill that might have reduced this imbalance?

Assuming that we are looking at a plan of the complete stuff preparation system at Old Mill, a number of suggestions can be made at this point as to how this imbalance might have been rectified combining them variously instead of relying on a single one:–

(a) The half–stuff engines may have been dual purpose engines. Rees' "Cyclopaedia", for instance, states that "In small mills, where the supply of water is limited, they frequently have but one engine, and use it for both washing and beating".

(b) Even if not wholly dual purpose, the washers in this case may have been capable of beating weak rag furnishes for wrappers etc., In other words 1 rag washer serving the beater and the other washing and beating for some other kind of paper not requiring prolonged beating.

(c) The beater may have made up for its lower rate of output by beating round the clock. That this was practiced see Le Francois de Lalande Bib.23 Art.47.

(d) By using sharp tackle Whatman may have shortened his beating times and brought it more into line with that of the washers. This by itself would not have overcome the disparity, but it might have contributed to a solution of the problem.

(e) In the original assumptions made above various other factors that might have mitigated this disparity (or even increased it) have been left out of the equation. For example, the losses of material in the rag washer might have been significant or the half–stuff might have required a lengthy drainage period and much handling time.

With such a wide variety of unknown quantities it is impossible to–day to visualize what the operating procedure at Old Mill might have been for coping with this imbalance.

SUMMARY

The main interest in these drawings is that they represent a unique lifting of a curtain enabling us to catch a glimpse of the state of the beater and the beating of rags in the year 1771. Up to this point illustrations tend to represent an idealized system and here one has an indication of some of the practical realities that a contemporary paper maker had to cope with.

One cannot be absolutely certain from the two drawings that have survived whether they represent a completely new beater floor installation or merely the addition of a new Engine (singular) adapted to a facility already in place, though parts of the latter may have required modification to make it compatible with a new beater, e.g. possibly an additional row of cogs, if not there before, to drive the beater. Did it end there? One might ask why did Whatman go to the trouble of illustrating what was evidently a rag washer bed–plate and dating it 1771, clearly differentiating it from the beater and bed–plate shown opposite? It seems almost certain that a prior installation resembled the radial arrangement shown in Drawing No.2 and which may have dated back to 1733 when such arrangements were already well–known. This subject has been mooted previously in Chapter VI[118] on the grounds that the entirely new–built Old Mill incorporated the New Technology and had never possessed a "pourrissoir"; while Harris' first objective when rebuilding Turkey mill would have been to rid himself of one.

Drawing No.1 makes it clear that one is witnessing an experiment in using sharp beating tackle, clearly much sharper than Whatman had been accustomed to, though whether he had formerly used grooved bars is a matter for conjecture; the use of single bars grouped in "tiers" might have been one of the features first tried out at Old Mill when it was built. The drawing also illustrates two other important features, namely, that the angling of the bed–plate, though

---

[118] cf. Chap VI p.244.

not optimized yet,[119] was accepted without special comment; and that increased cell space had improved the circulation and a reduction in knots.

Drawing No.2 provides dimensions of a contemporary installation from which several statistics of interest might be derived, such as capacities, roll speed, cutting length and beating area, not examined here.

One of the most remarkable features of these drawings, however, is that one is seeing a *paper maker* actually making engineering drawings and methodical observations of his new Engine, implying that the younger Whatman was abreast of the times and interested in the progress of contemporary engineering. It was perhaps an early manifestation of what we know for certain later on was his interest in the most recent advances in technology and his contact with the leading engineers of his day, John Smeaton and John Rennie; his interest in bleaching[120] and thoughts on the use of steam for heating his vats and drying his paper. (Balston, T. Bib.1 125. Joshua Gilpin's account of his visit to Whatman in 1796: Journals, Public Records Division, Harrisburg, Pennsylvania.)

---

[119] It is not known when the practice of angling the bedplate was first introduced intentionally; it might have been discovered by accident. If the fly–bars and bedplate bars are exactly parallel to one another, the force necessary to shear fibrages is so great that the roll takes on a jumping action. Initially this may have been regarded as acceptable and perhaps somebody noticed that some rolls did not jump as much as others and looked for a reason. In theory parallel bars will produce the maximum drawing out effect on fibres, but practically no cutting action will be obtained. Hence, the angle at which the bedplate is offset to the roll is known as the cutting angle. If the bedplate is elbowed, as in certain modern beaters, there is a danger that stuff may be forced into the angle and remain uncut; a similar danger could arise from the cutting angle being too great, the fibrage being carried through along the bars without being cut. (cf. comment on the performance of the Maalbak, Chap.VI p.225).

[120] Balston, T. Bib.1 103. It seems that Whatman had been in possession of the various Treatises on chemical bleaching at some point between 1788 and 1791 and may have learnt of this subject either from James Watt or John Rennie.

THE PROVENANCE OF THE PAPERMAKING HAND FELT IN BRITAIN

The question was asked in Chapter V, who made the papermaking felts ? and where? This is a subject that still needs a great deal of research to find answers to these questions. The author has made extensive searches, but the result to date is effectively negative. It is possible that the requirement for this kind of cloth in the early stages of the industry was so small, relative to the total demand for woollen cloth, that this particular product escaped specific classification in inventories and other records.

As wool and woollen cloth were Britain's premier industries from the time the paper industry came into being in this country up to the early part of the 18th C., one presumes that the immigrant White paper makers of the 15th to mid–17th Cs. would have found a cloth that suited their needs locally; and, further, that the Brown paper mills would have managed likewise, not having any need for a special type of cloth, continuing to draw on local sources of supply as the industry continued to grow, probably right up to the 19th C. But towards the end of the 17th C. and in the early 18th the White paper industry not only increased in size but the quality of its products was also raised and one of the factors in achieving this improvement was the introduction of the special two–sided felt. Where did the White paper makers obtain these felts from ? Did they continue to use local sources of supply, instructing the manufacturers in their special needs ? or was there some centralised agency for this ? When one examines this subject it will be seen that there are signs that point in both directions.

For example, in the 1690 Dublin mill inventory, Dupin did not specify as he did in other instances that the "wrought white Cloath" had come from London or anywhere in particular. One might draw the conclusion that he obtained his felts in Ireland.[121] Similarly, the "haircloth" referred to in late 17th C. Scottish inventories presumably came from local sources also. And yet, the item for "Woollen and Hair Ruggs for Papermakers, Cloathworkers, Sadlers and Pouch Makers" offered at the Coffee House Auction, near Cornhill, in 1695[122] points to a more centralised source of supply. Were these made in Lancashire, perhaps, for sale in London ? The fact that the announcement specifies "for Papermakers" suggests that some makers at least may have bought their felts regularly in a London market supplied by manufacturers who specialised in this type of cloth.

"The position is complicated by the fact that we know that during the 19th C. the papermaking Felt industry became established in the North of England, mainly in Lancashire, and the question that arises from this is whether the industry was indigenous to that part of England or became concentrated there later for other reasons ?

[121] Miss P.M. Pollard (letter 16:May:1986) has kindly informed me that Dupin started making paper at his Dublin mill just before the Irish wool trade became so prosperous that the English clamped down on it, forbidding exports. However, she added that "woollen" shrouds were considered ca.1698, which doubtless may be taken that white woollen cloth would have been available.
[122] Bagford, J. Bib.22.

A case can be made out to support the former suggestion; but, on the other hand, continuity of this trade cannot on present evidence be demonstrated with any degree of certainty. Lowe has shown[123] that Lancashire had a substantial industry during the 16th C. manufacturing narrow width (27 inches)[124] "Rugs" and "Friezes", which were sold not only in London but in a wide variety of local markets such as East Anglian towns, Cambridge, Witney, Banbury, Newbury, Daventry etc.; and some of their products were exported, until the War with Spain put a stop to it, to Italy, Spain and the Netherlands. Lancashire then was obviously an important source for this type of cloth, but before the paper industry had established itself in this country. London, it seems, continued to be an important market during the 17th C., but it is impossible to say whether the rugs referred to in the coffee house auction at the end of that century came from Lancashire or one of the other centres of the woollen industry. From their description it would appear that both rugs and friezes could have been used as papermaking felts, white or coloured; both were rather coarse cloths and friezes were characterised by their nap.

The next link in this hypothetical chain is provided by the history of the Kenyon family,[125] the firm of James Kenyon & Son of Bury in Lancashire having had a long history in the manufacture of papermaking felts. In a letter to the author, dated 12th January 1977, Mr. J.C. Kenyon (formerly Chairman of the Company) wrote "We, as very old Lancashire Blanket Weavers, have records of making hand felts, as well as blankets, horse cloths, duffle coating, overcoating etc." A certain similarity in this statement to the item in the 1695 Auction may be noted. Unfortunately, the documentary evidence for their involvement in papermaking felt manufacture only dates back with certainty to 1827 and with one possible exception relates to felts for Paper Machines.[126] All one can say at the moment is that (a) because of their

---

[123] Lowe, Norman, Bib.71 3/4 and Chap.V.

[124] A width that would have more than covered most paper sizes up to the mid–18th C.

[125] Muir, Augustus, "The Kenyon Tradition, 1664–1964" Heffer, Cambridge (1964).

[126] Although the Kenyon family were connected with wool in one way or another for over 300 years, the first identifiable evidence of their connection with the manufacture of papermaking felts found so far occurs in their "Ledger of Weaver's Work, 1827–1830" held by the Lancashire Record Office (DDX 823/2/2). There is an earlier "Journal of Output. 1820–1829" (DDX 823/2/1) where the column relating to "Sorts" might yield a reference to papermaking felts if interpreted by an expert in this business; the entries seen are, for example, "Broad", "Narrow", "flan" and "Machine" and might really refer to sorts used for a wide variety of purposes other than papermaking. A puzzling feature about this second Journal is that according to Muir's history of the Kenyon family, James Kenyon only set up his wool manufacturing business in Clerke Street, Bury, in 1826, having received his training quite independently in Robert Battersby's Wool business of which we know nothing. A possible answer to this problem is that the Journal of Output might refer to Crimble mill which had been in the family's hands during the last quarter of the 18th C. and which they turned over increasingly to cotton early in the 19th C. though wool continued to be processed there for a time. During the 1820's Crimble was sold to Charles Stott. The Journal of Output might thus refer to operations at Crimble and the Ledger to Clerke Street; and it is in the Ledger that references are found to felts being supplied to paper mills, mostly it would seem for paper machines.

The Kenyon family history is too complicated to consider here; one of James' sisters married John Battersby, the son of Robert who had the wool business in Bury and this might account for James Kenyon training there. It is possible that in founding his Clerke Street business James had picked up the papermaking felt trade from Battersby's and entered this field via the paper machine market rather than the hand felt one. By 1827 the market for papermachine clothing would have been expanding rapidly and have shown great promise for any prospective felt manufacturer, particularly as at that time machines would have required dry felts as well as wet. So, although the Kenyons' connection with wool was a very longstanding one, in the absence of documentary evidence any claim to have been involved in the manufacture of hand felts at an earlier date can only be seen as traditional.

It must be said, however, that Kenyons may well have supplied the paper industry with hand felts through, say, the London market in the form of rugs and friezes from a much earlier date. Because of the small size of this outlet it would

(continued...)

300-year history in the Lancashire wool industry (John Kenyon was organising the spinning and weaving of woollen cloth and fulling it on the River Roch in the 1660's), there is a strong probability that they would have been engaged in making and selling the same sort of cloth as had been made there in the 16th C., a cloth which by the sound of it could have served as hand felts; and (b) some 120-140 years later the family is still there, in the wool business, and at a time when many other firms, whom we know also supplied the paper industry with hand felts, were establishing themselves.

A third element in forming this picture of Lancashire as the principal source of the early paper industry's hand felts is suggested by the significant expansion of the paper industry in both Lancashire and Yorkshire during the 18th C. as the Table below, using Shorter as a source,[127] indicates:−

	Lancashire	Yorkshire
1701–1725	7	9
1726–1750	10	12
1751–1775	15	24
1776–1799	23	31

(These figures represent *very* approximately the total number of paper mills, all types, and must only be taken as an indication of the order of growth.)

Shorter[128] has suggested that a connection existed between many of these mills and the woollen industry. For example, some of them may have come into being to make Press papers used in the finishing of the woollen cloth. It is conceivable, therefore, that a certain reciprocity existed in that some weavers specialized in making felts for the paper makers and perhaps established a name for themselves in this way as Felt manufacturers.

Certainly, early in the 19th C. the papermaking felt industry was already well entrenched in Lancashire. Firms like Messrs. R.R. Whitehead & Bros. of Saddleworth, Oldham, founded in 1770, were later important suppliers of felts to the paper industry. Messrs. Porritt & Chadwick of Bury were founded in 1808 (later Porritt & Sons and more recently Porritts & Spencer) and were certainly engaged in making papermaking felts by 1832; and, moreover, survived into the 20th C. as one of the leading felt manufacturers. Other companies that one finds

---

[126](...continued)

not have warranted them specializing in it, but as the industry expanded it might have contributed to Lancashire becoming a centre, as indeed it did, for this product of which Kenyon's formed a part. The Kenyon tradition would certainly fit in with this picture. Lowe (Bib.71 Chap.VI), for instance, commented on the fact that towards the end of the 16th C., in spite of Government efforts to stop this, some of the Lancashire town industries, especially wool and linen, were moving out into rural areas for a variety of reasons, in search of water power for fulling; the ability to couple agricultural activities with spinning and weaving; and to escape the restrictions of the town craft guilds. The Kenyons were already established in the countryside as yeoman farmers and ca.1660 John Kenyon reversed this trend, left his farm in the hills and set up a wool business in Castleton, Lancashire, and to judge from their expanding activities in the 18th C. they prospered. So it is not difficult to envisage their involvement in supplying rugs and friezes, horsecloths and blankets to London, some of these finding their way into the concentrations of paper mills in the south.

(It would be appropriate to acknowledge at this point the very considerable help the author received in 1976/7 from Mr. J.C. Kenyon of Slaidburn, Lancashire; and Mr. J.K. Bishop, County Archivist at the Lancashire Record Office.)

[127] Shorter, A. Bib.12.

[128] Shorter, A. Bib.9 63.

references to were Kay Grundy of Bury; Hardmans and no doubt there were many others whose names and activities have been lost. At the beginning of this section it was said that much research remains to be done and a fruitful field for this would probably be found among the records held by the various Lancashire Institutions providing a much more substantial picture of the origins of the felt industry than it has been possible to give here.[129]

Turning now to the pointers in the other direction, during the 17th C. one has a picture of a wool industry that was dispersed all over the country and at a time when the small paper industry would still have been using the unmodified blanket–like cloth mostly for making Brown paper. Paper makers in these circumstances would undoubtedly have relied on local sources for their felts. Later, and turning to White paper, it has been shown that people like Eustace Burnaby and the COMPANY must have used an improved form of hand felt. As the most important centres of White paper manufacture were still close to London, did this become a central source of supply of these; or did paper makers still rely on local sources; or did they crop the cloth themselves? If the truth were known, it seems probable that all three sources were used.

Although in rapid decline now Kent would still have had a wool industry that could have met these needs and may well have been eager to do so. To the west of London the wool industry was still thriving. As in Lancashire, so in Oxfordshire the 300–year old Witney Blanket Industry would have been in a position to have supplied local paper mills with this material or a centralised market like Blackwell Hall in London.[130] It is unfortunate that no records have been discovered yet that would confirm any of these ideas.[131] The quantities involved may have been too small.[132]

For the early period then the pointers do not point in any specific direction; the improved felts that we are concerned with could have come from Lancashire or, equally, from one or the other centres of woollen manufacture; alternatively, they might have been bought locally or

---

[129] It should be noted that many of the better known firms, those that survived into the 20th C., not only took over many of the smaller firms but they, in turn, have since been taken over themselves. In most cases the records of these firms have been either lost or destroyed as a result of these mergers, so that it is not easy to come across information concerning their early history. It has been suggested that in addition to such records as may be held by local authorities and libraries, the various textile journals might also be useful sources of information.

[130] The Act of 1678 regulating the operation of the various halls in London handling woollen goods (Blackwell, Leadenhall and Welch Hall) mentions "Haircloth" amongst many other articles, but does not describe it. Since so much of the cloth appears to have been disposed of privately, depriving Christ's Hospital of Hallage, it is possible that a search amongst the Hospital Records of the fines imposed on Wool Traders breaking the regulations might reveal more information on this type of cloth and its origin.

[131] In addition to Mr. J.C. Kenyon and Mr. Bishop, mentioned earlier, I am greatly indebted also to Messrs. Porritts & Spencer (Lancashire); Mr. David Smith and Mr. Richard Early of the Witney Blanket Industry; the Oxfordshire County Archivist; the Archivist of the Oxford University Press; the Libraries of Oldham and Bury; Dr. D.T. Jenkins of the University of York; Mr. Paul Hindle of Salford; Mr. K.G. Ponting, Pasold Research; and Mr. Denis Richards through the Wool Industries Research Association for the help they have given me in this search, regrettably to no avail. In addition many contemporary sources, including inventories, have been examined; enquiries made to many Museums, etc. If early evidence is to be found, it seems likely that it will be found by chance among local records.

[132] A *very rough* calculation (but probably of the right order of magnitude) based on the younger Whatman's felt purchases in the 1780's produces a figure of some 3500–6500 yards p.a. of cloth (37 to 54 inches in width) as a requirement for the known White mills of this period, and obviously a great deal less than this at the beginning of the century. This is probably too insignificant a quantity to be traceable in company records.

from London or some other market. Is there any later evidence that points to supplies coming from areas other than Lancashire ?

The next mention seems to be that contained in the Whatman Ledger for the years 1780–1787 (PSM). Here various entries are to be found for different lengths and widths of "Felten" and "Flannel" all bought from one John Balcombe. Some of the details of these purchases will be found in Thomas Balston's biography of the Whatmans.[133] But, so far, it has not been possible to identify John Balcombe or discover where he lived. It is a name commonly found in Kent and at the beginning of the century at least one Balcombe was a Clothier of Frittenden, not far from Cranbrook, once the centre of the Kent Broadcloth industry but virtually defunct by the 1780's (according to Hasted in 1790 there were no clothiers left in Cranbrook), although there were still Balcombs in the district but it has not been possible to identify their occupations.[134]

The evidence provided by Whatman's method of payment does not tell us anything conclusive either; if anything, it suggests that Balcombe might have been a Factor in London, but no trace of him has been found in any of the London Directories confirming this.[135] Besides, the next piece of evidence suggests that in one way or another he could possibly have been either a felt manufacturer or finisher himself or the agent for one working somewhere in Kent perhaps.

Whereas it has been shown earlier that in the 19th C. the Felt industry was establishing itself in Lancashire, there is also evidence that for several years at the beginning of the century there was a "Felt Factory" in Maidstone itself.

A statement in the "Original Papermakers Recorder"[136] stated that a Dr. Charles had a Felt Factory in Sandling, nr. Maidstone, ca.1826 and that he was "practically the *only* maker of Felts for Papermakers. Subsequently, we believe, the goodwill and plant of his business was disposed to Messrs. Whitehead of Oldham". Details of Dr. William Charles and his family will be found in Appendix III and these show that he was described as early as 1794 in the Universal British Director, under Maidstone, as a "Felt Manufacturer". Where his premises were at that time is not known; but he appears in the Boxley Ratebook as early as 1784 and was assessed for a fulling mill at Sandling in 1800 and in the same year acquired Chillington House in Maidstone where he set up a "business of felting and blanket cleaning" in partnership with a Robert Harris. The business was sold in 1840, but no records, no ledgers or any other information, to confirm the transaction with Whiteheads, for instance, have come

---

[133] Balston, T. Bib.1 63.

[134] The Index of Apprentices of Great Britain (Bk.8 fo.77 Soc.Genealogists) shows that in 1720 Balcombe, Richard s.o. Richard of Frittenden, Clothier, was apprenticed to James Fuggle of do. Kent, Butcher. A John Balcombe was also assessed in the Sissinghurst Ward of Cranbrook in 1780. Another John Balcom*b* has even been found in Maidstone (1800), but the assessment for his Stone Street premises (3/–) is so low that it is clearly not that of our felt supplier.

[135] Whatman found it convenient to pay many of his London accounts through his Stationers, who acted as Bankers for him (Balston, T. Bib.1 53). The Ledger shows that he paid Balcombe through Wright, Gill & Co. (George Gill's grandson) for the years 1780–84, though in 1781 there is an entry of a payment made through Fourdrinier & Co. to Bloxam & Co. for a "piece of Felten" £4.4.0 Sept.28th and this charge does not appear under Balcombe elsewhere in the Ledger. In 1785 the entries under Balcombe show that he was paid "Cash". This was also the case in some other years and they could have been made while Whatman was visiting London. In 1786 Balcombe was paid through James Woodmason; and in 1787 through Fourdrinier & Co.

[136] April 1902, quoting the Paper Trade Review.

to light. (Messrs. Whiteheads were absorbed by Porritts & Spencer during the 20th C. and none of their records appear to have survived). William Charles' enterprise can be seen however as a pointer to the fact that the paper industry, at least in Kent, may have depended in earlier periods on local suppliers for their felts. The paper industry at Maidstone at the beginning of the 19th C. was dominated by its White paper manufacture and, if we are to believe the statements made about the Felt Factory there, then for it to have lasted for so long points to the fact that some of these White paper makers were buying their felts from William Charles.

Nothing conclusive has emerged yet from either the suggestion that the Lancashire frieze or the Yorkshire woollen industries had been the prime source of the improved felt; or that London was the focal point for their distribution, though there are signs that favour a Lancashire source. There are no clear pointers either to sources found in the vicinity of the White mills around London though, once again, there are hints that this might have been the case. But if one takes into account the more dispersed sites of White papermaking such as Dublin and Scotland being supplied with "wrought white Cloath" and "haircloth" there does seem to be a case for the local sources of supply perhaps under instruction from the paper makers of those districts.

The dearth of information on this subject may be due to the small quantities of cloth involved, at least initially; or the possibility that the early paper makers may have cropped their own unmodified felts. But there is evidence that towards the end of the 18th C. the manufacture of papermaking hand felts, if not located in one area, had by this time led to a centralized source for distribution. Support for this notion comes from a surprising quarter. Voorn,[137] in answer to a question from the author, pointed out that "The Dutch paper makers sometimes bought their felts locally, but more often from well-known Textile Centres, such as Leiden, Haarlem and also Amsterdam. Felts were expensive and most mills had only a limited quantity available". (He gave instances of 18th C. prices). He then added:–

"During the French Occupation of Holland [1801] after the French Revolution, the mills of Messrs. Van Gelder reported that they now use felts made at Leiden and Hilversum, *but that formerly they imported their Felts from England.* The Dutch Felts were not as good as the English were, but they were improving".

---

[137] Voorn, Henk, Bib.28 PAHIS 761652 3:Nov:1976.

A NOTE ON INTERNAL SIZING IN RELATION TO THE CONDITION OF 17TH, 18TH AND 19TH C. RAG FURNISHES

It is important for the reader of this section to bear in mind constantly the distinction that is made here between what may be called, loosely, the "surface" sizing of the paper with a film of gelatine and the sizing of those regions within the cellulose matrix where the gelatine has failed to penetrate and alternative methods have to be used to size them. The absence of sizing in the latter manifests itself when the surface of the paper is disturbed as, for instance, following the erasure of marks, writing etc. and the paper is then re-used. But it was also quite common for the surface sizing of the paper with gelatine to fail too, resulting in absorbent areas and leading, for example, to what was often called "colour sinking". As will be seen in this section the condition of the rags and their subsequent treatment could in certain circumstances either have contributed to or alleviated faults in the finished sheet of paper arising from the sizing process.

It will be recalled that in Chapter V the subject of Internal and Engine Sizing was raised in connection with Hagar's Patent of 1682, in which he claimed that he was able to size paper by adding some substance to the rags *in the mortars*. It was also pointed out there that gelatine, as used in the conventional sizing process, sometimes failed to penetrate the less accessible regions of the fibrous matrix of the paper and at some point in the history of papermaking to overcome this difficulty more mobile solutions were used containing water-repelling substances which could be precipitated onto the fibre *in situ* and thus impart a degree of sizing to the surface of the finer capillaries. When these substances were used, usually added during the beating stage, as an auxiliary sizing agent, the process was known as *Internal Sizing*. One of the commonest agents employed in more recent times was soap precipitated onto the fibre surface with alum to form there a coating of insoluble aluminium soap.

The origins of internal sizing as an intentional process are not known; or, put another way, it is not known precisely when the need for this first arose. A clue, but not necessarily the answer to this question, could be provided by the condition of the rags, and in turn the dispersed and beaten fibres, used to make a sheet of paper. In the course of laundering textiles insoluble calcium and magnesium soaps become cumulatively incorporated within the cloth (and, in the end, the rags) originating from those areas where common soap and hard water had been the principal cleansing agents;[138] and since soaps are derived from vegetable and animal fats, these insoluble deposits could impart an element of water-repellency, fractional though it may be, to the fibres, thereby conferring a degree of internal sizing on the waterleaf paper. In addition rags, discarded textiles, would have been in an unwashed state soiled with

---

[138] Substances known as fatty acids can be separated from vegetable and animal fats and combined with alkalis to form sodium or potassium soaps which are soluble in pure water. Because hard water contains calcium and magnesium salts these react with soluble soap to form insoluble calcium and magnesium soaps, the scum that may be observed when using hard water for washing. The greater part of this material will be rinsed out of the cloth, but a small proportion will have become deposited on the fibres. This will accumulate with repeated laundering unless steps are taken to remove it chemically.

unsaponified fatty matter[139] from human bodies or other sources of greasy material when they arrived at the mill. The question is how much of this water–repellent material would have survived the varying treatment the rags received as changes in process technology took place ?

Working backwards in time one can see three principal stages where modifications to the *papermaking process* could have affected the presence or otherwise of the fatty residues in question, that is for the moment not taking the origin of the rags into account, a subject that will be considered later. Once regular chemical treatment of rags had been introduced into the paper mill in the 19th C. (during or after the third decade in this particular case), fatty matter would have been effectively saponified and insoluble soaps hydrolyzed by boiling rags with alkali under pressure and subsequently removing the waste products by washing with *soft* water. Consequently, to retain any internal sizing effect it would be necessary to replace in a controlled manner the impure water–repellent substances that had been removed with pure materials. Hence the increasing interest in methods for internal sizing in the 19th C. There were two main methods of doing this and both used alum to precipitate a water–repellent film onto the fibre, in the one case to form an insoluble aluminium soap and in the other a finely dispersed hydrophobic coating from rosin, a gum residue derived from pine trees and consisting mainly of abietic acid, previously water solubilized with alkali. Moritz Illig, the first paper maker to publish a theoretical explanation of sizing (ca.1807), had invented the rosin form of internal sizing by 1800 though his work was not widely known until much later and his method only coming into general use after 1830. His method was really designed to replace the conventional gelatine sizing after–process and because it was carried out on the pulp in the beater it became known as *Engine Sizing*. This was a 19th C. invention so we are not concerned with it here.

As to the other method, just how early soap and alum were used for internal sizing can only be guessed at;[140] this process was also carried out in the Engine. There is documentary evidence to show that it was being investigated in 1815 and there are also signs that other aids to sizing had been considered earlier than this, some possibly much earlier and certainly *before the practice of boiling rags had been introduced into the paper mill* which, if effective, created a need for it.[141] In fact it looks as if the interest in this subject had arisen from some

---

[139] The unsaponified material referred to here is the naturally occurring vegetable or animal fat which contains glycerides from which the fatty acids used in soap making, for example, may be separated by saponification.

[140] This question is made especially difficult because papermaking became progressively divided during the 19th C. between the top qualities of White paper made in small mills and by hand and where the use of internal sizing using soap and alum would have been employed and papers made on machines and mainly used for printing where internal sizing would have been unnecessary and sizing with rosin more usual. As the volume of machine made papers rapidly outstripped the top qualities of White, the chances of finding records of internal sizing using soap and alum are corrrespondingly slender.

[141] Apart from evidence of interest in the use of soap and alum and other additives and unexplained purchases no positive evidence of the actual application of any of these substances to internal sizing has come to light yet. Nevertheless the evidence that these methods were being investigated may point to their use elsewhere for which no records have survived. The very fact that experiments were made must point to some reason that prompted them and we know later that at some unknown stage they came into and continued in use into the 20th C. Taking the items in question in chronological order we have: (i) In the Whatman Ledger (1780–1787 PSM) there are a number of entries showing that several hundredweight of soap were purchased annually from John Magrah & Co., Church Row, Aldgate. There is no indication as to what this was used for. It might have been used for washing felts or something else, but the fact that soap was bought at all may be noted; (ii) A sheet of Notes written by William Balston (undated but probably written before 1805 as the paper is marked "CRISPE & NEWMAN" whose partnership was dissolved in 1801/2) contains the following, "to try the effect

(continued...)

as yet unidentified cause;[142] or, alternatively, that a need for it had survived from a much earlier period when some remedy had been introduced to counteract the imperfections that were always likely to occur in the very difficult gelatine sizing operation. We will return to this subject again when considering the third and earliest stage.

The second or middle stage covers the transition between the abandonment of the Old Technology and the beginning of the chemical era in papermaking, that is approximately between 1730–1780/1790's; during this period the rags were no longer fermented in the more progressive mills. In spite of the fact that only the best and whitest of rags would have been used for White writing paper, most of them would have contained an accumulation of insoluble soaps together with a certain quantity of unsaponified greasy matter. After sorting they would have *received no intermediate treatment* before being subjected to a cold water wash in the rag–breaker during the stuff preparation process. In other words this would have left these water–repellent materials largely untouched and, as a consequence, one would have expected the papers made from these rags to have had a certain degree of in–built internal sizing.

However, not all rags arrived at the paper mill in the same state. There is some quite positive evidence that white rags imported from different parts of Europe had originated from cloth that had been laundered using totally different methods, those from northern Europe, including England, containing water–repellent materials and those from southern France, Spain and Italy largely free from these materials.[143] During this period it seems likely that the best classes of domestic writing paper would have been made from a mixture of British rags and those imported from Germany[144] and would thus have retained a significant element of water–

---

[141](...continued)
  of a boiled solution of wheat flour, or Rice in Plate papers either in the Engine or by way of size"; (iii) The Gaussen/William Balston correspondence (1815–1847 PSM) shows the year 1815 was one of considerable experimentation at Springfield mill, Maidstone, over a relatively wide field within the papermaking process and this activity appears to have been going on for some time. Gaussen's letter (9:June:1815) lists the current price of chemicals like oxalic acid, oxymuriatic acid, sulphuric acid and "resin about 3^d per lb. Starch, I believe, for rendering paper stiff if mixed with size. I have tried it with waterleaf and it gives stiffness even without size – for second papers this should answer". In his next letter (10:June) "I find that the precipitate from yellow soap yellow, but from resin white". Again in his letter (23:June) "In regard to the 2/3^d of alum being useless viz. the solution of alumine, I am convinced of being wrong as Gottling [?] says the solution combines with that of soap to make an earthy whence the weight cannot be lost".

[142] The introduction of the steel pen nib, for example. See Chap.V p.195.

[143] An example of this difference is found in a letter from Anthony Poggi, the engraver, to the younger Whatman written in the 1790's (Bib.39). In the following extract he makes a distinction between "All linen that becomes rags in the Northern provinces of France, in England, Holland and Germany having been washed by *dint of soap alone* does retain an inflexible quality and substance, which render it unfit for our purpose" (engraving) and the fact that "the greatest part of the French, Spanish and Italian Linen is consumed and brought to rags by a process totally different from English, Dutch and German". He goes on to say that "they take in those countries [that is the southern ones] a quantity of wood ashes [predominantly potassium carbonate] in proportion to the linen they have to wash, which ashes they boil in water for some hours until it becomes a strong lees, when they pour it hot over their linen, which are put in a tub and covered with a double coarse cloth that serves as a strainer to retain the ashes from the linen. Thus it remains for eight to ten hours, when these first lees are drained off and others added in the same manner as before. By this process the linen *becomes almost clean before any soap is used*, which I should add is only for the finest linen". He concludes this account by saying that "the small quantity of Soap that is used in washing in those parts of the Continent can have little or no effect towards *leaving an oily nourishment behind it....*" Further details of the effect of alkalis on cellulose may be found in this App. "Textile to Rag".

[144] See Chap.III note 78. It was pointed out there that although 75% of rags used by the British paper industry came from domestic sources, the balance imported mostly from Germany (but cf. note 157) would have gone into the best qualities of paper.

repellent deposits. In theory then there may have been no great need for sizing the fine papers of this period internally, even if it were ever contemplated. Although the fact remains that towards the end of the 18th C., if not earlier than this, Whatman was buying considerable quantities of soap which could have been used for this purpose.

One has to realise in this comparison between the first and second stages described above that the transition from the earlier of these periods to the chemical era in papermaking was not a sharply demarcated one. In spite of it being used earlier for textiles, the chemical bleaching of papermaking rags in the paper mill with chlorine, the first major chemical innovation in the papermaking process, was only being experimented with and introduced gradually from ca.1790 onwards.[145] This, by itself, would have had no really significant effect on the greasy materials beyond chlorinating them; it would still have required an alkaline extraction to dissolve and eliminate the chloro–compounds so formed. The effective removal of these fatty residues would thus have had to wait for the introduction of alkali treatment in the paper mill.

Joshua Gilpin, the American paper maker, described a visit to Polton mill, on the River North Esk, Midlothian ca.1795 and related how William Simpson had found that boiling his rags in lye (potassium carbonate) prior to bleaching with chlorine gas was very effective, though in his case the object of the boiling was to destroy the hard element of flax in the coarse linens.[146] All the same, it may have helped reduce the fatty content as well; but the object of mentioning this is merely to show that the idea of boiling rags at all in the paper mill was now a possible new line of approach to treating them. At the other end of this bracket it seems improbable that the boiling of rags with alkali under pressure was practised before the 1830's, when Donkin's rotary rag boiler was introduced.

Intermediate between these two events rags were treated in wire cages immersed in open "square" boilers, at first with hot water[147] and at a later stage with caustic soda. It has to be remembered that although it had been used earlier, caustic soda only came into general use after the repeal of Salt Duty in 1823. Obviously, in any undertaking larger than pilot plant scale, to be effective this hot treatment of rags would have been dependent on the introduction of steam generating plant into the paper mill (early instances included Howard & Houghton,

---

[145] Balston, T. Bib.1 103–108.

[146] Edelstein, S.M. "Papermaker Joshua Gilpin" (The Paper Maker, Hercules Powder Company 30, 1961 No.2 p.8). Gilpin's reference to the rag boiling at Polton mill can only refer to a practice not widely used at that time, because Johnson, writing in 1794 makes no mention of treating rags between sorting and beating, no hot water wash or boil with lye. (Johnson, B. jnr. "The Papermakers Guide, or the Art of Paper Making" privately printed, Ludgate Hill, London). Although rags at this date were not processed in this way, it is interesting to note that chemical processing of raw materials for paper was undertaken on an industrial scale as early as ca. 1800/01 by Matthias Koops implementing parts of his Patent (2392) for recovering waste printed paper and using another Patent (2481) for such diverse materials as flax waste, straw, hay and other cellulosic materials including shredded wood. The materials were steeped in limewater (calcium hydroxide) and potash (thereby liberating strong alkali) and then either boiled in this liquor or clean water followed by draining, pressing and beating. (For an extended account of Koops' activities see Hills, R.L., "Notes on Matthias Koops' Papers", The Quarterly [BAPH], No. 2, Aug. 1990, 3–5). In the event Koops was bankrupted. (cf. this App. note 2). John Rennie was involved in designing some of the machinery (Coleman, D.C. Bib.15 172/3).
The boiling of rags in alkali was probably delayed as a result of the salt tax. After its repeal in 1823 alkali works were quickly established in Lancashire and the way lay open then for such applications.

[147] Martin, Thomas (alias John Farey who also wrote the account of papermaking in Rees' "Cyclopaedia" in 1813) in "Circle of the Mechanical Arts" London (1813) stated that "Rags, if very dirty, are washed in hot water by a fulling mill, as is used by Dyers for washing cloth". (Le Francois de Lalande claims that some rags in France were washed before sale; see below p.256.

Hull, 1786; Koops, London, ca.1800; and Brown & Chalmers, Aberdeen, 1802). William Balston, to take another example, had this facility from 1807 onwards. Even so it seems doubtful, assuming that he was treating his rags with hot water or soda ash, that either of these would have been the sole cause of introducing soap and alum internal sizing. In any case there is no evidence that William Balston was treating his rags in this way at such an early date;[148] Farey, who must have visited Springfield mill sometime before 1813 only refers to hot water washing of the dirtiest of rags. So the conclusion must be that, if soap and alum were being contemplated in 1815, it cannot have been influenced by this embryonic new approach to the treatment of rags in the paper mill. One either has to see it as a survival from a much earlier period introduced, perhaps, as a remedy for deficient surface sizing; or to deficiencies in internal sizing shown up by the increasing use of the steel nib; or, possibly, to the growing use of cotton rags though why this should have created this need is not immediately apparent.

One can now revert to the third and earliest stage which covers papermaking in the British Isles *before* the adoption of the New Technology i.e. at a time when rags for White paper were still being fermented. When one goes back to this period there are two issues to consider, issues which may or may not have been related to the condition of the rags. In one case the interest lies in finding an alternative to the method of sizing paper with gelatine: and, in the other, a method of correcting some deficiency in the gelatine sizing process as it affected the paper. It has been demonstrated above that in the 19th C. both of these issues interested paper makers.

The position at the end of the 17th C., the time of Hagar's Patent, led one to ask the questions, was Hagar looking for a replacement process, or one of improvement ? The conclusion was reached (in Chap.V) that in his case it is very doubtful whether there was any substance that would have given the White paper, which his Patent covered, the durability required of paper in those times other than gelatine. If he did, in fact, employ a starch mucilage, one might conceivably see it being used to increase the stiffness of papers made from weak rags, bolstering up their strength. Otherwise one has to assume that his Patent was directed towards rectifying some deficiency in the Paper stemming from the sizing process such as poor surface sizing or inadequate internal sizing.

Even as late as the 1760's (in fact right up to the 20th C.) it was a common experience for scriveners and draughtsmen using pen or wash to find the surface sizing of paper deficient; as Lalande had noted "Le collage du papier manque souvent" and he lists a number of reasons for this.[149] The faults arose from the use of weak or decomposing gelatine, insufficient quantities of alum, premature gelling, unsuitable drying conditions ("the direct rays of the sun must not come nigh it till it is dry, for otherwise the size would become drawn out") and so on. All these faults, however, stemmed from either poor quality gelatine or incorrect procedure in its application and were, therefore, correctable without having to resort to the use of additives. Moreover, deficiencies in surface sizing would obviously overshadow any

---

[148] William Balston's Rag Stock Lists of the 1830's, though not necessarily complete, show for the early part of this decade that he lists amongst other items casks of Potass. and in 1837, as well as Potass., casks of Soda.

[149] Le Francois de Lalande, J.J. Bib.23 Art.107. The whole of his account of size and sizing appears to be concerned with surface sizing only; no reference has been noticed commenting on any deficiencies in the sizing of the interior of the sheet of paper.

that might arise from the failure of the gelatine to penetrate the interior of the cellulose matrix. One might add to this the question, would a deficiency in internal sizing have been noticed by anybody at that time ? Were the criteria stringent enough for its absence to have been a matter for concern, e.g to someone like Hagar? In terms of surface disturbance, as in the case of erasure, this seems very doubtful; but in terms of weaknesses, rather than total failure, in surface sizing one could ask whether these were compounded by the state of the rags. Did they lack that *"oily nourishment"* mentioned by Poggi?[150] In the middle period most rags probably contained sufficient fatty residue to have provided a modicum of sizing; but then there was nothing in the process to remove these residues. The use of soap and alum, or a starch mucilage, does provide a measure of sizing in itself. So, does one see in Hagar's Patent a means of abating what might have been a common nuisance, namely, imperfect surface sizing ?

Without experimental evidence, or a much more detailed account of the rag fermentation process as carried out in paper mills, one can only speculate as to whether the fatty residues were effectively removed from the rags or not. First of all it is clear that Lalande was aware of the presence of greasy matter in rags and this is the subject of several comments, which in substance may date back to M. Desbillettes' time at the end of the 17th C. Lalande refers, for example, to an experiment[151] in which rags and clay were mixed in the mortars producing "une pâte parfaitement dégraissée"; but it suffered from the disadvantage of making the paper greyish in colour. Nevertheless it demonstrates that the greasiness of rags was a matter of some concern. He also mentions the benefits derived by some paper mills from using rags washed with lye (lessiver) before being sold.[152] In another instance he compares the results of an experiment involving paper made from unfermented versus fermented rags.[153] The former resulted in stuff tending to clot in a viscous form and preventing it from settling uniformly on the mould cover during sheetmaking; he concludes by saying "cela prouveroit que le pourrissoir aide encore à dégraisser le chiffon". Though not directly connected with the above Lalande also refers to the use of lime to accelerate the operation in the pourrissoir.[154] One can only presume that the lime was used unwittingly to replace the fermentation process. In short where the hemicellulose content of the linen rags was low fermentation would have proceeded very slowly, too slowly for the paper maker's liking. So he resorted to adding lime. Lime would have neutralised the acid conditions favouring fermentation; and, though it may have substituted for its effects, it would have been coupled with much more damaging degradation of the fibre due to oxidation; moreover, it would have effectively retained any fatty residues present as insoluble calcium soaps. In France there were regulations forbidding completely the use of lime, although its use may have continued in some places. As there appears to have been no corresponding regulation in Britain covering papermaking,[155] it may have been practiced here as well, although the addition of lime was widely recognized as damaging to the strength of cloth and probably would not have been employed by reputable paper manufacturers of White paper.

---

[150]   See note 143.
[151]   Le Francois de Lalande, J.J. Bib.23 Art.11.
[152]   loc.cit. Art.10 & 11.
[153]   loc.cit. Art.20.
[154]   loc.cit. Art.18.
[155]   Strict laws were passed in the early 18th C. in England forbidding the use of lime in the bleaching of textiles.

These comments have been cited here merely to illustrate that fatty matter in rags, amongst other things, was a feature that some French paper makers were evidently preoccupied with, but should not be regarded as typical. In fact some of them complicate our concern here with the possible effects of the fermentation process on the fatty residues in rags; and other instances appear to contradict the observations made by Poggi on the treatment of French rags during laundering, where repeated steeping in hot lye would have reduced the chances of insoluble calcium soaps ever being deposited in their cloth, let alone being removed in the pourrissoir as claimed by Lalande. But there is one point in all this that is worth mentioning as possibly having a bearing on this issue. In all the different methods of fermenting rags described by Lalande[156] it is clear that the piles of rags were turned at frequent intervals, "on les retourne, & le centre vient à la surface pour faciliter la fermentation". With these points in mind, and assuming what would appear to have been a typical routine in the pourrissoir, one can now turn to some of the theoretical aspects of the fermentation process having a degreasing action on rags.

Though there may have been other impurities of this nature in the rags, attention is focused here on (a) natural fats from the body in the form of glycerides; and (b) insoluble calcium or magnesium soaps deposited on the fibres as a result of laundering textiles with soap in hard water, which appears to have been a characteristic of British and German washing conditions, typical but, of course, with exceptions. To reduce these impurities to a water soluble state, in other words where they could be removed by washing, one has to postulate a two stage reaction. In the case of the natural fats it is thought that lipase enzymes would almost certainly have been present in the fermenting mass of rags and these would have hydrolyzed the fats, splitting them into their components, glycerol (miscible in water) and the long chain fatty acids (insoluble in water). Hydrolysis of the insoluble calcium fatty acid soaps by the hot acid conditions of the fermentation might have been a more difficult matter and much would have depended on heat generated in the mass of rags, this according to some accounts could be very considerable. In any case one would end with further quantities of long chain fatty acids insoluble in water and not being removed by a later cold water wash. To break these down into water–soluble compounds, it is necessary to postulate the β–oxidation of these acids brought about by lipolytic bacteria, a process well–known in animal fat digestion. In the context of the fermentation of rags Dr. Gascoigne (Messrs. Cross and Bevan letter 8:May:86) has suggested that β–oxidation might well occur in this process too. But its effectiveness would have depended on prolonged fermentation and frequent turning of the rags to provide the necessary aeration; the lypolytic bacteria, he considers, would certainly be active under conditions well on the acid side of neutral, a condition considered to have been essential for a fermentation of this kind to take place. Theoretically then it should have been possible to eliminate the fatty impurities of the rags; but, as said, in the absence of experimental evidence this remains an hypothesis. We have only Lalande's comments that this was so. However, the method by which this could have been achieved, given certain circumstances, can be seen as credible and supporting his claim, albeit our concern is with the rags used by British paper makers rather than the French.

If one accepts the assumptions made above, then one can visualize a waterleaf paper made from fermented rags as being deficient in any sizing effect arising from the fatty impurities

---

[156] Le Francois de Lalande, J.J. Bib.23 Art.16 & 17.

and thus being more than usually disposed towards exposing and heightening any imperfections that might have occurred in the normal surface sizing process of the finished paper. Obviously, the underlying causes for such a condition would not have been known to Hagar or his contemporaries. For them the normal surface sizing process with gelatine may have been a source of many complaints and anything that might have improved this state of affairs would have been welcomed.

In this note the general condition (there must always have been exceptions to this) of the rags used in the earlier period for conversion into papermaking stuff has been examined with particular attention being paid to the fatty impurities in them and any sizing effect these may have exerted on the finished paper, whether it was in the form of internal sizing or as an insurance against imperfect surface sizing.

It has been shown that certain preparative processes in the 19th C. would have resulted in the removal of these impurities and because of increasingly stringent requirements by users of aqueuous graphic media there was evidently a demand for improvements in the sized properties of the interior of the paper.

A clear distinction can be made between these conditions and those obtaining for a period of 60 years or more in the 18th C. when rags were used without any pre-treatment. These rags, which would have been subjected only to a cold water wash in the Hollander rag-washer and which were probably selected with more than usual care for White papers during rag sorting, would have retained, therefore, any fatty impurities they may have collected during their lifetime. The fatty matter content, as Poggi has indicated, would have depended however on the provenance of the rag and the type of laundering treatment the parent textile had received. As a consequence one would expect that most British papers, if made from "northern" rags,[157] would have possessed an in-built form of sizing, quite possibly not required for internal sizing but nevertheless serving as a second line protection against partial failure of surface sizing.

There is, however, a blurred interface between these two periods during which new processes were being tried out, particularly in the pre-treatment given to rags before converting them into papermaking stuff. Apart from isolated experiments where the use of alkali may have been used, ostensibly for another purpose, no positive evidence has come down to us of hot alkaline treatment given to *rags* as a regular routine with the object of removing the impurities in them, amongst these the fatty impurities. There were hints that methods were being investigated, which we know later were certainly used to restore the sizing properties that the fatty materials imparted to the interior of the cellulose matrix.

---

[157] This assertion requires some qualification. The chances are that for most of the 18th C. (1725–1785) the best qualities of White paper would have been made from "northern" rags. Coleman (Bib.15 p.107 Table VI) shows that Germany and Flanders were the principal sources of imported rags, particularly after 1746. But the imports from Italy and the Straights rose significantly, if only temporarily, between 1786–1795 (between 1791–95 German 60% Southern 23% against a more usual 5%). It would be difficult to try to determine how these "southern" rags were distributed and for what papers they were used; but it might help explain some of Poggi's more obscure observations on the characteristics of certain late 18th C. Plate papers.

Finally, on a much more speculative basis, the condition of rags resulting from the biochemical treatment they received during the fermentation process used in the Old Technology has been considered. Given certain conditions, seemingly quite consistent with the description of the procedure that has come down to us, the conclusion was reached that many, if not all, of the fatty impurities would have been removed from the rags during this process; but the operative basis for this conclusion is that the fermentation was long enough, hot enough and aerated enough to allow the lipolytic bacteria to reduce the fatty materials to water–soluble compounds.

Besides the many changes introduced into the papermaking process during these three periods, all of which would have had their own characteristic effect on the properties of the finished paper, one can now add to these the effects arising from differences in the state of the rags ready for the beater, differences stemming not only from certain impurities associated with the textiles from which they were derived but also from the effect on these resulting from any pre–treatment the rags themselves may have received.

ADDENDUM ON SURFACE SIZING

Although the function of sizing paper and the method used during the 17th/18th Cs. have been referred to many times in this work in *general terms*, clearly there would have been variations in procedure from place to place. These have not been touched on partly because the intention has been to present a typical picture of the process rather than confuse it with an account of local practices;[158] and partly because there was no fundamental change in sizing technology until the introduction of steam heating of the drying lofts in the 19th C. and, later, the change to the use of aluminium sulphate in place of the potash alum used formerly.[159] (Drying lofts were heated towards the end of the 18th C. by means of stoves,[160] but it seems unlikely that these had quite the potential for accelerating drying that the later use of steam pipes had).

All the same adverse comments had been made on the sizing of 18th C. English papers and the reader may wonder, for instance, what the significance was of the stipulations laid down by the Council of Antiquaries in a memorandum prepared for them in April 1775 by Basire and sent to the younger Whatman regarding the sizing of his newly created Antiquarian paper.[161] The first of these was to ensure that the paper was uniform in substance etc.

The second stipulated that the paper "was not to be sized with Parchment, nor any Allum used for Whitening it". The third stated that the size was to be made from "Kid–Leather" and "the outward surface to approach, as near as may be, to that of the French"; and, finally, an

---

[158] The methods used for preparing gelatine size, the various materials available; types of additive including the addition of indigo to correct the yellowish colour of the size (cf. this App. notes 61, 78); common faults in sizing; procedures adopted in hot weather etc. are covered in some detail in Le Francois de Lalande, Bib.23 Art.97–109 for those who wish to pursue this subject further.

[159] Peter Spence (1806–83) patented its production in 1845 (No.10970) following this with a second Patent (No.13335) in 1850. Thomas Balston, however, claimed that it was not used in high class papers until 1880.

[160] See Chap.V, p.164 (item 15).

[161] For further details see Balston, T. Bib.1 31.

experiment was to be made on a few sheets of smaller paper with kid leather size before sizing the Antiquarian.

Size prepared from Parchment would have been darker in colour though much stronger than that prepared from kid leather (hides). Since the paper was to be used for an engraving the question of strength was secondary in importance (a) to the colour of the size;[162] and (b) to the acceptance of ink by the surface of the paper.[163]

With regard to the stipulation that "no Allum should be used *for Whitening it*" this would not have referred to the addition of potash alum to the gelatine to be used in the size bath, but to the practice of using other additives to improve the colour and clarity of dark and murky gelatine. It will be recalled that in Mr. Million's day[164] white vitriol (zinc sulphate) was added to the size to clarify and "bleach" gelatine that was otherwise dark and, perhaps, cloudy (copperas or ferrous sulphate was sometimes used in its place). It was held that these materials were unsuitable for papers on which ink was to be used since they reacted to produce a muddy effect.[165] Hence the proviso in the memorandum for their exclusion.

---

[162] It is known that the younger Whatman normally stocked two classes of alum, using some 55 cwt. per annum of a cheap grade and small quantities of a much more expensive quality. Similarly, he used two classes of scrowls and pieces, some 13 tons p.a. in total.

[163] See Chap.VI pp.240/241 and notes 75,76 re the hardness of the sized surface of contemporary English papers.

[164] See Chap.I p.40 and note 86.

[165] Le Francois de Lalande, Bib.23 Art.103 wrote "Outre l'alun qu'on met dans la colle, lorsqu'elle est clarifiée, certains Fabriquants y ajoutent un peu de couperose ou vitriol verd, d'autres du vitriol blanc, environ la dixieme partie de l'alun; cependant il y a des personnes qui prétendent que ce mêlange n'est point favorable pour l'écriture, & produit une espece de boue en se mêlant avec l'encre".

FURTHER DATA OBTAINED FROM THE EXAMINATION OF A SAMPLE OF EARLY PAPERS FOUND IN THE RECORDS OF ALL SAINTS' PARISH CHURCH, MAIDSTONE (1694–1732) KAO P 241 12/4–12/14

The examination of Paper (Bills) dating from 1694–1732 and found among the Parish Records of All Saints', Maidstone, Kent (KAO P 241 12/4–12/14) was carried out primarily to determine the incidence of Felt Hair Marks, causing "crazing", but some other observations were included and are given below.

1. *The Incidence of the Felt Hair Marks* (for analysis see Chap.VI Table VI)

TABLE XXIV

		(1)	(2)	(3)	(4)	
Date Range	Piece No.	Not Visible	very slight	slight	marked to severe	Sample Total
1694–99	12/4	1	–	–	6	7
1701–04	12/5	33	6	13	63	115
1705–07	12/6	41	3	13	34	91
1708–09	12/7	54	9	6	10	79
1710–12	12/8	64	14	7	14	99
1713–16	12/9	24	7	1	4	36
1717–19	12/10	Documents in a very fragile state				
1720–23	12/11	Examined for Countermarks only				ca.150
1724–26	12/12	65	11	3	4	83
1727–29	12/13	25	7	1	5	38
1730–32	12/14	113	20	8	8	149
TOTAL EXAMINED						697

1. Among the papers examined were 4 pieces of Whited–Brown paper; the results for these have not been included in the Table above.
2. In addition to piece No. 12/11 Pieces 12/15–20, covering a period 1733–70, were examined for Countermarks.

2. Wiremarks

Fifty–one samples were examined for the spacing of Chainwire marks and the number of Laid Wires per cm. In general the wiremarks were regular; a few were noticeably irregular, some slightly wavy. One example of "double" chainwire marks was noted.[166] The majority of examples (see below) had "conventional" spacings with chainwire marks ranging from 23–26 mm. apart and laid wire counts of (8)9–12(13) wires per cm. An analysis of these findings is given below.

Date Bracket	Number	Chainwire Spacings (mm).	No. Laid Wires (cm).
1694–1732	32	23–26	(8)9 – 12(13)
1699	1	34–36	10
1707–1713	7	19–21	13 – 14(14)[167]
1694–1699	4	19–22	10–11
1713	1	18–20	11–12
	6	near average	near average
	51		

[166] One of these marks is a rib and not a chainwire mark. (See Chap.V, note 36; and Loeber, E.G. Bib.45 p.43 btm).
[167] 14 Laid wires per cm. is a fine, but not the finest, type of laid cover; the count can go up to 16–17.

3. *Cleanliness*[168]

The Sample examined in this case (306) was White paper. A scale of 1–5 was used in the assessment (purely subjective value judgements) to indicate the state of cleanliness of the examples. The value 5 would have referred to an exceptionally clean sheet of paper, unlikely to have been found before the end of the 18th C., perhaps much later than this. The distribution of the results is shown below. It is emphasised that these results are based on value judgements and no particularly fine distinction is claimed here. All the same it is clear that the bulk of the sample lies in the middle of the range. It should be added that no particular pattern could be discerned in these findings e.g. a higher proportion of clean paper in any one period.

Scale of Cleanliness	Whited Brown.	1–2	2	2–3	3	3–4	4	
	4	3	19	66	163	44	7	Total 306

4. *The Victoria & Albert Museum Library Samples* (Examined 30:Apr:1981)

For comments see Chap.V, p.204.

TABLE XXV

Wiremarks	Nothing unusual	
	Chainwire Spacings (mm.)	23–25
	Laid Wires (w.p.cm)	10–12
Cleanliness	No differences observed between the English and the Foreign samples	3–4 or 4

[168] "Cleanliness" as used above refers to foreign bodies, coloured fibres, shive and other dirt lodged in the matrix of the sheet and these blemishes really have to be seen through a magnifying glass. It must not be confused with the current appearance of the paper which may have acquired superficial dirt accumulated over the centuries. The dirt embedded in the matrix of the sheet gives a rough indication of the kind of conditions that existed in a paper mill of that period, the care that they might have taken to see that their equipment was clean, their furnishes of high quality and their sorting exacting; it is also indicative of the general cleanliness of the linen from which the rags were derived.

COUNTERMARKS IN EARLY PAPERS FROM ALL SS'., MAIDSTONE. (KAO P 24L 12/4–12/14)

Figure 13

1.  d.o.ds. 1702, 1705, 1715, 1729 (?) approx. 15 mm. ht. x 14 mm. (According to Voorn [Bib.28 Letter 20:Aug:1985] "H" is very common in Holland from 1674 onward. Wessanen used it. He considers that it is not a countermark – see also Text).

2.  d.o.ds. 1706, 1710 (with Amsterdam Arms). Illustrated Churchill (Bib.3 No.25 with Amsterdam Arms d.o.d.1690) and Shorter (Bib.9 248c) who suggests George Gill. For discussion see Text.

3.  d.o.d. 1707 (see also No.18 below). unidentified

4.  d.o.d. 1703 (another 1705 ?). Gravell (Bib.16 Letter 4:May:1981) put forward 2 other similar unidentified marks. (right)

5. 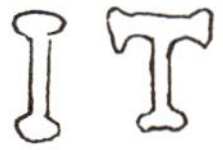 d.o.ds. 1694, 1702, 1705. Could also be "TI". unidentified.

6.  d.o.d. 1702 unidentified.

7. 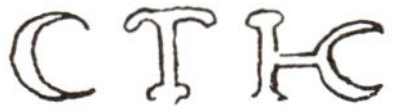 d.o.d. 1701 unidentified.

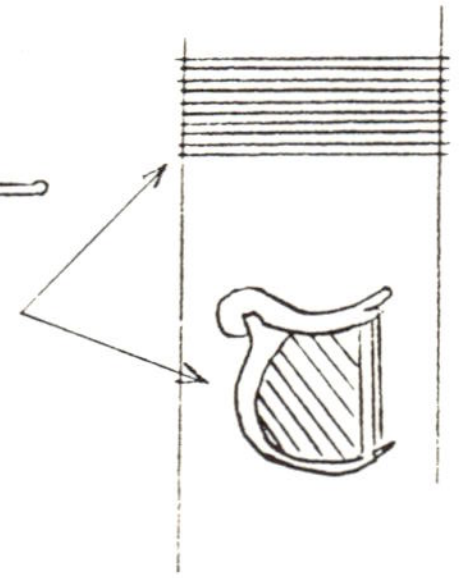

8.  d.o.d. 1699 Unidentified.

9.  d.o.d. 1705 in association with a Harp unidentified; see correspondence with Miss M. Pollard (formerly Trinity College Library Dublin).

10.  d.o.d. 1705 unidentified.

11. 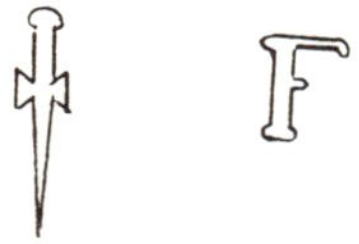 d.o.d. 1705 (possibly another "F" 1708) unidentified.

Many other truncated or indecipherable marks were observed

11.  d.o.d. 1707 — Voorn (Bib.28 Letter 20:Aug:1985) identifies this and No.12 below as the mark of Van Putten, an Amsterdam merchant (not Pieter de Vries as suggested originally).

12.  d.o.d. 1706 — As for No.11.

13. 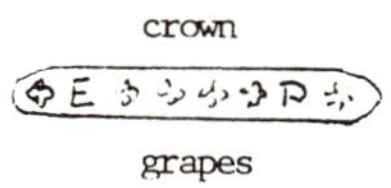 d.o.d. 1702 — unidentified (probably French). *crown / grapes*

14. 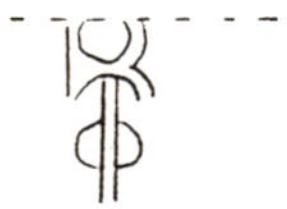 d.o.d. 1708 — unidentified

15.  d.o.ds. 1710, 1715 — unidentified

16.  d.o.d. 1707 — unidentified

17.  d.o.d. 1723 — unidentified

18.  d.o.d. 1710 — unidentified (see also No.3 above)

19.  d.o.d. 1712 — unidentified

20.  d.o.d. 1721 — unidentified

21.  d.o.d. 1710 — unidentified.

22. 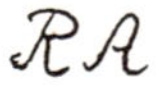 d.o.d. 1712 — suggested in text this might have belonged to Richard Archer (see App.IV also).

Dyce MS 43 "Odes & Satyrs in the reign of Charles II". (Date not known : 17th C.)

Vol.I (25.F.37)
Four Countermarks were found in this volume (p.117 numbered twice; pp.341–44 missing; p.375 omitted).

1. The bulk of the volume (pp.1–248 and 341–364) is made up of paper *Watermarked* with the Dutch Lion in a surround under a Crown.

Countermark	COMPANY
Chainwire	23–26 mm.
Laid w.p.cm.	11

2. In certain un–numbered pages (given here as pp.iii/iv, vii/viiim xi/xii) of The TABLE *no Watermarks* were noticed. The countermarks were very indistinct, set in good White paper, clean but on the thinnish side; the wiremarks were regular.

Countermark	"EB" (attributed to Eustace Burnaby)	*Illustrated below V&A No.1*
Chainwire sp.	23–26 mm.	
Laid w.p.cm.	10	

3. In pages 249/50–ca.311/312 *Watermark* The Arms of Amsterdam. Cleanliness good. No crazing.

Countermark	"GI" (?)	*Illustr. V&A No.2*
Chainwire sp.	23–26 mm.	
Laid w.p.cm.	11–12	

4. In pages ca.313/14–end of volume *Watermark* The Arms of Amsterdam. Cleanliness good. No crazing.

Countermark	Jacob Cornelisz Honig (Voorn, Henk Bib.28 20:Aug:1985).	*Illustr. V&A No.3*
Chainwire sp.	23–24 mm.	
Laid w.p.cm.	11	

Vol.II (25.F.38)
Three countermarks were found in this volume. (1) No.4 above pp.1–506 (2) No.3 above pp.507/8–600 and 648–720; (3) 5 below:–

5. In pages 600/647 and 721–end *Watermark* The Arms of Amsterdam

Countermark	"CS"	*Illustr. V&A No.4*
Chainwire sp.	24–25 mm.	
Laid w.p.cm.	10–11	

Dyce MS 44 (25.F.39) "Thirty Eight Poems by Constable" (Date not known : 17th C.)
In this small volume the paper was quite different; reasonable in appearance, but contained a lot of shive. The wiremarks were slightly wavy. There were several examples of the countermark, in which the "B" was quite clear; but only one where there seemed to be an "E" with an extended middle bar.

Countermark	"EB" (?) suggested as Eustace Burnaby (?)[1]	*Illustr. V&A No.5*
Chainwire sp.	20–21 mm.	
Laid w.p.cm.	8–9	

V&A No.1   No.2   No.3   No.4   No.5

[1] MS 43 appears to have been assembled from MSS spanning several years, making use of many different papers. It would not be surprising then to find Burnaby's papers made between 1675/7–82 cheek by jowl with COMPANY papers made in the 1690's. In the case of MS 44 the paper in contrast seems to have come from one source and because it was for a private collection (rather than for State Papers) Burnaby's papers could well have been used here also. The only other "EB" countermarks of this period, from ca. 1692 onwards, are those emanating from Elliston & Basket (the latter the King's Printer ?). The lettering of the latter's marks (Colin Tatman P.R.O. Conservation and T.L. Gravell Bib.16 16:May:90) are all similar in style to the example illustrated (right), quite distinct from the examples above, and usually accompanied by the London Arms.

(American doc. signed Wm.Penn
London 1705/6)   x 0.7

THE ENIGMA OF THE WHATMAN REAM STAMP

Plate 25

THE WHATMAN REAM STAMP

(photographed in a raking light : negative reversed)

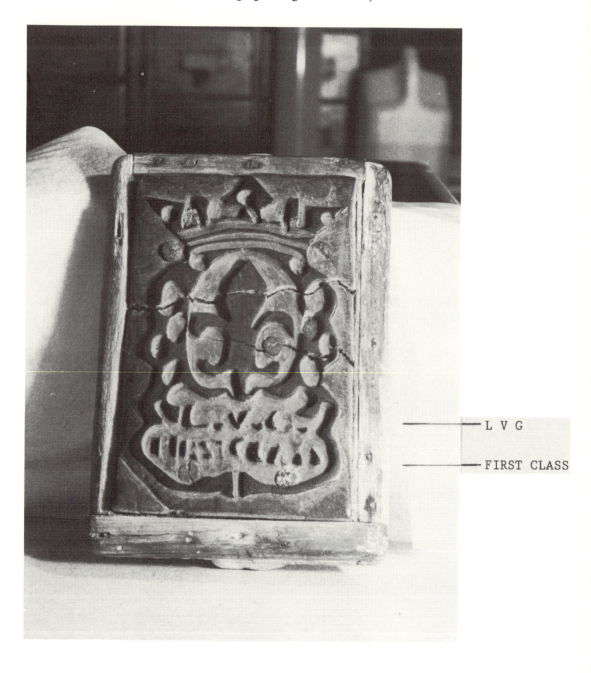

On loan to the Maidstone Museum.
Original photograph by Mr. R.A. Stutely, Museum Assistant, May 1982.

THE WHATMAN REAM STAMP

Reference has been made in the main text to this stamp.[169] The ream stamp, illustrated in PLATE 25 opposite, presents the investigator with a number of problems which have not been resolved so far. This particular specimen, which is now on loan to the Maidstone Museum, has been in the possession of the Balston family, as manufacturers of Whatman paper, for as long as anyone can remember and has always been known as "The Whatman Ream Stamp". But the question has now arisen, was it Whatman's Ream stamp ?

Recent investigations by the author have cast some doubt on this assumption, though it must be pointed out that no positive evidence has come to light that resolves this question one way or the other. Before proceeding further it would be as well to give a description of this specimen in case, as will be made plain later, other similar specimens as yet unrecognized are still in existence somewhere or other. The dimensions given below were kindly supplied to me by Mr. R.A. Stutely, a member of the staff of the Maidstone Museum who, at my suggestion, photographed the Ream Stamp in a raking light with the object of trying to decipher the bottom line on the block. Thanks to an excellent photograph by Mr. Stutely this question has now been answered.

The Ream Stamp is cut from a solid piece of wood and is secured to the backing frame by 7 screws. The wood is end–grain, possibly walnut. Its dimensions are 19 cm. (length) x 12.5 cm. (width) x approx. 2.5 cm. (thickness).

It is fixed to a backing frame with a wooden handle attached. The dimensions of the frame are 22.7 cm. (length) x 16.4 cm. (width).

The lettering is carved from the wood and until recently only "L V G" was decipherable below a Fleur de Lys under a Crown. But the recent photograph now shows that the bottom line has the two words "FIRST CLASS" in it.

The block is obviously a very ancient object, but its date is not known. The discovery of the words "FIRST CLASS" may, however, provide us with a date from which to work. An examination of the Excise Acts shows that it was not until the 1794 Act (34 Geo.III, c.20) that papers subject to Duty were described as "Papers of the First Class etc.". The original Excise Acts of Anne (10 Anne, c.9 and 12 Anne, c.9) of 1711 and 1713 respectively classified papers, omitting size of sheet as one consideration, in terms of "FINE" and "SECOND" qualities. The Act of 1781 (21 Geo.III, c.24) cancelled these Acts and in the three Acts of 1781, 1784 and 1787 papers are grouped into five Tables, the first of which covered Writing Papers, the second Writing or Copper Plate Printing etc. and it is not until the 1794 Act that FIRST CLASS is mentioned. On this basis then the block would appear to date from 1794 or after.

Blocks or Stamps were in use long before this date, so that one of the questions that has to be asked is whether an earlier block may not have had the same overall design with perhaps the word "FINE" used in place of "FIRST CLASS". This could well have been the case

---

[169] See Chap.III, p.103 + note 31.

because much earlier French and Dutch Ream Wrappers not only gave the paper maker's name but indicated quality as "PAPIER FIN" for example.

In our block there is nothing to indicate the paper maker's name; it could be argued that the imprint was so well known that it did not require further identification. The initials on the block "LVG" were quite widely pirated in the 18th C. Lubertus van Gerrevink, a Dutch paper maker of eminence, started making paper in 1691; but just when the mark was first used by other makers is not known, possibly towards the end of the 17th C.[170] The *sole* use of this mark was granted to Lubertus in 1726 by the States of Holland,[171] 7 years before the elder Whatman built his mill. But by this time the latter, through his contacts with the trade, could well have been alarmed by the large quantities of Pro Patria/Gerrevink marked paper being imported into England then, threatening the White paper makers of his district. The "LVG" mark was used by some English paper makers, but it is not known how widely or how early this practice was followed (see Note at the end of this Section). Whatman was certainly one of the first to make use of it, he may even have been the first English paper maker to do so; but whether we can legitimately link this with the incorporation of the "LVG" together with the Fleur de Lys into a Ream Stamp from this time is an open question. The problem that we are faced with here is how to explain the presence of an "LVG" mark in a Ream Stamp which we now believe may date from 1794 or after; one would have thought that by this date the "LVG" mark would have had little significance for anyone in the contemporary paper trade, mills, Excise or among Stationers unless it had some unusual personal connotations. By 1794 the status embodied in the use of countermarks, and presumably the same would have applied to ream stamps, was such that few paper makers of standing would have lost the opportunity of making themselves known by name, particularly under the heading of "First Class" papers.

This is not the only unsolved question. Whether an earlier block of the same design existed before 1794 and whether this may have originated with the Elder Whatman or not, there is another factor that complicates the issue surrounding this particular Ream Stamp. Is one to assume from the wording of the Act of 1787[172]:–

> In order "to prevent the Multiplication of Stamps upon such pieces of Vellum, Parchment or Paper, on which several Rates and Duties are by several Acts of Parliament imposed, it shall and may be lawful for the said Commissioners for managing the Duties....... instead of distinct Stamps, Dies or Marks directed to be provided to denote the several Duties......[be] charged therewith to cause *One new Stamp, Die or Mark to be provided* to denote the said several Duties on every such piece of Vellum, Parchment or Paper ......from time to time, as shall be by them thought needful, and *to repair, renew or alter the same*, as there shall be Occasion" (author's italics).

that the younger Whatman (and other paper makers) had been issued with Stamps such as these, or were the Stamps referred to different from our Stamp and retained by the Excise officers for their use only ? In 1794 there were approximately 400 licensed paper mills in England and Wales and it is difficult to believe that the Commissioners for Excise would have gone to the expense of handcarving 400 Ream Stamps; there would surely have been simpler ways of marking reams.

---

[170] Voorn, Henk, Bib.28 Letter 20:Aug:1985 writes: "...the imitations are most times too good to be sure". Lubertus acquired a second mill in 1696.

[171] The Patent may have been granted to protect his mark from that of a Lucas van Gerrevink (Churchill, W.A. Bib.3 40).

[172] I am indebted to Mr. D.J. Johnson, Deputy Clerk of the Records, The Record Office of the House of Lords for drawing my attention to the section on Stamps in 27 Geo.III, c.13.

As a consequence of the destruction of Paper Excise Records (Public Records Act of 1877) no information on the nature of these Stamps, Dies or Marks (referred to in the Act) or the method of indicating that Duty had been paid on a particular ream of paper is to be found today among the records of the Library and Museum of H.M. Customs and Excise[173] or elsewhere to the author's knowledge. On present evidence the Ream Stamp is an enigma. No other similar English Ream Stamp appears to have survived (?). Was it a one–off Stamp with personal connotations for Whatman or was it an Excise Stamp, in which case why did it not bear a "GR" under a crown ? On balance the former is considered to be the more likely solution and that the Stamp acted as a kind of trade mark well enough known among stationers for it to be recognized in the absence of Whatman's name appearing on the ream wrapper. If so, one might ask why was "LVG" used when he might have used his own cypher[174] ? The fact that the former rather than the latter was used suggests that the present Stamp is an up–dated version of a much more ancient implement. A final point, if the "LVG" had a more ancient usage, had it any special significance for the elder Whatman since his son seems to have discontinued its use by 1780 ?

A Note on the use of "LVG"

There does not seem to have been any systematic study made of paper makers who pirated Lubertus van Gerrevink's countermark. Voorn has suggested that the reason for this may be due to the difficulty for us to-day to distinguish the genuine from the pirated mark as the forgeries were so good. But it is known that both continental and British paper makers made use of this mark.

According to Heawood[175] "Much 'Pro Patria' paper came in from Holland, as did some with the 'Horn' mark and the name or intials of L.V. Gerrevink below" in the early part of the 18th C.; Heawood is not very specific about the date when this took place. Shorter[176] more or less repeated this statement in 1971. It has been claimed that this influx of Gerrevink's paper (forged or otherwise) led British paper makers to use his mark in an attempt to compete in the market place. But was this really the case? For example in a random sample of 697 pieces of paper from All SS', Maidstone Parish Documents, of which the greatest part must have been imported paper, only one "LVG" mark was found for the period 1694–1732[177] and this was found in a document dated 1728. That the mark was used by British paper makers we know, but by whom and how early is uncertain. It will be seen from the information given below that some claims give a misleading picture in that an early use is implied, when in fact the documents belong to a much later period, in one case 50 years after the period we are considering.

In the July 1938 copy (No.9) of "Signature" James Wardrop accused W.A. Churchill (author of "Watermarks in Paper" 1935 Bib.3) of a grave misstatement in attempting to associate certain British paper makers, namely Bates (act. from ca.1724); Portal (which Portal is not specified); and, Wardrop adds the comment that Churchill implies this, William Jubb, with Lubertus van Gerrevink in some sort of manufacturing agreement. Earlier Churchill (loc.cit.) had claimed quite inaccurately that Whatman had operated two paper mills, one in England and one in Holland and that papers made in the latter bore the marks "J WHATMAN" with "LVG", adding also that it was in Holland that Whatman had learnt the art of papermaking. Here we are concerned with the former issue;[178] Churchill was probably not aware that there were two Whatmans; the son certainly did visit Holland, twice,[179] and he also made

---

[173] Letter from Mr. T.G. Smith, Librarian and Archivist to the Library and Museum of H.M. Customs and Excise dated 22:June:82 ref. TGS 80/82.

[174] See PLATE 26.

[175] Bib.35 p.27

[176] Shorter, A Bib.9 47.

[177] KAO P 241 12/4–14

[178] As mentioned in the Text, on present evidence it seems most improbable that the elder Whatman ever visited Holland; it is not impossible, but unlikely. On the other hand a good case can be made for Richard Harris learning his papermaking there and having some kind of connection with Gerrevink at the time. (See Chap.IV pp.167/168).

[179] See this App. note 102.

use of the abbreviated "LVG" mark. But in the case of two of the other paper makers using this mark the documents Churchill exemplifies are dated 1775 for Bates (No.413) and 1780 for Portal (No.414), late in the 18th C. at a time when the younger Whatman, Thomas Balston has claimed,[180] was abandoning their use in paper. Apart from the fact that some French and Dutch paper makers appear to have pirated the "LVG" mark, sometimes in association with "IV", from a fairly early date, the question remains which British paper makers used "LVG" or variants in the earlier part of the century? and was Whatman the first to do this?

The Jubb family of Ewell mill, Surrey (act.1732-1796) were among the early users of Gerrevink's mark in Britain; there may have been others, but they have not been seen by the author. If, however, one sets an upper date limit of 1760 (quite an extended limit), one Jubb mark that pre-dates this makes use of "L V Gerrevink". The document in question is an American one dated 1754 (Gravell, T. Bib.68, 452). "Jubb" appears in a Bell together with the letter "S", indicating in all probability that it was Sarah Jubb's mark, widow of William snr. and her mark might have continued up to 1758 when William jnr. came of age.[181] There is another example of her mark illustrated in Shorter (Bib.12 No.88) but, unfortunately, no date is given for the document and the "Gerrevink" has been truncated so that it is impossible to see whether it is in an abbreviated or in a full form. The only certain "JUBB/LVG" mark known to the author is in a document dated 1775.[182] All this is in contrast to the elder Whatman's use of "LVG" as early as 1747 (Balston, T. Bib.1 158 fig.8). Balston (ibid. 145) states "when the elder Whatman used the traditional Crowned Shield watermarks (the Strasburg Arms, the Fleur-de-Lys and the Posthorn) he habitually gave them the initials "LVG" as an appendage". The general use of the "LVG" mark is not uncommon *after* 1760; earlier than this it was used by Whatman, possibly by other English paper makers (?), and unquestionably in imported papers, frequently in association with a Fleur-de-Lys in the other half of the sheet.[183]

Heawood (Bib.35) lists 21 typical examples of the use of "LVG" of which only 6 (with variations) have been found in early documents (before 1760); No.98 with Strasbourg Bend d.o.d.1748, which could have been a Whatman paper; No.159, also under a Strasbourg Bend d.o.d.1754, which is definitely Whatman; No.1808 under Fleur-de-Lys/Crown d.o.d. early 18th C. which could also be Whatman and two other similar uncertain ones, Nos.1809/10 (all conforming to the observation made by Thomas Balston). In addition there are No.1817 under Fleur-de-Lys + "IV" d.o.d. stated after 1707, possibly imported paper, and with this one might associate Nos.1826/7 with "I VILLEDARY" in full; No.1823 stated to be before 1731, again very probably imported paper closely resembling No.1829 d.o.d.1743 and found in Holland, both of these with Fleur-de-Lys; and, finally, Nos.2736/7 under Posthorn d.o.ds after 1714, which though found in England (MSS) might be linked with No.2739 found in Rotterdam. Churchill's few examples throw no further light on this subject.

Though examples of the use of "LVG" are to be found in British Documents before Whatman's time, marks which may have originated from Lubertus himself; or Lucas van Gerrevink (see Churchill, W.A. Bib.3 40); or have been pirated by other continental paper makers, we have no evidence as yet to show that the elder Whatman was not the first to use this mark among British Paper makers.

[180] Balston, T. Bib.1 145.

[181] William Jubb snr. died in 1739 (aged 42); his widow, Sarah, married a William Wells who insured Ewell mill in 1752 and 1759 (Hand-in-Hand Policies). William jnr. (b.1736) first insured the mill in 1766 (Wells d.1785). [Information kindly supplied by Prof. Alan Crocker, Univ. of Surrey]. It seems clear from the above and the appearance of "S" marks in association with "JUBB" that Sarah was determined to maintain the Jubb connection with this mill alive. For a somewhat similar use of "L V Gerrevink" see this App. p.279.

[182] Apart from his No.88 Shorter (Bib.12) only illustrates one other Jubb mark linked with Gerrevink (No.87) but no d.o.d. and no "S", probably post 1758. None of the others including a Fleur-de-Lys in a shield have any Gerrevink marks. Gravell No.454 "JUBB/LVG" was found in the document dated 1775; his only other example, 455, has no Gerrevink mark. (It may be noted that Gravell's No.453 is in no way associated with the Jubb/Gerrevink mark illustrated in No.452. The d.o.d.'s and the location [Old Swedes Accounts, Wilmington] are the same, but the elder Whatman's marks are found there also. Moreover, the chainwire spacings accord with the Whatman papers and are quite distinct from the Jubb example).

[183] Gravell, T. Bib.68. 409/410 illustrate an "IV/LVG" mark (d.o.d.1750) in which the Fleur-de-Lys above the "LVG" closely resembles those found in Hudson's exemplars 3, 4, 5, 6 & 7 (Bib.2) as indeed the chainwire spacings. Though this observation may be a coincidence, it could be said to be indicative of a continental paper rather than British.

NOTES ON "W" COUNTERMARKS

Plate 26 – EXAMPLES OF CYPHERS USED BY THE TWO JAMES WHATMANS

<u>The Elder Whatman</u>

<u>The Younger Whatman</u>

State Papers 1748

Also Bib.68. 714 d.o.d.1764

A typical example of his cypher
appended to a Post Paper watermark.

(taken from a London document with
a J WHATMAN   TURKEY MILL 1822
countermark & date).

State Papers 1747

State Papers 1754

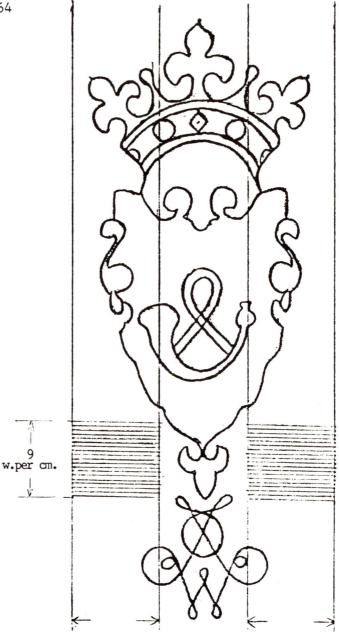

9
w.per cm.

23.5 mm. chainwire spacing

State Papers 1755

State Papers 1764

SOME NOTES ON "W" COUNTERMARKS

Reference has been made in Chapter III to the Elder Whatman's use of countermarks from the very start of his career as a paper maker and at a time when the occurrence of English countermarks was rare, possibly because there were few paper makers at that time whose White paper could match the quality of imported products and would therefore warrant the use of a countermark; or, maybe, because an increasing number of paper makers no longer had to compete directly with continental papers in limited markets like London, supplying instead their own localities now that restrictions on provincial printing had lapsed. Whatever the reason may have been, we find that the Elder Whatman, perhaps following a tradition initiated by George Gill, made extensive use of countermarks or cyphers, increasing in their variety as time went on. This practice was continued by his son.

It is probably true to say that in the majority of cases the Whatman countermarks are unmistakable, either taking the form of "I W" or "J W"; or the well–known "JW" cypher (see Plate 26 opposite); or "J WHATMAN" and variants, though in this last case, in the early part of the 19th C. the "J WHATMAN" mark came to be forged quite widely in several countries in Europe.[184] In addition to these marks, however, there are the less familiar single line "W" marks[185] used by both of the Whatmans as well as a wide variety of other "W" marks, usually in the form of double line block letters, many of which may not have originated from the Whatmans at all. The object of this section of Appendix V is to set out the findings, such as they are, regarding these less familiar marks.

---

[184] See Countermark section, Plate 16 (Preface to App.I & II, p.vi).
    An account of the countermarks and watermarks used by the Whatmans (with illustrated examples) will be found in Balston, T. Bib.1 App.V pp.156–164. In addition to these, countermarks and watermarks used by the successors to the Whatmans will also be found in Balston, T. Bib.74 App.IV 164–167; mention of some more recent findings is made in the countermark section referred to above.

[185] Thomas Balston (Bib.1 164) mentions the occurrence of both single and double line "W" countermarks in Writing papers used between "1740 and 1794" (implying, in the use of these dates, the span of the Whatmans' activities rather than the actual occurrence of these marks), but kept no record of them, which was unfortunate. Many other examples of these marks will also be found in Gravell, T.L. & Miller, G.E. Bib.68 21/22.

The Principal Types of "W" Countermark

Two types of "W" countermark have been referred to already, the single line and the double line block letter types. But during the 18th and early part of the 19th Cs. quite a variety of "W" countermarks have been found, some of which belong to these two classes and others which differ from these not only in style but in the position that they may occupy in the sheet. It is quite clear that some of these have no connection with either of the Whatmans. It is important, therefore, to make the reader aware of this subject and of the danger of attributing "W" marks indiscriminately to the Whatmans solely on the grounds that the papers, in which these marks have been found, could have been made by either of them during their lifetimes.

Types   (i) $\mathcal{W}$ (ii) $\bigvee\!\!\bigvee$ (iii) $\bigvee\!\!\bigvee$
                                                                       $\bigvee$

Broadly speaking three types of "W" countermark have been identified in British papers made during the 18th C.; and the use of some of them extends into the 19th C. as well. Though some of these marks can at times be identified with certain paper makers, there are a greater number of instances where the significance of these marks is not known. The types illustrated above are discussed briefly below, mainly with the intention of exposing this subject in the hope that it may lead to a more authoritative account of them than it is possible to give here.

Type (i) Cursive or Uncial "W" marks and others of an unusual character

Only two marks of the kind illustrated as (i) are known to the author. Both occur under a Posthorn Watermark. In the other half of the sheet the countermark and date are in one case "WILMOTT 1794" found in Philadelphia Customs House Records dated 1807;[186] and in the other case a "J WHATMAN 1805" (PSM) which has been discussed already in relation to William Balston's occupation of Eyehorne mill, Hollingbourne.[187] The dimensions of the two "W"s differ slightly, but both are formed in the same style. In the case of the Wilmott mark it could be argued quite convincingly that the "W" represented the name Wilmott, partly because as Shorter has pointed out[188] 1794 seems to have been the year in which William jnr. was in the process of handing over Sundridge paper mill to Charles Wilmott and that they preferred to use the surname only (other "WILMOTT" marks have been found for this year), hence the "W" only under the Posthorn; and partly because a "C WILMOTT 1817" mark has been found[189] in which a script "CW" monogram occupies the same position under the

---

[186] Gravell, T.L. Bib.16 letter 23:Oct:79 (not illustrated in Bib.68).
[187] See App.I, TABLE XVII.
[188] Shorter, A. Bib.12 183.
[189] Gravell, T.L. Bib.68 23 No.780/1 Massachusetts doc. 1823. Illustrated p.215.

TABLE XXVI   277

Illustrating a selection of Solitary Double Line Block Letter "W"s
found mainly in 18th C. Documents

SOURCE	d.o.d.	COUNTERMARK	WATERMARK	LOCATION & IDENTITY
J.N.B.	1725 1730's	W	Nil Nil	Leeds Village KAO P 222 5/8 Guildhall MS. 4655/13 Neither of them identified.
J.N.B.	1731/2	CAW	Nil	All SS'. Maidstone Document KAO P 241 12/14. Cornelis Adriaan Wessanen (act.1707-1733 Voorn, Henk).
Dr.Hudson	1740	IW	Uncertain; probably nil. If any, a Stereotype.	Bib.2 and discussed Chap.III The elder Whatman's mark.
T.Gravell	1765-1824	W		Bib.16 Letter 19:Feb:1980. Numerous; locations various. Many 20 x 11-13 mm. Some later ones on Wove paper.
E.Heawood	ca.1772	W	LVG under Strasbourg Bend " W " in other half.	Unspecified MS; and also in Blair's Chronology, BM Copy, (Title page 1768) includes Whatman papers ca.1770-1813. Probably Whatman (see Bib.1).
J.N.B.	1775	W	Nil	2 examples Leeds Village Docs. KAO P 222 5/11. Could be Whatman or Williams ?
J.N.B.	1776-1780	W	In surround; probably Britannia.	Leeds Village KAO P 222 12/34 Could be Robert Williams (?)
T.Gravell	1783	W	Under Crown with Dutch Lion Passant in Shield under Crown in other half	Bib.68 illustrated 719-720 Marbleton, N.Y. (William Wilmott jnr. ?)
T.Gravell	1788 1793	W	English Arms with Bell 1. GR under Crown r.	Bib.68 illustrated 721-723 (1788) Newburyport Mass. (1793) Kingston "Jamaique" Inf. Bib.16 Letter 19:Feb:1980
T.Gravell	1790	W	Within Wtmk. between Posthorn & Crown with GR under	Bib.68 illustrated 725 Univ. of Delaware, Clow.
E.Heawood	1792		In between Britannia (r) and GR under Crown (1)	Illustr. Library 1930-31 p.485 (?). referred to in letter to Bruce Bannerman 16:May:1931.
T.Gravell	1795	W	Nil	Bib.16 Letter 19:Feb:1980 Acct.Sheet Port au Prince Univ.Delaware D & W papers.
J.N.B.	1803	W	Britannia in Shield (r) 1801 under GR under Crwn " W " centre of sheet	Pencil Notes William Balston 26:July:1803 MPM Folder PSM. Balston & Hollingworth (?).
J.N.B.	(1812)	W	Britannia in Shield (r) 1812 under GR under Crwn " W " centre.	Pack 4 Springfield Collection PSM Balston (?).

Posthorn. In the case of the Whatman paper the "W" might also represent the Brand name here. But as will be seen below it is possible that by this date the marks may have had some other meaning; they might have referred, for example, to Writing Paper.

Two other unusual marks of this kind may be mentioned here. The first (illustrated left) is a script "W" which again has been found *under* a Posthorn watermark to which the usual (late form) "JW" cypher has been appended (see Bib.1 158 fig.9), Gravell found this mark in a Baltimore document dated 1795;[190] the paper was thus almost certainly made by the younger Whatman. There can surely be no doubt that in this case the added script "W" must have had some special meaning otherwise why should Whatman have placed this mark in the same half of the sheet as his cypher ? The other mark, slightly different to type (ii) mark (shown above), has been included here (illustrated left) because of its unusual form and, having a watermark date of "1795" underneath, was unlikely to have been a Whatman mark (to date no Whatman watermark dates for this year have been found).[191] This "W", yet again, lies between a Crown and Posthorn of a Posthorn watermark (illustr: Bib.68 p.200 No.724).[192] The other half of the sheet is unfortunately missing, so the maker remains unidentified. (It would be of interest to know, if other late 18th and early 19th C. marks of this kind are found, whether they are always associated with the Posthorn mark).

Type (ii)   Double Line Block Letter "W"s

These marks, a selection illustrated diagrammatically in Table XXVI, are far more numerous than the other two types and clearly cover a wide range of paper makers including makers of Dutch papers imported into this country early in the 18th C. and, possibly, some originating from English makers before the elder Whatman's time.[193] It is not intended to discuss these marks here, merely point out that searches to date indicate that the majority appear to belong to a period dating from ca.1765 onwards and up to the 1820's at least.

As the Table shows some of these "W" marks just pre–date the elder Whatman and are referred to again below. It is known for a fact that the elder Whatman used *solitary single line* "W" marks, but there is no unequivocal evidence that either of them used the block letter ones on their own though it is possible (see 1775 examples).[194] In view

[190]   Gravell, T.L. Bib.16 9:Feb:80 (not illustrated in Bib.68); confirmed in letter 1:Mar:80.

[191]   Although watermark dates in Whatman's Antiquarian paper have been discovered by the author as far back as 1784, in his ordinary papers watermark dates start with 1794 and are followed by 1797, 1799, 1801, 1804, 5, 6 etc.

[192]   Gravell, T.L. Bib.16 5:Aug:85 informed the author that two examples of this mark had been discovered. The reference in Bib.68 to the mark (d.o.d. 1790) is used in his text (p.22 No.724 Clow), whereas the other mark (d.o.d. 1797 Jeremie 1003 Bib.16 9:Feb:80) is used in the illustration (p.200 No.724). The marks should *not* be confused with his Bib.68 No.725 (illustrated in Table XXVI ex. Gravell 1790).

[193]   See note 199.

[194]   Edward Heawood noted (correspondence with the late Col. C.H. Balston in the 1930's) a double line block letter "W" in John Blair's "Chronology" of 1768, editions of which were clearly printed much later as they also contain papers with "J WHATMAN" (as in the earliest issue); "J WHATMAN & CO"; and "J WHATMAN & W BALSTON 1813". Such an association of marks might be taken as an indication of the younger Whatman making use of these type (ii) "W"s on

(continued...)

of the fact that the earliest Sundridge Land Tax Returns all bear "W" marks under a Crown (see also Table XXVI ex. Gravell 1783) it is probable that William Wilmott jnr. was the paper maker in this case;[195] but no marks yet can be associated with William snr., although both snr. and jnr. could be considered as candidates for any of the 1765-1824 examples illustrated in Table XXVI that fell within their active period as well as certain other marks (see below) and possibly some single line "W"s *without* loop; some of the latter, however, have been found linked to an "I" and clearly could not be assigned to any Wilmott.[196]

Whereas plausible explanations can be put forward for the appearance of some of the double line block letter "W"s e.g. used by one or both of the Whatmans (especially for small sheets of paper) and, later, by William Wilmott jnr., there are other examples for which there is no ready explanation. A case in point is a badly formed block letter "W" placed under "L V Gerrevink" in which the last 8 letters are single line and not very professionally formed either, these being underneath a Posthorn (facing left) in a shield under a Crown. This mark was found in a Montreal document of 1764.[197] The mark is very similar in appearance to *Sarah* Jubb's mark d.o.d. 1754 (see p.271), illustrated Gravell Bib. 68 No.452 except that there is no Bell only the type (ii) "W" placed under the capital "V".[198]

After 1775 the field of choice widens rapidly; and from 1790 onwards, especially during the 19th C., the "W" marks proliferate and might have come from anywhere, even referring in some cases to a client or representing some other function. Only a systematic collection and examination of this kind of mark is likely to provide any sort of solution.

Table XXVI illustrates a Cornelis Adriaan Wessanen (act.1707-1733) mark to show that the "W" differs from the two roughly contemporary unidentified "W"s. If not a Dutch mark, once again one is thrown back on remote English origins;[199] and there are problems in assigning them to a Dutch paper maker too.[200]

---

[194](...continued)
occasions. On the other hand there were books printed in the late 1780's that contain both Whatman and Wilmott papers. Likewise the two examples of type (ii) "W"s found by the author in Leeds village documents (see Table ex.1775) do not necessarily indicate that these papers were made by Whatman, then occupying Old Mill (1770-4), though the odds are that they were his since this mill lay within Leeds parish. It is just conceivable that these marks might have emanated from Eyehorne mill where Robert Williams had just started (1775). These "W"s are, however, different from the "W"s believed to have been used by William Wilmott jnr. (see note 195 below).

[195] See Chap.VII, note 156.
It will be noted that the "W"s under a Crown, suggested here as belonging to William Wilmott jnr. of Sundridge mill and found in the first 3 LT Returns for this parish, have an open top to the left arm of the left "V" as distinct from the Leeds village examples. Other examples of these "W"s under a Crown have been discovered by Mr. Colin Tatman (Paper Conservation, P.R.O. London) in the left half of sheets of Pott (rt. half Dutch Lion Passant as in Gravell Bib.68 No.719/20) used for a printed Register of Births ca.1789 (R.G.4/4275).

[196] Other less likely candidates for the double line "W"s include Robert Williams at Eyehorne mill, though he only started in 1775; Richard Ware (Harefield mill, Middlesex) possibly author of two "RW" marks (d.o.ds. 1747, 1767); and *much less likely* Robert Williams of Rickmansworth (act. ca. 1756-80) and Joseph Wells (Bagnor mills, Berkshire act. ca.1756-91) though Wells was followed by the Wickwars who did use countermarks.

[197] Gravell, T.L. Bib.16 19:Aug:1986 (9D series 10 Box) Peter Force Collection (Libr. of Congress MSS) d.o.d. 1764. Montreal, Moses Hagen.

[198] Very remote alternatives to a Sarah Jubb mark are (i) William Wells, William Jubb's step-father; (ii) William Wilmott snr.; (iii) The Warren family of Thornton-le-Dale, N. Yorks; (iv) William Wilkins, Upper Tovil (see Chap.VII p.288); or, finally, is one seeing here another of these mysterious dependent "W"s ? Of these perhaps William Wells is the most likely parallel to the Jubb "S" marks.

[199] The "W"s in question (d.o.ds 1725 and some time in the 1730's) have a resemblance to the "W"s in the first known "IW" countermarks for the year 1740 (see Chap.III), but short of postulating the unlikely event of Whatman trying a youthful hand at Gurney's mill one has to look first for paper makers who were active not too far from these dates. Though the chances are that they may have been Dutch marks a list of English makers whose papers could be considered to fall within this bracket is given below. (The Shorter Bib.12 p.no. is indicated on the left):

(249) Warren, William, Thornton-le-Dale, N.Yorks. 1680-1748
White paper was made here from 1680 and, it is claimed that another William Warren followed and, further, is stated to have rented the mill as late as 1778.

(214) Watkins, Thomas, Longford, Middlesex. 1723
Making White paper from ca.1713 to his bankruptcy in 1723; not known to have used countermarks.

(continued...)

Figure 14

Diagram of a typical specimen of a single line "W" with Loop placed
within a Pro Patria Watermark above the tail of the Lion.

8.5 Laid Wires
per cm.

9.0 Laid Wires
per cm.

← 25.5 mm. →

Reproduced here by kind permission of Mr. T.L. Gravell (U.S.A).

It should be noted that the illustration above is based only partly on a photographic print of an image obtained from a light sensitive proofing paper copy kindly supplied to the author by Mr. Gravell on 5:Jan:1980. The Pro Patria Watermark with the "W" countermark within it is in fact reproduced in Bib.68 No.713. Due to the fact that the author's print was not ideal for tracing and, in addition, parts were indistinct, the illustration above makes use of the Pro Patria / "AW" watermark reproduced on p.285 (and Bib.68 No.32) as a setting for the "W" countermark. It will be noticed, for instance, that the watermark illustrated above extends just beyond 4 chainwire spacings compared to the original (just within 4) taken from a single sheet of Accounts, Old Swedes Church, Wilmington, Delaware (d.o.d. 26:Aug:1758). However, in other places it has been altered as, for example, in lowering the Lion's Tail in order to accommodate the "W". Otherwise the details of the two watermarks are very similar as indeed the chainwire spacings and the number of Laid Wires per cm. The dimensions of the "W" in this particular example were 11 x 6.5 mm. It is regrettable that it has not proved possible to have had these countermarks copied by β–radiographic techniques; but the examples in question have been laminated preventing the use of this method.

Mention has also been made in App.1 on p.9 of the fact that two examples of single line "F" marks belonging to Thomas French have been found in identical positions to the "W" illustrated above and, moreover, with Pro Patria Watermarks containing the same features (see further comment in this App. p.287).

Type (iii)   Single Line "W" marks with and without loop

Two different types of this mark have been illustrated on p.276. One of them, the lower illustration, is derived from a relatively crudely bent piece of wire attached to the mould cover and, as will be shown below, can be identified with some certainty as a device used mainly by the elder James Whatman. The crudeness of these "W"s may have been due to the Mould Maker at that time having difficulty in forming small letters of this kind, the dimensions of these "W"s ranging from 8 x 5 mm. to 12.5 x 6.5 mm. in the case of the Pro Patria marks (Fig. 14) and probably much the same for letters found within the surrounds of other emblems like Britannia.[201]

The first document discovered to date with this single line "W" with a loop belongs to the Historical Society of Delaware's autograph collection and is in a sheet of Pott lying within a Pro Patria mark (situated just above the tail of the Lion) in one half and with "LVG" in the other; the document is dated November 1753 and the countermark was found by T.L. Gravell.[202] The paper must have been made, therefore, in 1752 or earlier. A whole series of these single line "W"s have been located by Mr. Gravell in various other Delaware documents covering a period of 12 years (d.o.ds. 1753–1765) with one freak occurrence in a document dated 1787.[203] But these small "W"s with a loop were not confined solely to Pro Patria marks, but have been found on their own in Wove Paper[204] (Capell's "Prolusions" 1759/60); within the surround of other watermarks such as Britannia[205] (d.o.d. 1770); or in

[199](...continued)
    (123 Watkins, Rice, Sutton Courtenay, Berks. 1672–1705 Made banknote paper. Cameron, W.J. Bib.18 p.12 and

    /4) Note 38 assigns a cursive "W" in the 1690 Paper–Sellers Petition "The Case and Circumstances of Paper–making in England" (BM Cup.645 b.11) to Rice Watkins on the grounds that in a later Petition "An Addition to the Case: without the letter "W" refers to "Mr. Watkins". Cameron's note gives Watkins dates incorrectly; he was in fact buried in 1705.

    (146) West, John, Horton mill, Bucks. late 17th C. Ancient Paper Maker (see App.IV).
         There are a few other "W" paper makers, but either their mills are known to have produced Brown paper, partly or entirely, or absolutely nothing is known about them, like John Washbourne of Littleton (Worcs.) act.1717–27 (Bib.12 248).

[200] Heawood, E. Bib.35, No.355 (d.o.d.1704) English doc.; and No.369 (d.o.d.1697) Dutch doc. illustrates two unidentified "RW" marks. Not only do the "W"s in these marks differ from the unidentified "W"s being considered here, but Voorn, H. Bib.28 Letter 20:Aug:85 states that the "RH"s illustrated by Heawood are not known in their collection at the Hague. (Later "RW" marks have been attributed to Richard Ware, Shorter, Bib.12 367 No.179).

[201] More detailed information on the dimensions of the Pro Patria "W"s and comments on various features of the PRO PATRIA designs enclosing them will be found in the Author's Bib.76.

[202] Gravell, T.L. Bib.16 Letter 17:Dec:1980; dimensions of "W" 8 x 5 mm. Paper dimensions 39.1 x 30 cm. (inf. supplied Mar./Apr.1981). The specimen was in the autograph collection of John Clayton jnr. The paper has been trimmed, but it seems more likely that the size was Pott rather than Foolscap.
    It is possible that earlier examples may turn up. The author has found a single line "W" with loop in an exterior position (not as in the others over the Lion's tail) to the left of a Pro Patria mark with the Maid of Holland facing right (in contrast to her facing left as in the others) in an All SS' Maidstone Parish Document (KAO P 241 12/17) dated Oct.1749. Unfortunately, this document has been torn immediately to the left of the "W" so that it will never be known whether it was preceded by an "I" or "J" or nothing at all.

[203] Gravell, T.L. Bib.16 Letter 17:Dec:1980; the mark was found in a letter signed Gunning Bedford jnr. dated 1787, St. George's Hundred, New Castle County, Delaware. As mentioned, it is most unfortunate that Thomas Balston kept no record of the "W"s that he had found so that we could see whether they covered the same sort of period as Mr. Gravell's examples.

[204] For full details see the Author's Bib.76.

[205] See App.I, TABLE XVI.

the English Arms under a Crown[206] (d.o.d. 1773, though judging from the Arms, which appear to be those of George I or II, the paper may have been made in the 1760's or earlier); and, of course, frequently associated with "JW" and in one case as "AW", the Ann Whatman countermark (see below). In general these small "W"s with loop appear to have been confined to the smaller sizes of sheet, Pott and Foolscap;[207] and it is suggested here, based on present findings, that the terminal date for the use of this kind of mark in moulds may lie somewhere in the 1760's, possibly the later examples to be seen as survivals from moulds used originally during the elder Whatman's régime (d.1759), the younger Whatman in due time adopting a new style.

In passing it may be noted that the author has discovered another series of much larger, but similar in style, single line "W"s with loop in some drawings dating (probably) between 1765–1775.[208] Typical dimensions of these specimens are: 17.5 x 9.5 mm. (in other words roughly double the size of the smaller ones) and they are placed centrally, on their own, between "GR" left (under Crown with decorations and encircled) and the Royal Coat of Arms right (in surround under a Crown and with Bell appended). It can only be assumed here, judging from their stylistic similarity to the smaller marks, that they are also Whatman countermarks; all the same this remains uncertain.[209] They do not appear to have been a great success practically because, being fragile and in an isolated position, many of them were found to be in a battered state, which may have been one reason for discontinuing them.

The other kind of single line "W" *without* loop (the upper one of type [iii] p.276) may belong to a later period. At present only five examples are known to the author, all discovered by Gravell in American documents dated 1765–1777.[210] The "W"s in these examples were clearly made by forming the letter round the sharp edges of a block; their more professional character may reflect a change in Whatman's mould maker, that is to say, if these marks belonged to Whatman. The issue is not a straightforward one, if another candidate is sought as the author of these marks. The question of attribution will be discussed in the next section; but it may be pointed out here that similar acute

[206] Gravell, T.L. Bib.68 pp.22 & 198 No.717 (illustrated).

[207] All the examples of the single line "W" marks (with loop) discovered both by Gravell and the author are associated with Watermark emblems that Thomas Balston claimed were found in papers of Pott or Foolscap size. The Pro Patria examples found by Mr. Gravell were in fact all in Pott or Foolscap papers confirming Thomas Balston's claim (Bib.1 161).

[208] These examples were found in an album in the Royal Institute of British Architects' Drawing Collection entitled "Original Architectural Plans" Thomas Hunt (1737–1808 +), leaves 16v, 47v, 55, 57v and 64. It may be noted that leaf No.62 contains the "J WHATMAN" countermark.

[209] As the dates of the documents containing these larger "W" marks is uncertain, William Wilmott jnr. could be considered as a possible alternative to the Whatmans. Otherwise one has to fall back on the suggestions already made in note 198, (see note 210 below).

[210] Two of these are illustrated in Gravell, T.L. Bib.68 Nos.716 provenance Dover, Delaware, d.o.d.1773 and Ipswich, Massachusetts d.o.d.1775 both with Bell appended to Royal Arms in Shield; another, No.718, provenance Philadelphia, d.o.d.1777 at the top of a surround to a Britannia watermark. The most recent find drawn to the author's attention (Bib.16 19:Aug:1986) is an acute angled "W" at the bottom of a circle cartouche with Royal Arms under Crown and Bell attached; d.o.d.1765, possibly Pott sized sheet and provenance Libr. of Congress, Peter Force Collection, Nassau Hall. It should be noted that these countermarks are all relatively small averaging ca.10 x 9 mm. The much earlier date of the last example clearly eliminates William Wilmott jnr. as a candidate and when associated with an "I" all the others (see note below).

I W

d.o.d.1773

angled single line "W"s like these have been found associated with an "I". An example is illustrated left discovered by Gravell in an American document of 1773.[211] If the paper was made anywhere near the date of the document then this is an unusually late use of "I" by the Whatmans who are generally thought to have switched to "JW" soon after 1750; to argue that the paper may have been much earlier produces a situation where one might have expected to find a "W" with a loop rather than an acute angled one.[212] Finally, this is not the only problem because some of these solitary acute angled "W"s are found in watermarks with an appended Bell. No absolutely unequivocal evidence of Whatman's use of a Bell is known to the author, but this is not to say that he might not have used them on occasions.[213]

[211] Gravell, T.L. Bib.16 31:Mar:1981 provenance Chester County, Philadelphia, d.o.d.1773. The mark is placed *not* in the surround but directly underneath the Royal Coat of Arms, seemingly those of Geo.I & II.

[212] Thomas Balston (Bib.1 163 fig.21) illustrates what appears to be an "IW" with an acute angled "W" over Britannia, but the illustration was taken from Churchill (Bib.3 No.22) where it is quite clear that the "W" has a small loop.

[213] Three of the acute angled "W"s referred to in note 210 have Bells attached to the Watermark. Not only are these "W"s similar in style to one without a Bell, but the dating of the last example taken together with the "IW" referred to in note 211 all point to Whatman as the author. If one adds to these the large single line solitary "W"s *with* loop and stylistically similar to the smaller "W"s with loop and takes into consideration that the accompanying watermarks have Bells attached, collectively these instances strongly suggest that one or other of the Whatmans used Bells on occasions in their watermarks.

Single Line "W" marks as Whatman Countermarks

In this section the two main types *only* of single line "W" marks, with and without loop, described in the previous pages and found in papers made in the 18th C. will be considered. The first kind can be identified with absolute certainty; the second is a more difficult matter.

Taking the more professionally formed acute angled "W"s first, there are at present too few examples to attempt a really positive attribution. To date, five examples of these have been discovered, all in America and with a d.o.d. range of 1765–1777. The paper used for the earliest of these documents was obviously made at least a year or two before 1765 and possibly even when the elder Whatman was still alive; or when moulds bearing his marks may still have been in use. The elder Whatman's longstanding use of solitary "W"s (see below) coupled with the fact that this kind of "W" has also been found associated with an "I" as "IW" and that all these appear in American documents, seemingly as exported paper, collectively point to one of the Whatmans, probably the elder and the continued use of his moulds, as a strong candidate for this countermark. Alternatives to this proposal have been considered, but when the "IW" example is taken into consideration as well as the export feature, one is in effect back to one or both of the Whatmans but with no absolute certainty.

In contrast to the above we have very definite evidence that the smaller single line "W"s with loop, and very probably the larger ones as well, can be attributed to the elder Whatman and that some of them may have been taken over by his son for as long as the original mould covers lasted. The most positive evidence we have that these countermarks belonged to Whatman is the appearance of one of these solitary type "W"s ca. 13 x 8 mm., very slightly larger than the Delaware series, in Edward Capell's "Prolusions" printed on Wove paper by Dryden Leach with a colophon date of 6th October 1759 and published by Tonson in 1760. It is known from a letter written by the younger Whatman on the 4th December 1768 that he had made this paper, though obviously in his father's lifetime.[214] Since Wove paper was manufactured at that time solely by the Whatmans, this piece of evidence is further confirmed by the presence of similar "W"s with loop in some of the paper used for Robert Dodsley's "Select Fables" printed by Baskerville on Wove paper and published in 1761. The papers in question were made on the same moulds, though the paper used for the latter was, in fact, made in two lots; part of this work was already in print by 1760 and the paper would, therefore, have been made in the elder Whatman's time; but more paper had to be made to complete the work, this was after 1760 i.e. during Ann Whatman's interregnum.[215]

As further support for the above, the same type of "W" in style and size is found in association with either "I" or "J" within the surrounds of several different emblems in the watermarks of Laid papers confidently attributed to the elder Whatman.[216]

---

[214] This is neither the place nor book to discuss the details of this evidence which has to be examined together with many other facts associated with it. It is, however, discussed fully in the author's work Bib.76.

[215] As in the note above there are several pieces of evidence, which it would be inappropriate to set out here, showing that the paper used for these two works was made on the same moulds. As to the "W" countermarks, in the case of Dodsley's "Select Fables" the single line ones with loop have been recorded by Thomas Balston (Bib.1 14) and Hanson, L.W. "Review and Bibliographical Notes", The Library 5th Series Vol.XV 1960, 141; but they were not seen in the only copy examined by the author (B.L.Shelfmark C.70.b.2). This may have been because they were too faint, trimmed off or because this copy had been made up from another batch of papers without marks.

[216] Balston, T. Bib.1 App.V.

Figure 15

Diagram of the single line "AW" (with Loop) countermark placed within a Pro Patria Watermark above the Gate to the Palisade.

8.5 Laid Wires per cm.

PRO PATRIA

A W

9.0 Laid Wires per cm.

← 25.5 mm. →

Reproduced here by kind permission of Mr. T.L. Gravell (U.S.A.)

The illustration above has been traced from an enlarged photographic print of an image obtained from a light sensitive proofing paper copy kindly supplied to the author by Mr. Gravell in 1979. (The watermark and countermark are reproduced in Bib.68 No.32). The dimensions of the "W" in this case were 8.5 x 5.5 mm., very similar to the solitary "W" of the November 1753 document (8 x 5 mm.) referred to in the text.

Thirdly, the same type of "W" has been found in local documents with dates ranging from 1749 (All SS'. Maidstone Parish document)[217] to 1766 and a solitary "W" of 1770 (Harrietsham documents).[218]

One really needs no further evidence to support the attribution of this type of solitary "W" with loop to the Whatmans at this time, more especially to the elder Whatman, so that one can say with some confidence that the papers carrying the smaller "W"s in the Delaware series, discovered by Mr. Gravell, were made by the elder Whatman.[219] They are precisely similar in style and have dimensions of the same order as those of the authenticated marks.

Finally, as a consequence of these findings it is not unreasonable to assume that the series of larger solitary "W"s with loop found in Thomas Hunt's Album (RIBA Drawings Collection) may also be attributed to one of the Whatmans. Likewise, on the same basis we may attribute the "AW" countermark[220] (Fig. 15) found within a similar PRO PATRIA watermark as the Delaware series by Mr. Gravell in a New Castle, Delaware document dated 1765 to Ann Whatman in paper which must have been made during her interregnum, 1759–1762. It must be emphasised here that in the case of this Ann Whatman countermark the attribution is not based solely on the similarity of the single line "W"s; there are many other features, such as chainwire spacings, the number of Laid Wires per cm., and various components of the Pro Patria watermark (like the diagonal bar to the gate in the palisade) that correspond sufficiently with some of the other Delaware marks to show that, in spite of the probability of more than one pair of moulds being in use, they must all have come from the same mill and mould maker.[221]

In conclusion we may say that the Whatmans are still good candidates for the acute angled solitary "W"s, but we cannot be as certain about this as we can in attributing the single line "W"s with a loop to them and more particularly to the father.

---

[217] KAO P 241 12/17.

[218] KAO P 173 5/2 Pt.I (see App.I Table XVI).

[219] An example is illustrated in Gravell, T.L. Bib.68 No.713 (see also Fig.14 in this App.). Other solitary single line "W"s with loop have been found elsewhere e.g. in the R.I.B.A. Drawings Collection Album L/9, William Newton (1730–98). Drawings date from 1763–80's and include several single line "JW"s (with loop) within Britannia watermarks; a block letter "W"; an early "J Whatman"; and one example of a single line "W" with loop (dwg.no.32[i]); this last was torn, but the presence of a "GR" orients the sheet in a way that confirms that it is a solitary "W" and not a truncated "JW".

[220] Illustrated Gravell, T.L. Bib.68 No.32.

[221] For detailed comparisons and measurements see author's Bib.76.

Single Line "F" countermarks within a Pro Patria Watermark

In passing it has been noted that single line "F" countermarks found within Pro Patria marks (having the same features as the Whatman watermarks) and placed in the same position as the solitary "W"s with loop have been found in one half of sheets of paper which bear "T FRENCH" in the other halves.[222] Thomas French, the son of Henry French snr. (see App.II for genealogy) was a contemporary of the younger Whatman and was active at Gurney's mill, Loose, from ca.1771–1795. Two explanations for the occurrence of this watermark and countermark may be advanced. One has already been discussed in App.I (p.9), where it was suggested that the Whatmans and the Frenchs may have shared a mould maker for a time and hence a similarity in their marks; but there is another. It is possible that Thomas French was making use of one of Whatman's cast–off moulds. That this is not as improbable as it might seem is supported by information kindly supplied by Gravell (Bib.16 letter 28:Feb:81) to the effect that in the ledger of Nathan Sellers,[223] the distinguished American mould maker, there is an entry recording an order to make alterations to "a pair of Whatman's Post Molds" in January 1789, demonstrating that his moulds evidently found their way into other paper makers' hands including American ones.

---

[222] 1781 Land Tax Loddington Ward (KAO Q/RP1 244); and Croydon Parish Doc.1781 (Letter from E. Heawood to Col. C.H. Balston 1931 : PSM).

[223] Nathan Seller's Ledger is held in the Library of the American Philosophical Society, Philadelphia.

Abbreviations

AP	Ancient Paper Maker
Arch.Cant.	Archaeologia Cantiana
BAPH	British Association of Paper Historians
EGL	Excise General List (or Letter)
GLRO	Greater London Record Office
IGI	(Mormon) International Genealogical Index
KAO	Kent Archives Office (redesignated Centre for Kentish Studies)
LT	Land Tax Returns
MPM	Master Paper Makers
PR	Parish Register
PSM	Papers Springfield Mill
RB	Ratebook
SFIP	Sun Fire Insurance Policy
~	Of the order of

Sources of Information and Bibliography

SOURCES OF INFORMATION

A variety of Sources have been used from which information has been obtained for inclusion in this book. Some of this information has been obtained from Printed Books and some from other sources such as Public and Private Archives, Correspondence and Personal Communications. Where frequent reference is made to Printed Works (or Collections of correspondence relating to a printed work in preparation), the works in question are listed below under the heading of "BIBLIOGRAPHY" and given a "Bib.No." for use in the text and footnotes. In the remaining cases the details of references are given in the text or in footnotes and only a general indication of the Sources is listed below under "OTHER SOURCES".

OTHER SOURCES

Kent County Archives (KAO), Maidstone.

Cathedral, City and Diocesan Record Office, Canterbury.

Guildhall Library MSS Department, London.

The Greater London Record Office.

The Record Offices of Bedfordshire, Buckinghamshire, Hampshire, Hertfordshire, Lancashire, Norfolk, Oxfordshire, Surrey and West Yorkshire.

The Public Record Office (Probate Division).

The Librarians of many different Libraries

Ordnance Survey Maps and Maps held in the Map Library, British Library.

The Royal Commission on Historical MSS., London.

Mr. Henk Voorn, Keeper, Paper Historical Dept., Koninklikje–Bibliotheek, The Hague.

All Saints', Maidstone Parish Registers.

Balston Archives (PSM, Papers from Springfield Mill, Kent).

BIBLIOGRAPHY

Bib.No.

1	*Balston*, Thomas	"James Whatman, Father and Son" (Methuen).	1957
2	*Hudson*, Frederick	"The Earliest Paper made by James Whatman, the Elder (1702–1759) and its significance in relation to G.F. Handel and John Walsh" (The Music Review. Vol.38 Feb. 1977 15–32).	1977
3	*Churchill*, W.A.	"Watermarks in Paper in the XVII and XVIII Centuries" (Amsterdam, 1935).	1935
4	*Harris*, John	"History of Kent"	1719
5	*Hazen*, A.T.	"Eustace Burnaby's manufacture of white paper in England" (Papers of the Bibliographical Soc. America. xlviii. 1954.329).	1954
6	*Hasted*, Edward	"History of Kent" (Vol.II). (See also under Bib.42, a later Edition, 1798).	1782
7	*Shorter*, Alfred	"Paper Mills in the Maidstone District" (The Paper–Maker & Brit. Paper Trade Jo.1960.Pts.I–IV Mar., Apr., May & July 1960).	1960
8	*Spain*, R.J.	"The Len Watermills" (Arch.Cant. LXXXII.1967.32).	1967
9	*Shorter,* Alfred	"Paper Making in the British Isles, an Historical and Geographical Study" (David & Charles).	1971
10	*Spain*, R.J.	"The Loose Watermills" (I) (Arch.Cant. LXXXVII.1972.Pt.1.43).	1972
11	*Spain*, R.J.	"The Loose Watermills" (II) (Arch.Cant. LXXXVIII.1973.159).	1973
12	*Shorter*, Alfred	"Paper Mills and Paper Makers in England" (Paper Publications Soc., Hilversum Series Monumenta Chartae Papyraceae Historiam Illustrantia.Vol.VI).	1957
13	*Fuller*, M.J.	"The Watermills of the East Malling Stream" (Privately printed, 1973).	1973
14	*Melling*, E.	"Some Roads and Bridges" Kentish Sources I. (Kent County Council, 1959).	1959
15	*Coleman*, D.C.	"The British Paper Industry, 1495–1860" (Oxford, Clarendon Press 1958).	1958
16	*Gravell*, Thomas L.	Letters 1979–1991 (See also Bib.68).	Current

17	*How*, Aspley	Aspley Guise Papers (Bedfordshire Record Office Acc. 809 MK42 9AP).	mid-18th C.
18	*Cameron*, W.J.	"The Company of White–Paper–Makers of England, 1686–1696" (Univ. of Auckland Bull. 68 Economic History Series 1).	1964
19	*Pollard*, M.	"White Paper–making in Ireland in the 1690's" (Proc.Roy.Irish Academy, Dublin 77. C. No.6.223).	1977
20	*Houghton*, John	"Husbandry and Trade Improved" London, 1691–1703 (Richard Bradley Edition 1727)	end 17th C.
21	*Desmarest*, Nicholas	"Art de fabriquer le Papier" Encycl. Méthodique, Arts et Métiers: Mécaniques. V. Paris).	1788
22	*Bagford*, John	Bagford Papers, Harley 5942 (Dept.Printed Books, British Library, Great Russell St., London).	ca. 1700
23	*Le Francois de Lalande*, J.J.	"Art de faire le Papier" (Descriptions des Arts et Métiers, Académie des Sciences, Paris).	1761
24	*Gachet*, Henri	"Monsieur Des Billettes, Historien du Papier" (Papiergeschichte Vol.VII No.4 July 1957.49).	1957
25	*Savary* des Bruslons, J.	Dictionnaire Universel de Commerce (1723–1730).	1723
26	*Imberdis*, S.J.	"Papyrus" (transl. Laughton, E. 1952).	1693
27	*Elliot*, Joseph Mosely	"Machinery for working wire and for the manufacture of Moulds for Papermaking" (British Patent No. 1959).	1793
28	*Voorn*, Henk	Letters 1976–1985.	Current
29	Calendar of Treasury Books	1672/5 BIV. 768.	1672
30	*Evelyn*, John	Diary (Edition William Bray, London 1879)	1678
31	*Fiennes*, Celia	"Journeys of Celia Fiennes" (Christopher Morris, Cresset Press 1947).	1697
32	*Steel*.	"The elements and practice of rigging and seamanship"	1794
33	*Clapperton*, R.H.	"Paper, an historical account of its making by hand from earliest times down to the present day" (Oxford, 1934).	1934
34	*Jenkins*, Rhys	"Papermaking in England, 1682–1714" (Library Assoc. Record III.1901.239).	1901
35	*Heawood*, Edward	"Watermarks, mainly of the 17th and 18th centuries" (Paper Publications Society: Mon. Chart.Paper Hist.I.Hilversum, 1950).	1950

36	*Labarre*, E.J.	"A Dictionary of Paper and Papermaking Terms" (Swets & Zeitlinger, Amsterdam, 1937).	1937
37	*Hunter*, Dard	"Papermaking, the History and Technique of an Ancient Craft" (Cresset Press, London 1957).	1957
38	*Gill*, C.	"The Rise of the Irish Linen Industry" (O.U.P. 1925).	1925
39	*Poggi*, A.C. de	Journals 1790–1809 (Vol.LVII) MS. Grp.58 Joshua & Thomas Gilpin Collection (Pennsylvania Historical and Museum Commission, Pa.17120).	1790's
40	*Green*, J.Barcham	"Paper Making by Hand" (Maidstone, 1967).	1967
41	*Russell*, J.M.	"The History of Maidstone" (William S. Vivish, Maidstone 1881).	1881
42	*Hasted*, Edward	"History of Kent" (see also Bib.6)	1798
43	*Scott*, W.R.	The Constitution and Finance of English, Scottish amd Irish Joint Stock Companies to 1720." (Cambridge University Press 1912: reprinted Peter Smith, New York, 1951).	1912
44	*Thomson*, A.G.	"The Paper Industry in Scotland, 1590–1861" (Scottish Academic Press, 1974)	1974
45	*Loeber*, E.G.	"Paper Mould and Mouldmaker" (Paper Publications Society Amsterdam, 1982).	1982
46	*Dunkin*, J.	"History and Antiquities of Dartford"	1844
47	*Coleman*, D.C.	"Premiums for paper : The Society and the early paper industry" (Jo.Roy.Soc.Arts 1959.Apr.362/3).	1959
48	*Voorn*, Henk	"Some curious early experiments in Dutch papermaking" (The Paper Maker, Hercules Powder Company. 32.1963.No.l 17).	1963
49	*Martin*, Thomas (alias John Farey)	"Circle of Mechanical Arts" (London, 1813).	1813
50	*Rees*, Abraham	"Cyclopaedia" (1802–1820) (Paper Section publ. 1813 probably written by John Farey, Engineer).	1813
51	*Voorn*, Henk	"Papermaking and Papermakers of long ago in Holland" (The Paper Maker, Hercules Powder Company. 20.1951.No.2.8).	1951
52	*Voorn*, Henk	"In search of new Raw Materials" (The Paper Maker, Hercules Powder Company. 21.1952.No.2 1–14).	1952
53	*Wescher*, H.	"Cotton in the Ancient World" "The beginnings of the Cotton Industry in Europe" "Cotton growing and Cotton Trade in the Orient during the Middle Ages"	1948

"Fustian weaving in South Germany from 14th to 16th Century"
"Schürlitz Weaving in Switzerland"
(CIBA Review 1948. No.64 2322–54).

| 54 | *Edwards*, Michael M. | "The growth of the British Cotton Trade, 1780–1815" (Manchester Univ.Press 1967). | 1967 |

| 55 | *Wadsworth*, A.P. & *Mann*, Julia de Lacy | "The Cotton Trade and Industrial Lancashire, 1600–1780" (Manchester, 1931). | 1931 |

| 56 | *Renouard*, A.A. | "Annales de l'Imprimerie des Aldes" (3rd Edition 1834) | 1834 |

| 57 | *Temming*, Heinz & *Grunert*, Heinz | "Temming Linters" (Glückstadt, Germany, 1966). | 1966 |

| 58 | *Beadle*, Clayton & *Stevens*, Henry P. | "Handmade Papers of different periods" (Jo.Roy.Soc.Arts 1909 26:Feb:293). | 1909 |

| 60 | *Eyre*, J.Vargas & *Nodder*, C.R. | "An Experimental Study of Flax Retting" (J.Text.Inst.1924.15. T.237). | 1924 |

| 61 | *Clapperton*, R.H. | "Rags and their preparation for Papermaking" (Proc.Tech.Sec.Papermakers Association of Great Britain & Ireland IX.1928.34). | 1928 |

| 62 | *Edelstein*, S.M. | "Two Scottish Physicians and the Bleaching Industry: The contributions of Home and Black" (Amer: Dyestuff Reporter 1954.35). | 1954 |

| 63 | *Urquhart*, A.R. & *Denton*, M.J. | "The Absorption of Liquids by Paper" (An unpublished monograph prepared for W. & R. Balston Ltd. Springfield Mill Maidstone, Kent, 1966). | 1966 |

| 64 | *Barkas*, W.W. | "The Swelling and Shrinking of Papermaking Fibres" (Proc.Tech.Sec. B.P.B.M.A. Vol.31 1950.III.471). | 1950 |

| 65 | Authors various (Ed. *Harte*, N.B. & *Ponting*, K.G.). | Textile History and Economic History "Essays in honour of Miss Julia de Lacy Mann" (Manchester University Press 1973). | 1973 |
| | Authors cited:– | (a) *Harte*, N.B. Protection and the English Linen Trade, 1690–1790", pp 74–112 (b) *Chapman*, S.D. "Industrial Capital before the Industrial Revolution", pp 113–137 | |

| 66 | *Overend*, G.H. | "Notes on the Earlier History of the Manufacture of Paper in England" (Proc. Hugenot Society of London Vol.VIII.1909.177–220). | 1909 |

| 67 | *Hyde–Price*, William | "The English Patents of Monopoly" (Harvard University Press, 1913). | 1913 |

68	*Gravell*, Thomas L. & *Miller*, George E	"A Catalogue of Foreign Watermarks found on Paper used in America, 1700–1835" (Garland Publishing Inc. New York & London, 1983)	1983
69	*Finerty*, E.T.	"History of Paper Mills in Hertfordshire" (Paper–Maker & Brit. Paper Trade Jo. 1957 Apr., May, June; 308, 422 & 510).	1957
70	*Stoker*, David	"The early history of papermaking in Norfolk" (Norfolk Archaelogy XXXVI Part III 1976 247 : and Correspondence).	1976
71	*Lowe*, Norman	"The Lancashire Textile Industry in the 16th Century" (Manchester, Cheetham Society, 1972).	1972
72	*Gaskell*, Philip	"Notes on Eighteenth–Century British Paper" (The Library, 1957.Mar.34–42).	1957
73	*Povey*, K. & *Foster*, I.J.C.	"Turned Chain–Lines" (The Library V. 1950–51.184–200).	1950
74	*Balston*, T.	"William Balston, Paper Maker, 1759–1849" (Methuen, 1954)	1954
75	*Clapperton*, R.H.	"The Paper–making Machine, its Invention, Evolution and Development" (Pergamon Press, Oxford, 1967)	1967
76	*Balston*, J.N.	"The Development of Early Wove Papers" (In preparation)	
77	*Pilkington*, Austin	"Frogmore and the first Fourdrinier" (The British Paper Company 1890–1990 publ: Laurence Viney Ltd.).	1990
78	*Hills*, Dr. Richard L.	"Papermaking in Britain 1488–1988". A short history (The Athlone Press)	1988

Fold–out Maps

1. 1864/5 Ordnance Survey Map (Scale 6 inches to 1 mile) Map 9 of the Parishes of HOLLINGBOURNE and LEEDS.

2. 1864/5 Ordnance Survey Map (Scale 6 inches to 1 mile) Map 10 of the Parishes of BOXLEY, AYLESFORD and DITTON.

INDEX OF PAPER MILLS

INDEX OF PAPER MAKERS

INDEX OF COUNTERMARKS

GENERAL INDEX

In order to simplify the process of information retrieval, particularly where terminology may be variable, this Index contains a number of Sections relating to certain activities or materials central to the subject of this book where references have been concentrated and are listed under appropriate subheadings. These Sections, for example, cover subjects such as Cellulose; Duty & Excise; Linen; Patents; Paper; Papermaking; Rags; Sizing; Water; and Wool. Individual items are still listed alphabetically but have the added direction to "see under" in many instances.